IONIC SOLIDS AT HIGH TEMPERATURES

SERIES ON DIRECTIONS IN CONDENSED MATTER PHYSICS

Published

Vol. 1 – Directions in Condensed Matter Physics – Memorial Volume in Honor of Shang-keng Ma
edited by G Grinstein & G Mazenko

Vol. 3 – Directions in Chaos (Vol. 1)
edited by Hao Bailin

Vol. 4 – Directions in Chaos (Vol. 2)
edited by Hao Bailin

Vol. 5 – Defect Processes Induced by Electronic Excitation in Insulators
edited by N Itoh

Forthcoming

Vol. 6 – Spin Glasses & Biology
edited by D Stein

Vol. 7 – Interaction of Electromagnetic Field with Condensed Matter
edited by N N Bogolubov, Jr, A S Shumovsky & V I Yukalov

Vol. 8 – Scattering and Localization of Classical Waves in Random Media
edited by P Sheng

Vol. 9 – Geometry in Condensed Matter Physics
edited by J F Sadoc

Vol. 10 – Directions in Chaos (Vol. 3)
edited by Hao Bailin

Directions in Condensed Matter Physics — Vol. 2

IONIC SOLIDS AT HIGH TEMPERATURES

Edited by

A. M. Stoneham

Materials Physics and Metallurgy Division
Harwell Laboratory

World Scientific
Singapore • New Jersey • London • Hong Kong

Published by

World Scientific Publishing Co. Pte. Ltd.,
P O Box 128, Farrer Road, Singapore 9128
USA office: 687 Hartwell Street, Teaneck, NJ 07666
UK office: 73 Lynton Mead, Totteridge, London N20 8DH

IONIC SOLIDS AT HIGH TEMPERATURES

ISBN 9971-50-335-2

Printed in Singapore by General Printing Services Pte. Ltd.

CONTENTS

ACKNOWLEDGEMENTS

We are especially indebted to those who have funded the research which is described in this volume. In particular, we should like to mention the UK Department of Energy's General Nuclear Safety Research programme (AMS, JHH, MTH, MJG), the UKAEA Underlying Programme of longer-term research (AMS, RB, JHH, MJG, MTH), the SERC (WH, RB), the US Department of Energy (RJ), NASA (RJ) and the US NSF (RJ).

The index was prepared by Nicola Stoneham, to whom we are most grateful.

We have benefited from discussions with many of our colleagues, and we should like to express our gratitude for their stimulus, encouragement and insight.

INTRODUCTION

Ionic solids at high temperatures

A M Stoneham

Collectively, the developments from recent studies of ionic solids have provided a much clearer idea of the wealth of phenomena at high temperatures. Experiments have probed structure, transport and dynamics by a range of powerful methods. The newer theoretical approaches offer routes to modelling and to predicting behaviour which promise to match previous striking successes at lower temperatures. The several components, both theoretical and experimental, have been brought together in this book so that the approaches can be related to one another, their significance identified, and their varied strengths and limitations identified.

Solid State Processes: The Technological Background

Many of the most fundamental of the scientific problems in condensed matter physics are low temperature phenomena. Examples include the many facets of liquid helium, some quantum tunnelling behaviour, and most aspects of superconductors. As far as is possible, the major technological processes (such as the photographic process, the Xerox process, or the processes in semiconductor devices) are designed to work at room temperature. Yet there remain plenty of important technological problems involving condensed matter at far higher temperatures. Sometimes this is because of special features of processing, like ensuring that a steel component is malleable, or enabling controlled crystallisation to occur, whether of a semiconductor like silicon or of diamond, where the stability regime is not easy to attain. It is, of course, marginally easier to heat a sample than to cool it, but usually high temperature methods are adopted only when there are no low temperature options. For example ceramics are of technical value because of their resistance to high temperatures, and usually this means that they need high temperatures for their processing. In geophysics (see, e.g. Chapter 5 by Jeanloz), the temperatures and pressures inside the Earth are not under human control. If one wishes to assess hypothetical reactor incidents, these may postulate nuclear fuels at extreme temperatures or pressures (cf. Chapters 2,3 and 4 by Harding, by Gillan, and by Hutchings and Hayes here). The more exotic processing methods, as for the laser- or electron beam- heating of special materials, involve local regions which have transients at very high temperatures. If an ionic conductor of above some chosen conductivity is wanted, possibly suitable as a solid electrolyte, it will need to operate at some minimum temperature (see eg Hayes and Stoneham 1985). This temperature is critical in any proposed practical use of the so-called superionic materials, and the principles which determine the lowest working temperatures and highest currents are linked to structural and other properties which will be discussed in later chapters. In cases such as these, a key issue is the intrinsic defect concentration, and here (fig 1) there are several possible consequences of increasing defect levels, including the formation of new phases instead of increases in ionic conductivity.

The aim of the theoretical studies to be discussed is the quantitative understanding of what happens at high temperatures. Implicit here is the more significant ability to predict behaviour in circumstances beyond that observed directly. The technical difficulty of the experiments described in Chapters 4 and 5 is very considerable, and shows a need for the combination of theory with experiment in interpretation, in extrapolation, and in confirming confidence in understanding. The drive to tackle the very difficult experiments of Chapter 4 came from two main needs with components in common. The first was the need for fast-ion conductors, with their range of possible applications like batteries, fuel cells and solid-state gas sensors. The second need was in the comprehensive analysis of safety issues of nuclear reactors, where the thermodynamic and transport properties at high temperatures are an essential part of the input data. These two technical requirements, together with other possible needs, as for laser host materials, have one component in common; the fluorite structure, shared both by UO_2 and by promising ionic conductors. Fluorite has, in fact, been of interest for many years, initially as a mineral which crystallised in a characteristically cubic form, and as a transparent solid whose colour came from radiation-induced intrinsic defect centres. It is natural science, rather than industrial science, which stimulated the experiments of Chapter 5, which concerns such fundamental issues as the nature of the interior of the Earth and the related questions of its origin.

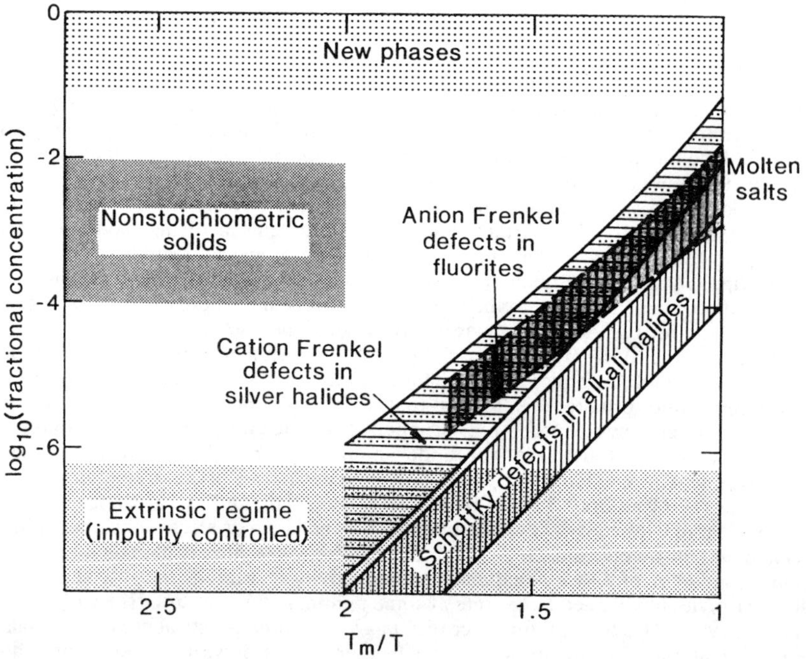

Figure 1 Fractional defect concentrations in thermal equilibrium : typical values for ionic crystals are shown as a function of reduced temperature. Here T_m is the melting temperature.

Theoretical Issues

The chapters in this book use a variety of models for interpretation, analysis and prediction. These models describe the motions of ions at widely different levels and apparently in very different ways. In this discussion of theoretical issues the relationships between the various descriptions are explored.

The low temperature behaviour of crystals exploits the periodicity of the lattice structure. In these *lattice-based* theories, the harmonic vibrations are described in terms of delocalised phonon modes with definite wave vectors. Generalisations to alloy systems involve the same ideas, using either a scattering description (in which the disorder leads to elastic scatter from modes with one wave vector to those with another) or a description in terms of the harmonic (but not necessarily delocalised) modes of the alloy lattice. At the harmonic level, the Hamiltonian can be written as the sum of three terms:

(1) The kinetic energy, T, which alone depends on the particle masses;
(2) Terms linear in the displacements, which vanish for equilibrium;
(3) Terms quadratic in the displacements from equilibrium, each such term being proportional to the second derivative of a force constant matrix.

In the harmonic approximation, there always exist modes which are independent dynamically. In real solids as the temperature is raised, anharmonic corrections of various sorts appear. However, the degrees of freedom recognised in lattice descriptions are only those which specifically associate atoms and sites: the diffusive degrees of freedom and other defect-related motions are not described directly, so that a more general description becomes necessary at sufficiently high temperature. One approach is to use *molecular dynamics* (as in Chapter 3), which does not explicitly discuss the phonon modes, the harmonic vibrations emerging simply through the equations of motion of the interacting ions. Yet provided the temperature and degree of disorder are not too high, the lattice-based description can do a remarkably good job too (cf Chapter 2). These two approaches form the basis of accurate quantitative predictions of ionic solids.

The Shell Model. The dominant interaction in ionic crystals is, of course, the Coulomb interaction which, with the short range repulsions postulated by Born many years ago, gives a respectable zeroth order picture of behaviour. Most molecular dynamics calculations still use models at this level, for it has acceptable accuracy and gives good value for computer money. Other calculations go somewhat further, and include electronic polarisation by having the individual ions polarisable. This sophistication enables the refractive index to be fitted too. However, the basis of almost all of the most accurate quantitative modelling of ionic crystals at lower temperatures has been the shell model of Dick and Overhauser (1958). This model (see Cochran 1971) generalises the Born model in two steps. The first step gives the way in which electronic polarisation is included, i.e. the representation of the way that individual ions develop dipoles in the fields of the others. In the shell model, each ion consists of a shell (corresponding to the outer electrons of the ion, and represented in effect by a point charge $Q_s|e|$ of small mass m) and a core (corresponding to the nucleus and core electrons, with charge $Q_i - Q_s = Q_c$ and virtually the whole ionic mass M); the electronic polarisation involves the relative displacement of shell and core, the two being assumed coupled harmonically. The second key step in the shell model comes from the realisation that the ionic polarisability depends on the ion's environment, because it is affected by the short-range repulsions.

The shell model has many clear virtues. The first is that it works well in practice; at the very least it provides a convenient framework for interpreting harmonic properties of crystals. Indeed, with minor generalisations, such as to include bond charges, the shell model offers a general framework at the harmonic level. Most lattice-based models assume additionally that the adiabatic and dipole approximations hold, and the shell model appears to succeed well in such systems, and indeed in many cases even where these basic assumptions might be questioned. A second virtue is that the shell model is physically sensible, giving a plausible description of the corrections to simple polarisable ion models and some sort of link to chemical descriptions. A third virtue is that the shell model parameters are often transferable with reasonable accuracy, both from one crystal to another and from solid to molecule (this transfer property should not be taken for granted; see Stoneham and Harding 1987). A fourth virtue is that the shell model may be used for certain types of electronic calculation, at least in cases where the energy of interest is dominated by polarisation and distortion (see Catlow and Mackrodt 1982; Stoneham 1987; Tasker and Harding 1989). Thus, for example, charge-transfer energies and the energies of small-polaron stabilisation and motion can be modelled.

The virtues of the shell model are achieved at a price, and this price has to be paid when there is a need to go beyond harmonic vibrations or purely static models. Thus the low mass m of the shell, which is needed to ensure the shell follows the core motion adiabatically (the Born-Oppenheimer approximation), can also mean that the shell vibrational amplitudes become unacceptably large. How this can be remedied is discussed in Chapter 1 , where the formal basis is given for using the shell model in self-consistent phonon models. Here the main idea is that the form of the Hamiltonian given above is still sufficient. However, in the third term listed above, no longer do the second derivatives of the potential appear: instead they are replaced by variational parameters. The hope is that the model harmonic hamiltonian, with this variational extension, will adequately reproduce the predictions of the full Hamiltonian.

Simpler Approximations. Molecular dynamics and lattice-based theories are not the only possible approaches, however. Two simpler descriptions are still used, namely the hard sphere model and the local potential model. The hard sphere model is commonly used for rare gas systems, where the short-range repulsion between atoms is dominant. This approximation is not especially good, nor is it often necessary. There is a wealth of empirical and other information that the short-range repulsion is rather accurately represented by the Born-Mayer form Aexp $(-r/\alpha)$, with α about 0.3Å; it is also very well known that Coulomb interactions in ionic systems favour strongly ionic arrangements which surround cations by anions, and vice versa. Since there are efficient, convenient and widely-used codes available (like the HADES and SHEOL codes of Chapter 2) for far more accurate descriptions than hard spheres, these codes should always be used for serious calculations. Nevertheless, there are occasions where the directness and simplicity of a hard-sphere model does help as a first orientation, and this is noted in some of the cluster models of Chapter 4.

The local potential descriptions are somewhat different. The idea is that each ion can be considered as vibrating independently in a potential which can be highly anharmonic (Lennard-Jones and Devonshire 1938; Ree and Holt 1970). Since there is only one particle, quite sophisticated calculations can be done; indeed, this local potential approach has been used to make workable some self-consistent phonon calculations (though not those of Chapter 1; see Dienes *et al* 1980; Shukla *et al* 1981). The advantage of this description (as well as of its generalisations, like itinerant-oscillator models) is especially evident for the fluorite structure, which is perhaps the most important crystal structure discussed in this volume. The anions (fluorine or oxygen) lie at sites of tetrahedral symmetry, and an important component of the anharmonic motion of the anions is describable by a tetrahedral term in the potential, proportional to (xyz/r^3). This term can be estimated from neutron scattering data (see chapter 4) and may be used for other calculations, like the prediction of an anharmonic component of the specific heat

(Hyland and Stoneham 1981). However, the status of the potential needs clarification. The tetrahedral term in, say, UO_2 clearly relates to O motion relative to U, just because of its symmetry, and this term will give an particularly good description when the O motion is well represented by an Einstein model (both because of the high mass of U relative to O and because there are strong short-range forces which favour the local potential description). But how should one interpret the harmonic terms in the local potential? These (see Hyland and Stoneham 1981) involve the in-phase motion of U and O, i.e. in a lattice description, the acoustic modes contribute as well, not just the optic modes. Thus the harmonic and anharmonic parts of the local potential do not correspond in the way implied by the local potential, so that any quantitative use of this picture is limited.

Static and Dynamical Methods. Chapters 2 and 3 illustrate two different, but complementary, approaches: static models (with their natural generalisation to look at harmonic vibrations) and molecular dynamic models. In static simulations, one has to decide which processes to calculate, but, having chosen, one can isolate a specific process very easily; in molecular dynamics, the processes are determined during the calculation (a great help if the important mechanism is not obvious) but are only isolated with difficulty. In static simulations, it is easy to get activation energies (i.e. the temperature dependence of a rate) but harder to get the absolute rate; in molecular dynamics, it is rather easier to predict the absolute rate than its temperature dependence. The shell modes is easily used in static approaches, but is often too expensive for molecular dynamics. There are no problems for molecular dynamics when the activation energy is small and so diffusion and vibration becomes hard to separate. The experimental quantities derived differ too, so that there are strong links of molecular dynamics with neutron scatter and of static methods with electronic properties, volume changes,etc. Molecular dynamics has its own generalisations to include electronic behaviour (Car and Parrinello 1986), with particular uses when the nature of the electron state (e.g. self-trapped or not) is not certain. The two approaches provide a strong basis for predictions for comparison with the wide range of experiments now done. In particular, both approaches have the great advantage of predicting what is observed: there is no need for a substantial re-interpretation of experimental data before a comparison can be made. Related to this is the way one can use theoretical methods like these to ask whether some model – a simple picture or qualitative argument – is acceptable, for even with the best theoretical approaches, such models provide a conceptual framework which has a value of its own.

Experimental Issues

The two experimental chapter highlight some of the recent work on ionic solids at high temperatures and, in doing so, show complementary features. Both give structural information by diffraction. Neutron scattering gives details of dynamics on an atomic scale: the degree and nature of the disorder, the way the lattice vibrations change, and the motions involved in diffusion processes. The laser-heated diamond cell can provide optical properties, macroscopic evidence for diffusion (including that in a temperature gradient), thermodynamic quantities, and phase diagram information over a wide range of temperatures and pressures. Table 1 compares some of the actual or potential observations by these important approaches.

There are, of course, experimental studies beyond those described here. Infrared emission is one example (e.g. Hisano and Tanaka 1986). The shock-wave work, and the information contained in the Hugoniot, is another instance, and is specifically important here as a test of the interatomic potentials used in the theory chapters (see also Stoneham and Harding 1987). Materials processing is another source of important information, though this is often too complex (as for slags, both in the liquid and solidified states, or as in some of the ceramic processing area) to analyse beyond a simple equilibrium thermodynamic description.

Table 1: Comparison of High Temperature Techniques.

	Neutron beam experiments	Laser-heated diamond cell
Structure	Crystal structure by diffraction; some defect structural information	Mainly X-ray data giving structure Melting versus pressure
Elastic properties	Measured from phonon dispersion curves	Possible by Brillouin scatter Stress relaxation can be monitored
Transport properties	Diffusion constants plus information on nature of atomic motions involved	Includes motion in a temperature gradient Electrical and thermal conduction possible
Spectroscopy	Crystal field splittings	Optical emission and absorption possible
Range of conditions	Usually ambient pressure with temperature limited by available furnace (e.g. below 3000K) Samples best if single crystal of reasonable size	Pulsed to >5000K; Continuous 5000 to 7000K Pressure to 160 GPa or higher.

The Present Achievement: Where are the Gaps?

Progress in understanding condensed matter at high temperatures, when anharmonic motion cannot be ignored, has touched many areas. The present volume is solely concerned with ionic solids, so it omits some of the important work in other areas: some of the applications to rare-gas solids, for instance, do not appear, nor do some of the path-integral methods (Barker 1979; Chandler and Wolynes 1981, Jacucci 1984, Sprik and Klein 1988) nor the techniques used to analyse plasmas and ion and electron solvation in liquids.

The present volume does not – and could not – describe all the rich variety of phenomena in ionic solids at high temperatures. Some of the gaps are intentional. Liquids and glasses are excluded by choice, to keep the volume coherent and well-focussed. Phase changes are a subject area in themselves and are likewise excluded. Perhaps the main gaps are studies of interfaces (including the phenomena relating to evaporation), where there has been substantial progress on segregation and allied phenomena, and where full free energy calculations may be expected soon. For mechanical properties like creep, and the modelling of plastic behaviour, atomistic descriptions are at an early stage. At a fundamental level, a full description of electronic processes still awaits a detailed theory. The methods used here have often been used to give approximate descriptions of some electronic processes (REDOX processes, small polaron conduction, optical charge transfer, etc) even though these stretch the shell model to its limits or even beyond. Such attempts are not covered here, though it is to be hoped that the next generation of theories will achieve a valid and practical approach to the key electronic processes.

It is useful to remark on some of the points arising from the several articles in this volume, since collectively they represent a substantial advance in the last few years. Richard Ball's chapter shows that quantum aspects cannot be ignored, unlike the harmonic case where the identity of the Heisenberg equations and the Lagrange equations means that many results are identical. In the shell model, however, when the shell mass tends to zero, the zero point amplitudes could easily become embarassing, and special techniques prove essential. These methods come from quite different areas of quantum physics, but allow a self-consistent phonons method to be derived for the shell model. John Harding's chapter shows the major changes occurring half a century after the classic paper of Mott and Littleton, and indeed mainly changes of the last decade. Here a major development has been the writing of well-tested, general-purpose computer codes, like HADES and SHEOL. The modelling of defect processes, using the basic ingredient of free energies, is now practical for a broad class of ionic systems; the accuracy is at least as good as experiment in favourable cases, and fully sufficient as a complement to experiment in other cases. Michael Gillan's chapter re-emphasises these points by quite a different method. This shows the insights that can be given by simulations in a form which allows a very direct modelling of the system and the experiment performed on it. Thus the approach covers cases where there are many types of defects, or when the processes are complicated, as when the effective activation energy is small, and not much larger than thermal or phonon energies.

Given this theoretical progress, one may ask what the future directions are. Some are clear from the omissions just mentioned: the interrelation of mechanical and electronic properties is an example. More generally, it is still very hard to handle highly defective systems, where I include polar semiconductors with high carrier densities as well as high densities of ionic defects. Yet it is these types of system, together with altogether more complex (and often less well-characterised) systems which are of especial industrial importance, and which may attract especial attention. It is in such systems that the link to experiment is both difficult and important. As regards experiment, the two papers by William Hayes and Michael Hutchings, and by Raymond Jeanloz, show what can be done in practice by two entirely different routes. No longer need theory proceed in isolation by speculating from ill-characterised experiments. There is a new generation of substantial, well-controlled and novel experiments which can be verified against lower temperature work.

References

J Barker 1979 *J Chem Phys* **70** 2914

R Car and M Parrinello 1986 *Phys Rev Lett* **55** 2471

C R A Catlow and W C Mackrodt (editors) 1982 *Computer Simulation of Solids* (Berlin: Springer)

D Chandler and P G Wolynes 1981 *J Chem Phys* **74** 7
W Cochran 1971 *Crit Rev Sol St* **2** 1 B G Dick and A Overhauser 1958 *Phys Rev* **112** 90
G J Dienes, D O Welch and A Paskin 1980 *J Phys Chem Sol* **41** 1373
W Hayes and A M Stoneham 1985 *Defects and Defect Processes in Non-Metallic Solids* (New York: John Wiley)
K Hisano H Tanaka 1986 *J Phys C* **19** 6311
G J Hyland and A M Stoneham 1981 *J Nucl Mat* **96** 1
G Jacucci 1984, in *Monte Carlo Methods in Quantum Problems* (edited M H Kolos; Reidel, Dordrecht)
M L Klein 1985 *Ann Rev Phys Chem* **36** 525
J E Lennard-Jones and A F Devonshire 1938 *Proc Roy Soc A* **163** 53, *A* **167** 1
M Parrinello and A Rahman 1984 *J Chem Phys* **80** 860
F H Ree and A C Holt 1970 *Phys Rev B* **8** 826
K Shukla, A Paskin, D O Welch and G J Dienes 1981 *Phys Rev B* **24** 724
M Sprik and M L Klein 1988 *Comp Phys Rep* **7** 149
A M Stoneham 1987 *Phys Chem Min* **14** 401
A M Stoneham and J H Harding 1987 *Ann Rev Phys Chem* **37** 53
P W Tasker and J H Harding 1989 *Computer Simulation of Ionic Materials* (Bristol, UK: Adam Hilger)

Chapter 1

ANHARMONIC LATTICE DYNAMICS

R.D. Ball

The Blackett Laboratory, Imperial College of Science, Technology and Medicine, Prince Consort Road, London SW7 2BZ, U.K.

State University of New York, at Stony Brook
Stony Brook, New York 11794, USA

and

Theoretical Physics Division, Harwell Laboratory,
Didcot, Oxfordshire OX11 0RA, U.K.

We review the theory and application of finite-temperature perturbation theory to phonons in a quantised crystal lattice with an anharmonic interatomic potential, and in particular the self-consistent phonon theory, designed to yield a perturbation expansion valid for strong anharmonicities, high temperatures, and even unstable potentials. The extension to nonperturbative and nonpolynomial field theories is kept in mind. Practical applications, and in particular the hard-core problem, phonon damping, self-consistent thermodynamics, and shell models are also considered.

Introduction

Traditional lattice dynamics [Born and Huang, 1954] is based on three fundamental approximations, namely

i) an adiabatic approximation; the dynamics of the electron clouds is treated adiabatically so that a potential may be derived which dictates the motion of the nuclei [Born and Oppenheimer, 1927],

ii) a two-body approximation; either one ignores n-body forces, $n \geq 3$, or uses a simple model such as the shell model, in which the interplay between short-range interactions and polarizability is incorporated in a dipole (effectively two-body) approximation [Cochran, 1959], and

iii) a harmonic approximation; the potential is approximated by its Taylor series about the equilibrium positions, truncated to second order in the atomic displacements.

When applying lattice dynamics to the study of real crystals, in many situations of physical interest the weakest point in the above scheme is the harmonic approximation. It relies on displacements being relatively small, the "smallness" being defined strictly by the magnitude of the next order corrections to the physical property under consideration (so that it may not be enough to require that the root mean square strain be much less than one).

We may identify several cases where the harmonic approximation may be inadequate by considering an Einstein oscillator with $<x^2>= (\frac{\hbar}{m\omega})(n+\frac{1}{2})$. In particular, consider the following situations:

a) The nuclear potential as constructed via the adiabatic approximation may not have a minimum at the equilibrium positions, in which case the frequency ω is not well defined (this is often the case in the lattice dynamics of the rare gases, for example bcc^3He, He, or fcc^{4}He, Ne, Ar, Kr, and is also found in certain defect problems).

b) The anharmonic energy may be particularly important at modest $<x^2>$, for example with a potential which $\sim x^4$ at small x. Here there have been model calculations of several sorts, notably linear chains and single oscillator models.

c) Low frequency resonances, $\omega \to 0$, provided $m\omega$ is not too large (i.e. a flat potential). Then $< x^2 >$ will be large, and anharmonic effects important.

d) Small mass, $m \to 0$, for example local modes of light impurities; then again $< x^2 >$ will be large.

e) Potentials with important hard core effects (large repulsions at very small nuclear separations) which will not be included in ω.

f) At high temperatures, i.e. when $< n + \frac{1}{2} > \gg \frac{1}{2}$. For $T \geq \frac{1}{3} T_m$, anharmonic effects are very important for all crystals, and are ultimately responsible for the melting of the crystal.

The various methods used to overcome the limitations of the simple harmonic approximation offer very varied levels of sophistication, and thus a wide range of possible areas of application. Here we will consider briefly a few of the analytical approaches current in the literature to assess some of their strengths and weaknesses.

Single Oscillator Models: a typical Hamiltonian is

$$\frac{1}{2m} p^2 + \frac{1}{2} m\omega^2 x^2 - cx^3 + bx^4$$

which is solved either variationally, or using thermodynamic perturbation theory. One may then deduce approximate values for Debye-Waller factors, specific heats, and, with caution, expansion coefficients [Lima and Tsallis, 1980]. Bounds on thermodynamic properties may also be devised [Witschel and Boymann, 1980].

Quasiharmonic Models: At the opposite extreme to the single particle approach, one may change the lattice parameter according to some externally imposed criterion (e.g. observed expansion coefficients) and continue to assume the same interionic interactions in calculating energies for static ions. There are thus no thermal pressure terms. The main problems are that one is still calculating an internal energy (rather than a free energy), there is no guarantee of stability even for sensible geometries, and there is an essential (and false) relationship implied between volume, pressure, and temperature derivatives of thermodynamic quantities [Catlow et al., 1981].

A more sophisticated approach [Born and Huang, 1954] is to include the vibrational energy explicitly, written as

$$f_i(\omega_i, T) = \tfrac{1}{2}\hbar\omega_i + kTln(1 - exp(\frac{-\hbar\omega_i}{kT}))$$

where $\omega_i(V)$ are the mode frequencies, calculated with the specified lattice parameter. Then the free energy

$$F(V,T) = U(V) + \Sigma_i f_i(\omega_i(V), T)$$

This may then be used to obtain an equation of state [Boyer, 1981]. Although most predictions will be improved, the quasiharmonic method still cannot handle any awkward instabilities, even if they are only artefacts.

Cell Models [Lennard-Jones and Devonshire (1938), Huckaby and Salzburg, 1970]: Here one calculates the self-consistent harmonic energy and mean square displacement of a central atom (coordinate $\underline{r}$) interacting with consistently adjusted neighbours. Thus for on operator Ω we define

$$< \Omega >= \int d^3\underline{r}\, \Omega e^{-\frac{1}{2}< E_{xx}> r^2/kT} \Big/ \int d^3\underline{r}\; e^{-\frac{1}{2}<E_{xx}>r^2/kT}$$

and calculate $< E_{xx} > \equiv < \partial^2 E/\partial x^2 >$ via an expansion of E in powers of (x, y, z), and a calculation of $< x^2 >$. It is assumed that $< xy >= 0$, and, since the distribution is Gaussian, $<x^4>= 3 <x^2>^2$, etc. The equations are iterated to self consistency. Then $< E_{xx} >$ may be thought of as a temperature dependent renormalized force constant. Even at this simple level, the technique has had some successes, both in thermodynamic and shock predictions [Welch, Dienes and Paskin, 1978, 1980] and adiabatic and isothermal elastic constants vs. temperature [Ree and Holt, 1970]. Its main advantage is that instabilities may be eliminated dynamically.

Thermodynamic Perturbation Theory: The Helmholtz free energy is evaluated as a perturbation series in terms of a specified interatomic potential $\varphi(r)$ and its derivatives, up to anharmonic third and fourth order terms [Kadanoff and Baym, 1962, Abrikosov et al., 1963, Fetter and Walecka, 1971, Barron and Klein, 1974]. Many simple calculations of thermodynamic properties have been performed in this way [for example Shukla, 1980, Shukla and McDonald, 1980, McDonald and McDonald, 1981]. However, the perturbation theory may underestimate both the specific heat at high temperature and the occurrence of large (but non-melting) atomic displacements [McDonald and Mountain, 1979]. Also, the method is incapable of dealing with instabilities.

Self consistent Phonon Theory [Choquard, 1967, Gillis et al., , 1968 Werthamer, 1970, Glyde and Klein, 1971, Horner, 1974]. This is an attempt to combine the self consistency ideas in the cell models with an extended version of thermodynamic perturbation theory, to give a theory which, though quasi-harmonic in the sense that the lattice dynamics are still described in terms of phonons of infinite lifetime (though this restriction may be relaxed by including the effects of residual effective interactions), the phonon frequencies are obtained self consistently, and vary with the lattice constant and temperature in a well defined way. In fact, one uses a thermally averaged potential energy

$$<\varphi(\underline{R})>= \frac{1}{((2\pi)^3 G)^{\frac{1}{2}}} \int d^3\underline{u}\, \varphi(\underline{R}+\underline{u}) e^{-\underline{\underline{G}}^{-1} : \underline{u}\,\underline{u}},$$

where $G \sim < uu >$ is the displacement - displacement correlation function. In this way one obtains a theory valid for all temperatures up to T_m, which is also capable of dealing with instabilities and (with a little modification) hard cores. It is to the exposition and application of self consistent phonon theory that this article will be devoted.

A drastic simplification of self consistent phonon theory, brought about by defining a single average phonon frequency, and neglecting correlations between the motions of different atoms, produces results directly related to those of the cell models [Shukla et al., 1981].

There are several other techniques, at various stages of development, which we will not have space to consider here in any detail. There is an alternative perturbation theory which claims to include phonons and short range forces simultaneously [Horner, 1970], as does the collective mode approach [Fredkin and Werthamer, 1965, Gillis and Werthamer, 1968]. The Jastrow theory [Jastrow, 1955, Morita and Hiroike, 1961, DeDominicis, 1962] deals with hard cores directly, and has been developed to a sophisticated level in the guise of the hypernetted chain method [see Ripka, 1979, and references therein]. There is also a self consistent field method [Kerr and Sjölander, 1970, Zubov, 1970-1981] and a 'localized crystal' method [Yukalov, 1979-81] which are essentially direct developments of cell model ideas.

This review will be organized as follows. Firstly we discuss the dynamics of a perfect crystal at finite temperature using a path integral representation and the proper-time formalism [Matsubara, 1955] - this section will also serve to set up most of the notation for what follows. Then, in Section I we consider the harmonic approximation, and develop the thermodynamic perturbation series in terms of phonon propagators (Greens functions) and interaction vertices. After considering the usual truncations of this series, and its thereby limited region of validity, we consider the evaluation of the free energy and phonon self energy.

Section II deals with the derivation of the self consistent phonon scheme, via the resummation of infinite sets of Feynman diagrams to obtain a new perturbation series with improved convergence properties. This is then made self consistent by the introduction of model force constants, suitably adjusted to eliminate self energy insertions.

In the next section we consider firstly some of the truncations which can be made to the self consistent phonon expansion to make calculations tractable. We then discuss (though only qualitatively) the convergence and stability of the self consistency iterations, and this will allow us to make a qualitative assessment of the schemes accuracy and domain of validity. We also look at the difficulties associated with short range repulsive potentials (the 'hard core problem') and their resolution.

Section IV deals with the application of self consistent phonons to the calculation of crystal properties. Firstly we look at the microscopic variables used to describe the system, and their relation to the thermodynamic quantities of interest. General expressions are then derived for these in various approximations. We also discuss time-dependent properties, and in particular phonon damping, useful for the interpretation of thermal neutron scattering data. Then in Section V we will show how to apply self consistent phonon techniques to real crystals, and in particular consider non-Bravais lattices, and the inclusion of n-body forces in ionic crystals by a generalization of the harmonic shell model. Finally we offer a discussion of some of the practical problems in the numerical evaluation and solution of the self consistent phonon equations.

Although primarily intended for the solid state physicist, it is also hoped that this article may also be of some interest to the particle physicist and field theorist. In particular, the infrared resummations performed here to obtain a self consistent particle propagator have many applications in field theory; there are close analogies with the gap equations of dynamical symmetry breakdown [Jackiw et al., 1973, 1974, Cornwall and Norton, 1973], and the Gaussian effective potential for $\lambda\varphi^4$ theory [Bardeen and Moshe, 1973, Stevenson, 1984, 1985]. An extension of the complete quantum mechanical scheme presented here to φ^3 in two dimensions has recently been completed by Ingermanson (1986). The summation of cocoon diagrams necessary for a phonon system with arbitrary interatomic potential is of direct relevance to nonpolynomial field theories, and in particular σ -models, massive gauge theories and quantum gravity [Salam et al., 1971, 1972, Honerkamp et al., 1971-1973], the full implications for which may not yet have been realized. There are also analogies to be made with infinite component or bilocal field theories. Finally, the quantum mechanics of phonons is an ideal place to address the problems of analytic continuations to real time inherent in finite temperature quantum field theory and (in a different guise) in any Euclidean approach to nonperturbative quantum field theory.

The Perfect Crystal

Consider a perfect lattice of atoms or ions, labelled by an index l, with positions $\underline{R}_l$. Then we use the notation

$$\begin{aligned} \rho &\equiv l' - l, \\ \text{so that} \qquad \underline{R}_\rho &\equiv \underline{R}_{l'} - \underline{R}_l \equiv \eta_\rho \underline{R}_l, \end{aligned} \tag{1}$$

where η_ρ is a displacement operator. Similarly, we let $\underline{u}_l$ be the displacement from equilibrium of atom l, and

$$\underline{u}_\rho \equiv \underline{u}_{l'} - \underline{u}_l \equiv \eta_\rho \underline{u}_l. \tag{2}$$

If the atoms form a Bravais lattice, we may define the usual Fourier decompositions

$$\begin{aligned} \underline{u}_{\underline{q}} &\equiv \frac{1}{\sqrt{N}} \sum_l e^{-i\underline{q}\cdot\underline{R}_l} \underline{u}_l, \\ \text{and} \qquad \underline{u}_l &= \frac{1}{\sqrt{N}} \sum_{\underline{q}} e^{i\underline{q}\cdot\underline{R}_l} \underline{u}_{\underline{q}}, \end{aligned} \tag{3}$$

where for an infinite lattice $\sum_{\underline{q}}$ denotes Brillouin zone integration. Then

$$\underline{u}_\rho = \underline{u}_{\ell'} - \underline{u}_\ell = \frac{1}{\sqrt{N}} \sum_{\underline{q}} \eta_\rho(\underline{q}) e^{i\underline{q}\cdot\underline{R}_\ell} \underline{u}_{\underline{q}}, \tag{4}$$

with

$$\eta_\rho(\underline{q}) \equiv e^{i\underline{q}\underline{R}_\rho} - 1. \tag{5}$$

If we have a set of basis vectors $\underline{e}^\lambda(\underline{q})$, with

$$\underline{e}^\lambda . \underline{e}^{\lambda' *} = \delta^{\lambda\lambda'}, \qquad \sum_\lambda \underline{e}^\lambda \underline{e}^{\lambda *} = \underline{\underline{1}}, \tag{6}$$

we may expand

$$\underline{u}_{\underline{q}} = \sum_\lambda \underline{e}^\lambda(\underline{q}) u_{\underline{q}\lambda}. \tag{7}$$

We will often use the notation $q \equiv (\underline{q}, \lambda)$, so that $\underline{u}_q \equiv \underline{u}_{\underline{q}\lambda}$ etc.

The generalization to non-Bravais lattices (and thus also to finite molecules) in straightforward but tedious, and will be deferred to §5.1; in what follows now we will always assume we are dealing with a Bravais lattice for notational simplicity.

Now to proceed to a dynamical description of a real solid, we will use the adiabatic approximation, that is we will assume that electron-phonon interactions are small [Born and Oppenheimer, 1927]. Then we may solve for the motion of the electrons, assuming the nuclei to be fixed, and use these solutions to supply an effective potential $V[u_\ell]$ for the motion of the nuclei [Meissner, 1970]. This approximation is generally a very good one as the electron mass is much less than the nuclear mass [Moshinsky and Kittel, 1968].

The nuclear Hamiltonian is thus

$$H[\underline{u}_\ell] = \frac{1}{2M} \sum_\ell \underline{p}_\ell^2 + V[\underline{u}_\ell], \tag{8}$$

where the conjugate momenta $\underline{p}_\ell \equiv M\partial \underline{u}_\ell / \partial t$. The dynamics of the crystal is then summarized by the real time Greens functions

$$G_n(t_1....t_n) \equiv Tr\, e^{-\beta H} T_c u(t_1)...u(t_n),$$

where $\beta \equiv 1/kT$, the functional trace is a sum over the complete space of states $|u(t)>$, T_c denotes time ordering, and we have dropped the suffix l for convenience. These Greens functions may be generated by the functional

$$Z[J] = Tr\ e^{-\beta H} T_c exp\left(\frac{i}{\hbar}\int_{-T}^{T} dt\ J(t)u(t)\right),$$

where $-T < t_1, t_2, ...t_n < T$, and $J(t)$ are external sources. This may in turn be conveniently expressed as a path integral [Feynman, 1948, 1953, 1965, 1973];

$$Z[J] = \int [\mathcal{D}u] < u(0) \mid e^{-\beta H} T_c exp\left(\frac{i}{\hbar}\int_{-T}^{T} dt\ J(t)u(t)\right) \mid u(0) > .$$

Now the states $\mid u(t)>$ evolve in time according to the Schrödinger equation

$$i\hbar\frac{\partial}{\partial t} \mid u(t) >= H \mid u(t)>$$

so that $\mid u(t) >= e^{-iHt/\hbar} \mid u(0) >$, for a time independent Hamiltonian. It is thus useful to analytically continue the space of states to the complex plane; we then have in particular that $\mid u(-i\beta\hbar) >= e^{-\beta H} \mid u(0) >$, so that generalizing the procedure adopted at T=0 [Abers and Lee, 1973], we may write [Niemi and Semenoff, 1984]

$$Z[J] = \int [\mathcal{D}u] < u(-i\beta\hbar) \mid T_c exp(\frac{i}{\hbar}\int_{-T}^{T} dt\ J(t)u(t)) \mid u(0)>$$

$$= \int [\mathcal{D}u]_C\ exp\ \frac{i}{\hbar}\int_C dt(L[u(t)] + J(t)u(t)), \qquad (9)$$

where the Lagrangian

$$L[u(t)] = \frac{1}{2}M\left(\frac{\partial u}{\partial t}\right)^2 + V[u(t)], \qquad (10)$$

the field $u(t)$ is periodic

$$u(t) = u(t - i\beta\hbar),$$

and the contour C goes from $-T$ to $-T - i\beta\hbar$, passing through the points $t_1, t_2, ...t_n$ on the real axis [Mills, 1969].

Now for static situations at finite temperature, it is sufficient to take the contour C from the origin to -$i\beta\hbar$ down the imaginary axis, since Green's functions with real time agreements are not required [Matsubara, 1955]. Letting $\tau = -it$ (τ is known as the "proper-time" or "imaginary time"), equations (9) and (10) become

$$Z[J] = \int [Du]_\beta \, exp\left(-\tfrac{1}{\hbar}S[u] + J \cdot u\right), \tag{11}$$

where the action

$$S[u] \equiv \int_0^{\beta\hbar} d\tau \left(\tfrac{1}{2}M\dot{u}(\tau)^2 + V[u(\tau)]\right), \tag{12}$$

$$J \cdot u \equiv \int_0^{\beta\hbar} d\tau J(\tau)u(\tau),$$

and the path integral measure is now the product of integrations over all paths $\{u_\ell(\tau) : 0 \leq \tau \leq \beta\hbar, u_\ell(\tau) = u_\ell(\tau + \beta\hbar)\}$:

$$\int [\mathcal{D}u]_\beta \equiv \prod_\ell \int ds_\ell \int [\mathcal{D}u_\ell(\tau)] \,|_{u_\ell(0)=u_\ell(\beta\hbar)=s_\ell} \,.$$

In principle the representation (11), (12) may be used to determine the generating functional $Z[J]$ for all values of t by analytic continuation from the imaginary axis [Baym and Mermin, 1961, Abrikosov et al., 1963]. In practice however this analytic continuation may be very difficult to realize, and the proper-time formalism is generally restricted to static situations (we will return to this in §4.3).

The temperature Green's functions $G_n(\tau_1....\tau_n)$ are defined in analogy with the real time Greens functions $G_n(t_1....t_n)$ by

$$\begin{aligned} G_n(\tau_1...\tau_n) &= Tre^{-\beta H} T_c u(\tau_1)....u(\tau_n) \\ &\equiv \langle u(\tau_1)...u(\tau_n) \rangle \\ &= \prod_{i=1}^{n} \frac{\delta}{\delta J(\tau_i)} Z[J] \,|_{J=0}, \end{aligned} \tag{13}$$

where we have used the notation

$$<f[u]>\equiv \int [\mathcal{D}u]_\beta exp(-\tfrac{1}{\hbar}S[u])f[u].$$

The temperature Green's functions together contain (at least in principle) all information about the dynamics of the crystal, and in particular phonon scattering amplitudes. For most thermodynamical applications, however, it is sufficient to evaluate the free energy

$$F \equiv -\beta^{-1}\ln Z[0]. \tag{14}$$

It is also useful to define the connected Greens functions $G_n^c(\tau_1....\tau_n)$ by

$$G_n^c(\tau_1....\tau_n) \equiv \prod_{i=1}^{n}\left(\frac{\delta}{\delta J(\tau_i)}\right)\ln Z[J]\,|_{J=0}, \tag{15}$$

so that if

$$\begin{aligned} < f[u]>_C &\equiv \frac{1}{Z[0]}\int [\mathcal{D}u]_\beta exp(-\tfrac{1}{\hbar}S[u])f[u], \\ G_n^c(\tau_1....\tau_n) &= < u(\tau_1)...u(\tau_n) >_C . \end{aligned}$$

In perturbation theory it is easy to see that G_n^c contain only connected diagrams: if W[J] generates the connected diagrams then since

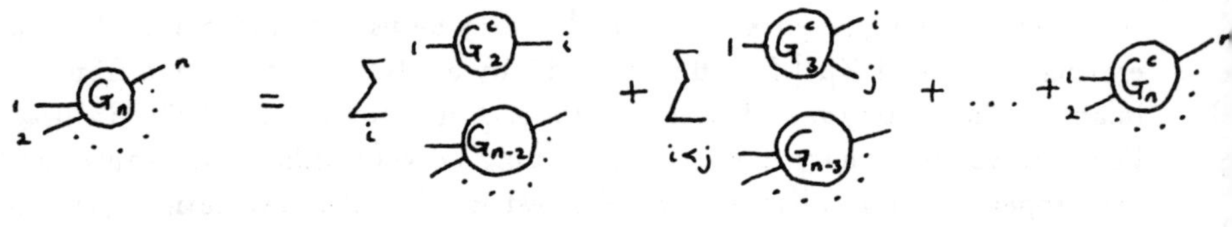

summing from $n = 1\ldots\infty$

$$\frac{\delta Z[J]}{\delta J(x_1)} = Z[J]\frac{\delta W[J]}{\delta J(x_1)},$$

whence $$W[J] = \ln Z[J].$$

Note that $$F = -\beta^{-1}\ln Z[0] = -\beta^{-1}G_o^c.$$

Finally we introduce the generating functional $\Gamma[\bar{u}]$ defined by a Legendre transformation

$$\Gamma[\bar{u}] \equiv W[J] - \bar{u} \cdot J,$$

so that
$$\frac{\delta W}{\delta J} = \bar{u} \quad \text{and} \quad \frac{\delta \Gamma}{\delta \bar{u}} = J. \tag{16}$$

$\bar{u} = <u>$ may be thought of as the classical value of the field u in an external source J. $\Gamma[\bar{u}]$ is the effective action functional for this field. It is easy to see that

$$e^{-\Gamma[\bar{u}]} = \frac{1}{Z[0]} \int [\mathcal{D}u]_\beta e^{-\frac{1}{\hbar} S[\bar{u}+u]}. \tag{17}$$

We define the proper Greens functions to be

$$G_n^P(\tau_1....\tau_n) \equiv \prod_{i=1}^n \frac{\delta}{\delta \bar{u}(\tau_i)} \Gamma[\bar{u}] \mid_{\bar{u}=o} . \tag{18}$$

In perturbation theory, one may show that G_n^P contain only proper diagrams - diagrams that cannot be cut into two pieces by cutting an internal line. This is essentially because

$$\frac{\delta}{\delta \bar{u}} = \frac{\delta J}{\delta \bar{u}} \frac{\delta}{\delta J} = -\frac{\delta^2 \Gamma}{\delta \bar{u} \delta \bar{u}} \frac{\delta}{\delta J} = \left(\frac{\delta \bar{u}}{\delta J}\right)^{-1} \frac{\delta}{\delta J} = \left(\frac{\delta^2 W}{\delta J \delta J}\right)^{-1} \frac{\delta}{\delta J} = G_2^{c-1} \frac{\delta}{\delta J}.$$

So $\delta/\delta\bar{u}$ acting on Γ adds an external line and removes a propagator G_2^c from this line. This continual amputation of propagators is what keeps the diagrams proper.

1 Weak Anharmonicity

1.1 Free Bare Phonons

Consider a harmonic potential

$$V_0[u(\tau)] = \tfrac{1}{2}\sum_{\rho\ell} \underline{u}_\rho(\tau)^T.\underline{\underline{\mathcal{D}}}_\rho.\underline{u}_\rho(\tau), \tag{1.1}$$

so that the harmonic action is (in momentum space)

$$S_0[u(\tau)] = \sum_q \int_0^\beta d\tau \left[\tfrac{1}{2}M\underline{\dot{u}}_q(\tau).\underline{\dot{u}}_q(\tau) + \tfrac{1}{2}M\underline{u}_q(\tau)^T.\underline{\underline{\mathcal{D}}}_q.\underline{u}_{-q}(\tau)\right], \tag{1.2}$$

where the dynamical matrix

$$\underline{\underline{\mathcal{D}}}_q = \frac{1}{M}\sum_\rho \tfrac{1}{2}\eta_\rho(\underline{q})\eta_\rho(-\underline{q})\underline{\underline{\mathcal{D}}}_\rho.$$

The classical equation of motion is then

$$-\underline{\ddot{u}}_q + \underline{\underline{\mathcal{D}}}_q.\underline{u}_q = 0. \tag{1.3}$$

If we diagonalize the dynamical matrix by a suitable choice of $\underline{e}^\lambda(\underline{q})$ so that

$$\underline{\underline{\mathcal{D}}}_q.\underline{e}^\lambda(\underline{q}) = \omega_{\underline{q}\lambda}^2\underline{e}^\lambda(\underline{q}) \tag{1.4}$$

then using the expansion (7),

$$S_0[u(\tau)] = \sum_q \int_0^\beta d\tau M\left[\tfrac{1}{2}\dot{u}_q\dot{u}_{-q} + \tfrac{1}{2}\omega_q^2 u_q u_{-q}\right], \tag{1.5}$$

and the equation of motion takes on the simple form

$$-\ddot{u}_q + \omega_q^2 u_q = 0, \tag{1.6}$$

for each mode independently.

The partition function Z_0 may now be evaluated by (functional) Gaussian integration (11)): setting initially the sources J to zero we have

$$Z_0[0] = \int [\mathcal{D}u] exp\left(-\sum_q \int_0^\beta d\tau M \tfrac{1}{2} u_q(-\partial_\tau^2 + \omega_q^2) u_{-q}\right)$$

$$= \prod_q \left[\frac{det(-\partial_\tau^2 + \omega_q^2)}{det(-\partial_\tau^2)}\right]^{-\frac{1}{2}} \tag{1.7}$$

The division by $det(-\partial_\tau^2)$ ensures that the functional integration is well defined, with $Z_0 = 1$ when $\omega_q \equiv 0$; zero modes are ignored in defining the functional determinant. To evaluate (1.7), we use a Fourier series representation

$$u_q(\tau) = \sum_\nu u_{q\nu} e^{i\Omega_\nu \tau \hbar}, \tag{1.8}$$

where

$$\Omega_\nu \equiv \frac{2\pi\nu}{\hbar\beta}. \tag{1.9}$$

Then

$$S_0[u] = \beta \sum_\nu \sum_q \tfrac{1}{2} u_{q\nu}(\Omega_\nu^2 + \omega_q^2) u_{-q\nu}, \tag{1.10}$$

so

$$\begin{aligned} Z_0[0] &= \prod_q \left\{ \prod_\nu (\Omega_\nu^2 + \omega_q^2)^{-\frac{1}{2}} / \prod_{\nu \neq o} (\Omega_\nu^2)^{-\frac{1}{2}} \right\} \\ &= \prod_q \omega_q^{-1} \prod_{\nu > o} \left(1 + \frac{\omega_q^2}{\Omega_\nu^2}\right)^{-1} \\ &= \prod_q (2 \sinh \tfrac{1}{2}\beta_q)^{-1}, \end{aligned} \tag{1.11}$$

where

$$\beta_q \equiv \beta\hbar\omega_q, \tag{1.12}$$

and we have used the representation

$$\sinh x = x \prod_{k=1}^{\infty} \left(1 + \frac{x^2}{k^2 \pi^2} \right) . \tag{1.13}$$

The free energy is thus given by taking the logarithm of (1.12):

$$\begin{aligned} F_0 &= \beta^{-1} \sum_{2} \ln(2 \sinh \tfrac{1}{2} \beta_q) \\ &= \beta^{-1} \sum_{q} \left(\tfrac{1}{2} \beta_q + \ln(1 - e^{-\beta_q}) \right) , \end{aligned} \tag{1.14}$$

the first term being that due to zero point motion.

The full partition function $Z_0[J]$ may now be evaluated by a change of variables in the (functional) Gaussian integration, to give

$$Z_0[J] = Z_0[0] exp \left(\tfrac{1}{2} \sum_{q\nu} \frac{1}{M} J_{q\nu} G_{q\nu} J_{-q\nu} \right) , \tag{1.15}$$

where $J_{q\nu}$ are the Fourier transforms of the sources $J_\rho(\tau)$, and $G_{q\nu}$ is essentially the inverse of $\Omega_\nu^2 + \omega_q^2$:

$$G_{q\nu} = \beta^{-1} (\Omega_\nu^2 + \omega_q^2)^{-1} . \tag{1.16}$$

Thus

$$G_q(\tau) = \sum_{\nu} G_{q\nu} e^{i\Omega_\nu \tau \hbar} \tag{1.17}$$

is the Green's function for the operator $(-\partial_\tau^2 + \omega_q^2)$:

$$(-\partial_\tau^2 + \omega_q^2) G_q(\tau) = \delta(\tau) . \tag{1.18}$$

The free propagator $G_q(\tau)$ is then the only non-trivial connected Green's function of the harmonic theory (compare (1.15) with (15)), and

$$G_q(\tau - \tau') = < u_q(\tau) u_q(\tau') >_0 , \tag{1.19}$$

where $<>_0$ denotes a (connected) expectation value taken with the harmonic action S_0.

Performing the summation (1.17) explicitly, we get

$$G_q(\tau) = \frac{\hbar}{2\omega_q} \frac{\cosh(\tfrac{1}{2}\beta - |\tau|)\hbar\omega_q}{\sinh \tfrac{1}{2}\beta_q} , \tag{1.20}$$

and in particular

$$G_q(0) = \frac{\hbar}{2\omega_q} \coth \tfrac{1}{2}\beta_q. \tag{1.21}$$

It will prove useful later to define the position space Green's functions (or 'linking correlation functions')

$$\begin{aligned} \underline{\underline{G}}^{\sigma}_{\rho\rho'}(\tau - \tau') &\equiv < \eta_\rho \underline{u}_\ell(\tau) \eta_{\rho'} \underline{u}_{\ell'}(\tau') >_0 . \\ &= < (\underline{u}_{\ell''}(\tau) - \underline{u}_\ell(\tau))(\underline{u}_{\ell'''}(\tau') - u_{\ell'}(\tau')) >_0 . \\ &= \frac{1}{MN} \sum_{\underline{q}\nu} e^{-i\Omega_\nu \tau \hbar} e^{i\underline{q}.\underline{R}_\sigma} \eta_\rho(\underline{q}) \eta_{\rho'}(-\underline{q}) \underline{\underline{G}}_{\underline{q}\nu}, \end{aligned} \tag{1.22}$$

where $\sigma \equiv \ell - \ell'$, and

$$\underline{\underline{G}}_{\underline{q}\nu} = \beta^{-1}(\underline{\underline{1}}\Omega^2_\nu + \underline{\underline{D}}_{\underline{q}})^{-1} = \sum_\lambda \underline{e}^\lambda(\underline{q}) \underline{e}^\lambda(-\underline{q}) G_{q\nu}. \tag{1.23}$$

We may also define the 'self linked correlation functions'

$$\begin{aligned} \underline{\underline{G}}_\rho \equiv \underline{\underline{G}}^o_{\rho\rho}(0) &= < \eta_\rho \underline{u}_\ell \eta_\rho \underline{u}_\ell >_0 \\ &= < (\underline{u}_{\ell'} - \underline{u}_\ell)(\underline{u}_{\ell'} - \underline{u}_\ell) >_0 \\ &= \frac{1}{MN} \sum_{\underline{q}\nu} \eta_\rho(\underline{q}) \eta_\rho(-\underline{q}) \underline{\underline{G}}_{\underline{q}\nu} \\ &= \frac{2}{MN} \sum_{\underline{q}\nu} (1 - cos \underline{q}.\underline{R}_\rho) \underline{\underline{G}}_{\underline{q}\nu}. \end{aligned} \tag{1.24}$$

Now the discussion offered above for the harmonic action (1.2), may be usefully extended to a more general harmonic action in which the dynamical matrix depends on the proper time τ, viz

$$\begin{aligned} S_0[u] &= \sum_{\underline{q}} \Big\{ \int_o^\beta d\tau \tfrac{1}{2} M \dot{\underline{u}}_{\underline{q}}(\tau).\dot{\underline{u}}_{-\underline{q}}(\tau) \\ &+ \beta^{-1} \int_o^\beta d\tau \int_o^\beta d\tau' \, \tfrac{1}{2} \underline{u}_{\underline{q}}(\tau).\underline{\underline{D}}_{\underline{q}}(\tau - \tau').\underline{u}_{-\underline{q}}(\tau') \Big\}. \end{aligned} \tag{1.25}$$

We now have

$$Z_0[0] = \prod_q \omega^{-1}_{qo} \prod_{\nu > o} (1 + \frac{\omega^2_{q\nu}}{\Omega^2_\nu})^{-1}$$

where

$$\underline{\underline{\mathcal{D}}}_{q\nu}(\tau) = \sum_{\nu} \underline{\underline{\mathcal{D}}}_{q\nu} e^{i\Omega_\nu \tau \hbar},$$

and $\omega^2_{q\nu}$ are the eigenvalues of $\underline{\underline{\mathcal{D}}}_{q\nu}$. Then the free energy

$$F_0 = -\beta^{-1} \ln Z_0[0] = \tfrac{1}{2}\beta^{-1} \sum_{q\nu} \ln(\Omega^2_\nu + \omega^2_{q\nu}) + const., \tag{1.26}$$

where the (infinite) constant is $-3N\beta^{-1} \sum_{\nu>o} \ln \Omega^2_\nu$ and renders the expression (1.26) finite. The propagator is now

$$G_{q\nu} = \beta^{-1}(\Omega^2_\nu + \omega^2_{q\nu})^{-1} \tag{1.27}$$

so that (1.26) may be written

$$F_0 = -\tfrac{1}{2}\beta^{-1} \sum_{q\nu} tr \ln \beta \underline{\underline{G}}_{q\nu} + const. \tag{1.28}$$

Note that now that since $\underline{\underline{\mathcal{D}}}_q$ is non-local (ie depends on $\tau - \tau'$) we may no longer perform explicitly the summations over ν to obtain results in closed form analogous to (1.14), (1.20) and (1.21). However in the limits of high temperature (the 'classical limit') and low temperature (the 'quantum limit') we have

$$G_q(\tau) \sim \begin{cases} \frac{1}{\beta\omega^2_{qo}} & \frac{\hbar\omega_{q\nu}}{kT} \ll 1 \quad \text{classical} \\ \frac{\hbar}{2\omega_{qo}} e^{-\tau\hbar\omega_{qo}} & \frac{\hbar\omega_{q\nu}}{kT} \gg 1 \quad \text{quantum} \end{cases}$$

1.2 Perturbation Theory

Consider a weakly anharmonic potential of the form

$$V[u(\tau)] = V_0[u(\tau)] + V_1[u(\tau)] \tag{1.29}$$

where for an appropriate measure the perturbation $| V_1 | \ll | V_0 |$, at least for small u. We may thus expect the perturbation theory that follows to be useful at low temperatures in weakly anharmonic crystals.

Now (II) reads

$$\begin{aligned} Z[J] &= \int [\mathcal{D}u]\, exp\{-(S_0[u]+S_1[u])+J.u\} \\ &= exp(-S_1[\frac{\delta}{\delta J}]) \int [\mathcal{D}u]\, exp(-S_0[u]+J.u) \\ &= exp(-S_1[\frac{\delta}{\delta J}]) Z_0[J], \end{aligned}$$

since

$$\int [\mathcal{D}u] f[u] g[u] e^{J\cdot u} = f[\frac{\delta}{\delta J}] \int [\mathcal{D}u] g[u] e^{J\cdot u}.$$

Making use of the identity

$$F[-\frac{\delta}{\delta x}] G[x] = G[-\frac{\delta}{\delta y}] F[y] e^{-xy} \mid_{y\equiv 0},$$

we have

$$Z[J] = Z_0[-\frac{\delta}{\delta u}] exp(-S_1[u] - J\cdot u) \mid_{u=0},$$

and we can now use (1.15) to give

$$\begin{aligned} Z[J] &= Z_0[0] exp\left(\tfrac{1}{2}\sum_{q\nu} G_{q\nu} \frac{\delta}{\delta u_{q\nu}} \frac{\delta}{\delta u_{q\nu}}\right) exp\left(-\int_o^\beta d\tau V_1[u] + J\cdot u\right) \\ &= Z_0[0] exp\left(\sum_{\rho\rho'\sigma} \beta^{-2} \int_o^\beta d\tau \int_o^\beta d\tau' \underline{\underline{G}}^\sigma_{\rho\rho'}(\tau-\tau') : \tfrac{1}{2} \frac{\delta}{\delta \underline{u}_\rho(\tau)} \frac{\delta}{\delta \underline{u}_{\rho'(\tau')}}\right) \\ & exp\left(-\int_o^\beta d\tau V_1[u_\rho(\tau)] + \sum_\rho \int_o^\beta d\tau J_\rho(\tau) u_\rho(\tau)\right) \qquad (1.30) \end{aligned}$$

$$= Z_0[0]\Big(\sum_{n=o}^{\infty} \frac{1}{n!}\Big\{\sum_{\rho\rho'\sigma} \beta^{-2} \int_o^\beta d\tau \int_o^\beta d\tau' \underline{\underline{G}}^\sigma_{\rho\rho'}(\tau-\tau') : \tfrac{1}{2} \frac{\delta}{\delta \underline{u}_\rho(\tau)} \frac{\delta}{\delta \underline{u}_{\rho'}(\tau')}\Big\}^n\Big)$$

$$\begin{aligned} &\Big(\sum_{m=o}^{\infty} \frac{1}{m!}\Big\{-\int_o^\beta d\tau \sum_{r=o}^{\infty} (\frac{1}{r!} \sum_\rho \underline{u}_\rho \ldots \underline{u}_\rho \vdots \underline{\nabla} \ldots \underline{\nabla} v_\rho \mid_{u=0}) \\ &+ \sum_\rho \int_o^\beta d\tau J_\rho(\tau) u_\rho(\tau)\Big\}^m\Big), \qquad (1.31) \end{aligned}$$

where we have assumed for simplicity that V_1 is a pairwise potential, which may in turn be expanded in a Taylor series about $u = 0$;

$$V_1[u] = \sum_\rho v_\rho(\underline{R}_\rho + \underline{u}_\rho) = \sum_\rho \sum_{r=0}^{\infty} \frac{1}{r!} \underline{u}_\rho \ldots \underline{u}_\rho \vdots \underline{\nabla} \ldots \underline{\nabla} v_\rho \, |_{u=0} \, . \qquad (1.32)$$

It is now straightforward to read off from (1.31) the "Feynman rules" for phonon perturbation theory in position space:

To evaluate a particular (connected) Green's function $G^c_n(\tau_1 \ldots \tau_n)$ draw all (connected) diagrams with n external lines which are topologically distinct. Then to evaluate the contribution of each diagram we assign factors as follows:

a) Internal phonon lines

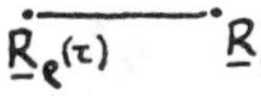

$\frac{1}{2}\underline{\underline{G}}^\sigma_{\rho\rho'}(\tau - \tau') = \underline{\nabla}\,\underline{\nabla}'$

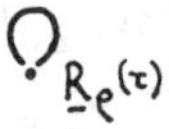

$$\tfrac{1}{2}\underline{\underline{G}}_\rho : \underline{\nabla}\,\underline{\nabla} \qquad (1.33)$$

b) Vertices

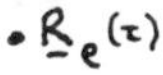

$\beta^{-1} \sum_\rho \int_o^\beta d\tau v_\rho$

c) Each diagram has a symmetry factor of n_s^{-1}, where n_s is the order of its symmetry group (the number of ways of permuting its internal propagators and vertices)

To calculate to a given order in the perturbation V_1, one must include all diagrams with not more than a given number of vertices. This is sometimes called the "cumulant expansion".

Taking the logarithm of (1.31) and setting J to zero, we have the free energy

$$F = F_o + \mathcal{H} = -\beta \ln Z_0[0] - \beta \sum_{m=1}^{\infty} \frac{(-\beta)^m}{m!} \chi_m, \qquad (1.34)$$

where χ_m are the usual cumulants of the cumulant expansion, and consist of all connected diagrams with m vertices, and no external legs ('bubble diagrams', or 'vacuum diagrams'). Since the connected Green's function may be obtained by functional differentiation of the logarithm of (1.31), they also have a cumulant expansion;

$$G^c_n(\tau_1 \ldots \tau_n) = \sum_{m=1}^{\infty} \frac{(-\beta)^m}{m!} \frac{\delta^n}{\delta J(\tau_1) \ldots \delta J(\tau_n)} \chi_m[J]. \qquad (1.35)$$

We may also obtain Feynman rules in momentum space by rewriting (1.31) in the form

$$Z[J] = Z_0[0](\sum_{m=o}^{\infty}\frac{1}{n!}\{\sum_{q\nu} G_{q\nu}\tfrac{1}{2}\frac{\delta}{\delta u_{q\nu}}\frac{\delta}{\delta u_{q\nu}}\}^n)(\sum_{m=0}^{\infty}\{-\beta\sum_{r=0}^{\infty}\frac{1}{r!}$$

$$\Big(\sum_{\substack{q_1\ldots q_r\\ \nu_1\ldots\nu_r}}\Delta(\underline{q}_1+\ldots+\underline{q}_r)\delta_{\nu_1+\ldots+\nu_1}u_{q_1\nu_1}\ldots u_{q_r\nu_r}V_r(q_1\ldots q_r)\Big)+\sum_{q\nu}J_{q\nu}U_{q\nu}\}^m) \qquad (1.36)$$

where

$$\Delta(\underline{q}_2+\ldots\underline{q}_r) = \begin{cases} 1 & \text{if } \underline{q}_1+\ldots\underline{q}_r = \underline{0} \\ 0 & \text{otherwise,} \end{cases}$$

$$\delta_{\nu_1+\ldots\nu_r} = \begin{cases} 1 & \text{if } \nu_1+\ldots+\nu_r = 0 \\ 0 & \text{otherwise.} \end{cases}$$

These factors come from the overall integral over τ and the sum over l in the second factor of (1.31), and express energy and momentum conservation at each vertex. The factor $V_r(q_1\ldots q_r)$ is defined as

$$V_r(q_1\ldots q_r) \equiv \sum_{\rho}\underline{e}^{\lambda_1}(\underline{q}_1)\eta_\rho(\underline{q}_1)\ldots\underline{e}^{\lambda_r}(\underline{q}_r)\eta_\rho(\underline{q}_r)\vdots\underline{\nabla}\ldots\underline{\nabla}v_\rho\,|_{u=o}\,. \qquad (1.37)$$

Thus the momentum space Feynman rules take the form

a) Internal phonon lines 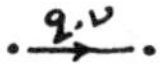 $\sum_{q\nu}\frac{1}{2}G_{q\nu}$

b) Vertices  $\Delta(\underline{q}_1+\ldots+\underline{q}_r)\delta_{\nu_1+\ldots+\nu_r}V_r(q_1,\ldots q_r)$

c) Symmetry factors as before.

(1.38)

Notice that we can use the energy-momentum conservation at each vertex to perform some of the $(\underline{q},\nu)$ sums. The remaining sums over ν may be performed by contour integrations of the form

$$\beta^{-1}\sum_{\nu}f(i\Omega_\nu) = \frac{1}{2\pi i}\int_c\frac{f(\omega)d\omega}{e^{\beta\omega}-1} \qquad (1.39)$$

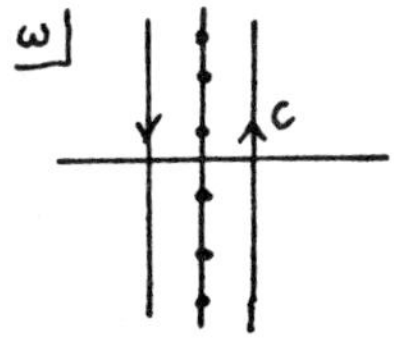

Although this is useful for providing simple expressions for low order diagrams [Cowley, 1971, Klein et al., 1969, Tripathi and Pathak 1974], for numerical calculations it is quicker to use the position space representation [Van Doren, 1973, Chattopadhyay et al., 1977]. This will also be more useful when we begin to sum over classes of diagrams in the next chapter.

For lattices in which each site is a centre of symmetry (and in particular a Bravais lattice), the propagator vanishes for $\underline{q} = \underline{0}$, and so all diagrams with $\underline{q} = \underline{0}$ lines vanish. In particular

 = 0.

We will assume throughout what follows that this is indeed the case.

The restriction to pairwise potentials is for convenience only, to avoid having to consider non-local vertices; if desired the results given above could be generalized to include more complicated potentials incorporating n-body forces.

1.3 Interacting Bare Phonons

Since the perturbation series involves an infinite set of terms (or Feynman diagrams), it is necessary to truncate it in some way in order to perform explicit calculations. The following procedures are those usually adopted:

A1) The pairwise potential

$$V = V_0 + V_1 = \sum_{\rho \ell} \phi_\rho, \tag{1.40}$$

is expanded in a Taylor series;

$$\phi_\rho(\underline{R}_\rho + \underline{u}_\rho) = \sum_{r=0}^{\infty} \frac{1}{r!} \underline{u}_\rho \ldots \underline{u}_\rho \vdots \underline{\nabla} \ldots \underline{\nabla} \phi_\rho \,|_{u=0} \,. \tag{1.41}$$

Then we make the identification

$$\begin{aligned} \underline{\underline{\mathcal{D}}}_\rho &\equiv \underline{\nabla}\,\underline{\nabla}\phi_\rho \\ v_\rho &= \phi_\rho - \tfrac{1}{2}\underline{u}_\rho . \underline{\underline{\mathcal{D}}}_\rho . \underline{u}_\rho, \end{aligned} \tag{1.42}$$

so that V_1 contains terms cubic in u and higher;

$$V_1 = \sum_{\rho\ell} v_\rho = \sum_{\rho\ell} \sum_{r=3}^{\infty} \frac{1}{r!} \underline{u}_\rho \cdots \underline{u}_\rho \vdots \underline{\nabla} \cdots \underline{\nabla} \phi_\rho . \tag{1.43}$$

We have assumed here that the static potential may be ignored, that the first derivative term vanishes (the equilibrium condition for the bare crystal is $\sum_\rho \underline{\nabla}\phi_\rho \mid_{u=0} = 0$), and that $\underline{\underline{D}}_\rho$ as defined in (1.42) is indeed positive definite, so that the equilibrium is stable and the phonon frequencies ω_q are real.

A2) The cumulant expansions, like (1.34) and (1.35), are truncated : we assume V_1 is small ($| V_1 | \ll | V_o |$) so that we only need consider diagrams with a small number of vertices.

A3) The Taylor expansion (1.43) is itself truncated, usually so as to include only three and four-point interactions;

$$V_1^t = \sum_{\rho\ell} \left\{ \tfrac{1}{3!} \underline{u}_\rho \underline{u}_\rho \underline{u}_\rho \vdots \underline{\nabla}\,\underline{\nabla}\,\underline{\nabla} \phi_\rho + \tfrac{1}{4!} \underline{u}_\rho \underline{u}_\rho \underline{u}_\rho \underline{u}_\rho \vdots \underline{\nabla}\,\underline{\nabla}\,\underline{\nabla}\,\underline{\nabla} \phi_\rho \right\} \tag{1.44}$$

In other words, the potential V is assumed to be very nearly harmonic, at least for small u.

In the next section we will show how all three of these approximations may be relaxed, which will lead us quite naturally to the self consistent phonon schemes.

Peierls Classification

For a small amplitude of vibration δ (so that $\lambda \equiv \delta/R \ll 1$, where R is such that $(\nabla)^n \phi \sim \phi R^{-n}$), we regard each r-point vertex of a connected diagram as $O(\lambda^{r-2})$. So a cubic term in V_1 is regarded as first order in the anharmonicity, a quartic term as second order, etc. A given diagram with p internal lines (propagators), n vertices, and e external lines is then of order $\lambda^{2(p-n)+e}$. Now such a diagram has $p - n + 1$ closed loops and, in the classical limit ($kT \gg \hbar\omega$), has temperature dependence $(\beta)^{-(p-n+1)}$. So in particular free energy diagrams of order λ^{2m} require $m + 1$ momentum space integrations, and have temperature dependence β^{-m+1}. This means that the

Peierls expansion (or loop expansion) in powers of λ will only be valid at low temperatures.

Free Energy Corrections to order λ^4

To lowest order (*ie* λ^2) the diagrams contributing to the free energy are

(a) (b)

which, using the position space Feynman rules (1.33) give contributions

$$\begin{aligned} \Delta F^{(a)} &= \tfrac{1}{2}\sum_{\rho}(\tfrac{1}{2}\underline{\underline{G}}_{\rho} : \underline{\nabla}\,\underline{\nabla})^2\phi_{\rho} \\ \Delta F^{(b)} &= -\frac{1}{3!}\sum_{\rho\rho'\sigma}\int_{o}^{\beta} d\tau(\tfrac{1}{2}\underline{\underline{G}}^{\sigma}_{\rho\rho'} : \underline{\nabla}\,\underline{\nabla}')\phi_{\rho}\phi_{\rho'}, \end{aligned} \tag{1.45}$$

or in momentum space (after performing the sums over ν)

$$\begin{aligned} \Delta F^{(a)} &= \tfrac{1}{8}\sum_{q_1 q_2} V_4(q_1, -q_1, q_2, -q_2)(2\bar{n}_1 + 1)(2\bar{n}_2 + 1), \\ \Delta F^{(b)} &= -\frac{1}{3!}\sum_{q_1 q_2 q_3} \Delta(q_1 + q_2 + q_3) \mid V_3(q_1, q_2, q_3) \mid^2 \end{aligned} \tag{1.46}$$

$$[\frac{(\bar{n}_1 + 1)(\bar{n}_2 + \bar{n}_3 + 1) + \bar{n}_2\bar{n}_3}{\omega_1 + \omega_2 + \omega_3} + 3\frac{\bar{n}_1\bar{n}_2 + \bar{n}_1\bar{n}_3 - \bar{n}_2\bar{n}_3 + \bar{n}_1}{\omega_2 + \omega_3 - \omega_1}],$$

where $\bar{n}_i \equiv e^{\beta\hbar\omega_i} - 1, \omega_i \equiv \omega_{q_i}$. These expressions have been evaluated explicitly for $NaCl$ [Cowley, 1971].

The next order (λ^4) diagrams are

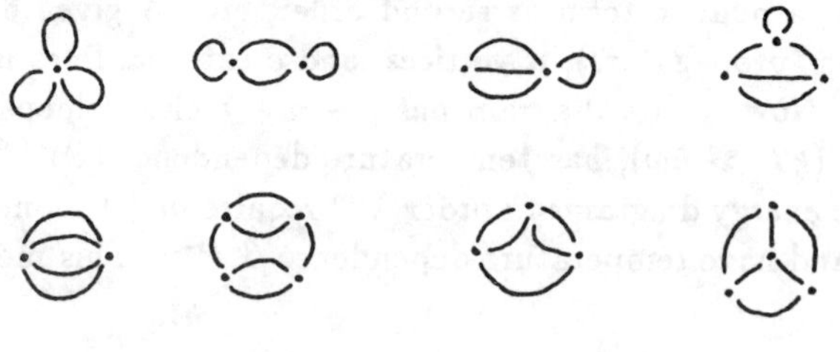

Expressions for these diagrams and an estimation of their relative magnitudes show that the cancellation between the diagrams is rather high, and their total contribution is as large as that of the order λ^2 diagrams at temperatures only $\frac{1}{3}$ of the melting temperature, even for quite weak anharmonicity [Shukla and Cowley, 1971].

So the Peierls expansion can only be expected to give reasonable results when

1. $\lambda \ll 1$ (very weak anharmonicity)
2. $kT \gg \hbar\omega$, so that quantum effects are small (for many quantum crystals $\mathcal{D}$ is no longer positive definite and convergence is nonexistent)
3. $T \ll T_m$, where T_m is the melting temperature.

Even when all of these conditions are met, the series generally converges rather slowly.

1.4 The Phonon Self Energy

Consider the full (renormalized) phonon propagator, or connected two-point function

$$\mathbf{G}(\tau - \tau') \equiv G_2^c(\tau, \tau') = \frac{\delta}{\delta J(\tau)} \frac{\delta}{\delta J(\tau')} \ln Z[J] \mid_{J=0} \tag{1.47}$$

This is a useful quantity to know when we are looking at real phonon propagation, for example in thermal neutron scattering. It will also be central to the setting up of the self consistent theory in the next section.

To calculate $\mathbf{G}$ in perturbation theory we need consider only proper diagrams (diagrams which cannot be separated into two pieces by cutting one internal phonon line). These are generated by $\Gamma[\bar{u}]$. We thus define the self energy as the sum of all proper diagrams with two external legs:

$$-\beta\Sigma(\tau - \tau') \equiv G_2^P(\tau, \tau') = \frac{\delta}{\delta \bar{u}(\tau)} \frac{\delta}{\delta \bar{u}(\tau')} \Gamma[\bar{u}] \mid_{\bar{u}=0} . \tag{1.48}$$

Then

$$\mathbf{G}_{q\nu} = G_{q\nu} - \beta G_{q\nu} \Sigma_{q\nu} \mathbf{G}_{q\nu} \tag{1.49}$$

or diagramatically

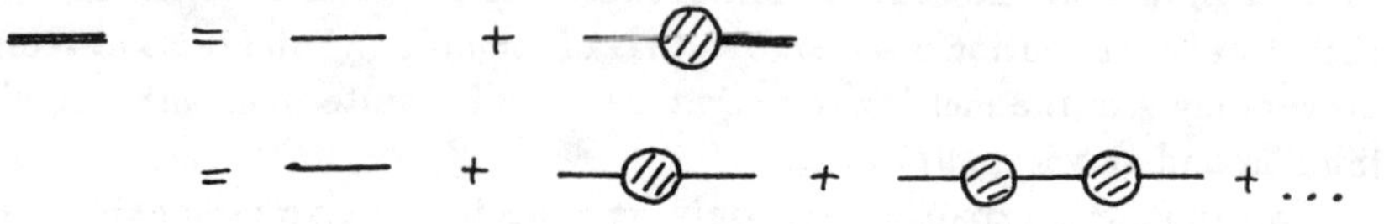

This is the Dyson equation for the full propagator in terms of the bare propagator and the self energy insertion: its solution is

$$\mathbf{G}_{q\nu} = \{G_{q\nu}^{-1} + \beta\Sigma_{q\nu}\}^{-1} = \beta^{-1}\{\mathcal{D}_q + \Sigma_{q\nu} + \Omega_\nu^2\}^{-1}, \tag{1.50}$$

So $\Sigma_{q\nu}$ is no more than a correction to the dynamical matrix $\mathcal{D}_q$, generated by the interactions.

To order λ^2 the proper diagrams contributing to $\Sigma_{q\nu}$ are

(a)

(b)

corresponding to the free energy graphs a) and b) (just cut one internal line), with contributions (in position space)

$$\begin{aligned} \underline{\underline{\Sigma}}^a_\rho &= \left(\tfrac{1}{2}\underline{\underline{G}}_\rho : \underline{\nabla}\,\underline{\nabla}\right)\underline{\nabla}\,\underline{\nabla}\phi_\rho, \\ (\underline{\underline{\Sigma}}^b)^\sigma_{\rho\rho'} &= \tfrac{1}{2}\left(\tfrac{1}{2}\underline{\underline{G}}^\sigma_{\rho\rho'} : \underline{\nabla}\,\underline{\nabla}'\right)^2 \underline{\nabla}\phi_\rho\underline{\nabla}'\phi_{\rho'}, \end{aligned} \tag{1.51}$$

or in momentum space

$$\Sigma^a_q = \tfrac{1}{2}\sum_{q\nu} V(q,-q,q_1,-q_1)(2\bar{n}_1+1), \tag{1.52}$$

$$\Sigma^b_{q\nu} = \tfrac{1}{2}\sum_{q_1 q_2} \mid V(q,q_1,q_2)\mid^2 \left\{\frac{\bar{n}_1+\bar{n}_2+1}{\omega_1+\omega_2+i\Omega_\nu} + \frac{\bar{n}_1-\bar{n}_2}{\omega_1-\omega_2+i\Omega_\nu} + \text{c.c.}\right\}$$

We shall consider the analytic continuation of these expressions to real frequencies in §4.3. Here we simply state that local diagrams of type a) are real, and simply shift the frequency spectrum ω_q, whereas non-local diagrams of type b) also develop an imaginary part when $i\Omega_\nu \to \omega + i\epsilon$, which damps the phonons, giving them a finite lifetime. Thus it is useful to separate all self energy insertions into two classes, local and non-local; those diagrams in which the external legs couple to the same and different vertices respectively.

To order λ^4 the self energy insertions are

Local:

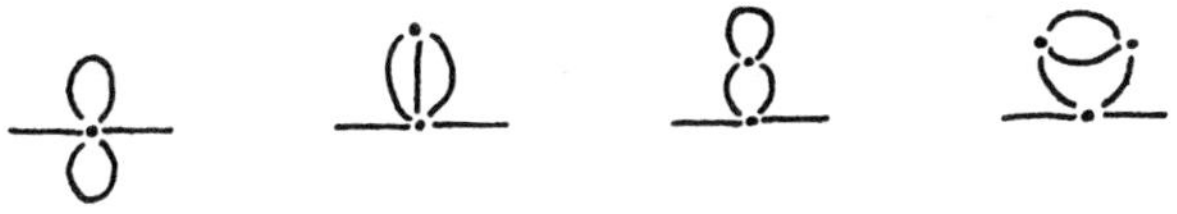

Non-local:

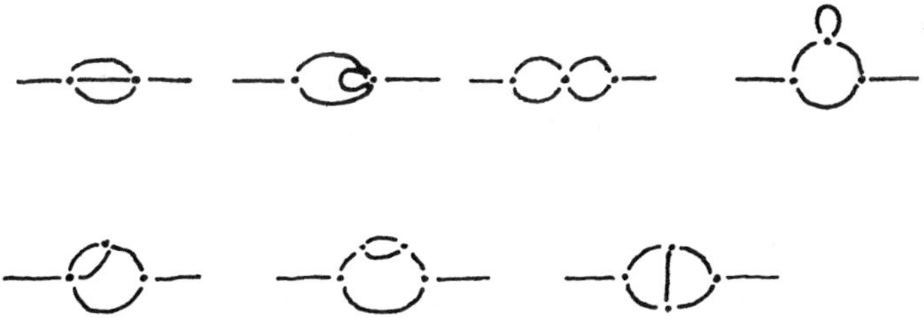

Explicit algebraic expressions for these diagrams may be found in [Tripathi and Pathak, 1974].

2 Strong Anharmonicity

In order to improve on the severe limitations of the 'weak anharmonicity' perturbation series described in the previous section, we will now explain how one can develop a new perturbation series which will

1. be equally valid at all temperatures below the melting point,
2. converge more rapidly, and
3. make sense even for quantum crystals where $\underline{\nabla}\,\underline{\nabla}\varphi$ is not necessarily positive definite.

We will do this in the following three separate stages (corresponding to the progressive relaxation of the assumptions A3, A2 and A1 respectively of § 1.3):

1. We sum in closed form all the diagrams contained in a particular cumulant (i.e. all diagrams with a fixed number of vertices but an arbitrarily large number of propagators - 'cocoon diagrams'). The truncation (1.44)(and indeed the Taylor expansion (1.41)) is thus unnecessary and the evaluation of each cumulant is valid at all temperatures.
2. To improve the convergence of the cumulant expansion, we sum up the class of diagrams which is formally the most divergent (the two particle reducible diagrams, or 'ring diagrams'). This may be regarded as a Dyson summation, or equivalently as a Legendre transformation.
3. We replace the ansatz (1.42) by introducing a set of model force constants which are then adjusted so that the low lying spectrum of the model harmonic system approximates as closely as possible the corresponding spectrum of the real system. This optimizes the ring summation by making the propagator self consistent; the modified perturbation theory with the model force constants has vanishing self energy insertions.

The resulting theory (the self consistent phonon expansion) contains as a first order approximation the more familiar renormalized harmonic and anharmonic theories (often called "self consistent phonons" and "improved self consistent phonons") and the derivation presented here is intended to clarify the connections between the various derivations in the literature [Horton, 1958, Boccarma and Sarma, 1965, Horner, 1967, Choquard, 1967, Koehler, 1966-68, Werthamer, 1970], and offer more insight into the nature of the expansion and the approximations involved in it.

2.1 Summing the Tadpoles and Cocoons

i) The first cumulant consists of all (connected) diagrams with only one vertex:

$\chi_1 \equiv$  $\equiv$  $+$  $\ldots \equiv \sum_{n\geq 2}$ 

So using the Feynman rules in position space,

$$\begin{aligned}
\chi_1 &= \tfrac{1}{2}\sum_{\rho}\sum_{n\geq 2}\tfrac{1}{n!}(\tfrac{1}{2}\underline{\underline{G}}_{\rho} : \underline{\nabla}\,\underline{\nabla})^n v_{\rho} \\
&= \tfrac{1}{2}\sum_{\rho}(exp(\tfrac{1}{2}\underline{\underline{G}}_{\rho} : \underline{\nabla}\,\underline{\nabla}) - 1 - \tfrac{1}{2}\underline{\underline{G}}_{\rho} : \underline{\nabla}\,\underline{\nabla})v_{\rho} \\
&\equiv \tfrac{1}{2}\sum_{\rho}\varepsilon_2(\tfrac{1}{2}\underline{\underline{G}}_{\rho} : \underline{\nabla}\,\underline{\nabla})v_{\rho} \\
&\equiv \tfrac{1}{2}\sum_{\rho}(< v_{\rho} > - v_{\rho} - \tfrac{1}{2}\underline{\underline{G}}_{\rho} : \underline{\nabla}\,\underline{\nabla} v_{\rho}),
\end{aligned} \tag{2.1}$$

where

$$\varepsilon_n(x) \equiv e^x - \sum_{r=o}^{n-1}\tfrac{1}{r!}x^r, \tag{2.2}$$

and the smeared potential

$$< v_{\rho} > \equiv exp(\tfrac{1}{2}\underline{\underline{G}}_{\rho} : \underline{\nabla}\,\underline{\nabla})v_{\rho}. \tag{2.3}$$

Using the Fourier transformation

$$\frac{1}{(2\pi)^3}\int d^3\underline{x}\int d^3\underline{k}\, e^{i\underline{k}.\underline{x}} = 1,$$

this becomes

$$< v_\rho > = \frac{1}{((2\pi)^3 det\underline{\underline{G}}_\rho)^{\frac{1}{2}}} \int d^0\underline{u}\; v(\underline{R}_\rho + \underline{u}) e^{-\frac{1}{2}\underline{u}.\underline{\underline{G}}_\rho^{-1}.\underline{u}}, \qquad (2.4)$$

which may be evaluated explicitly for most soft potentials (we shall consider potentials with hard cores later).

ii) Similarly the second cumulant is

$\chi_2 =$ 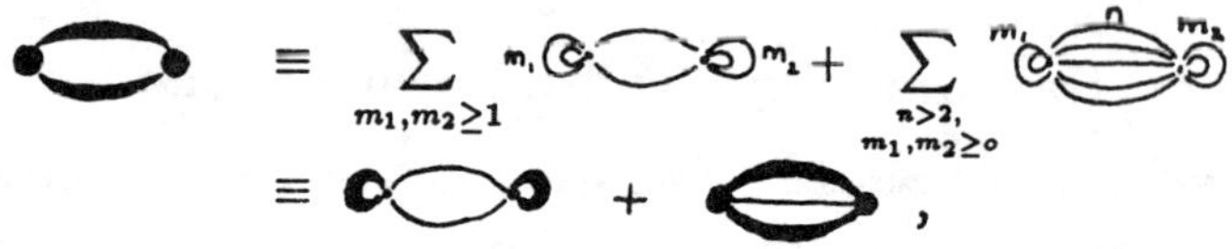

so that using position space Feynman rules again

$$\begin{aligned} \chi_2 &= -\sum_{\rho\rho'\sigma} \beta^{-1} \int_0^\beta d\tau \{ \tfrac{1}{2}(\underline{\underline{G}}^\rho_{\rho\rho'}(\tau) : \underline{\nabla}\,\underline{\nabla}')^2 (< v_\rho > - v_\rho)(< v_{\rho'} > - v_{\rho'}) \\ &+ \varepsilon_3(\underline{\underline{G}}^\rho_{\rho\rho'}(\tau) : \underline{\nabla}\,\underline{\nabla}') < v_\rho >< v_{\rho'} > \} \qquad (2.5) \end{aligned}$$

Again we may evaluate this by Fourier transforming, so that, for example,

$$exp(\tfrac{1}{2}\underline{\underline{G}}^\sigma_{\rho\rho'} : \underline{\nabla}\,\underline{\nabla}') < v_\rho >< v_{\rho'} > = \frac{1}{((2\pi)^6 det\mathcal{G})^{\frac{1}{2}}} \int d^3\underline{u} \int d^3\underline{u}'$$

$$v_\rho(\underline{R}_\rho + \underline{u}) v_\rho(\underline{R}_{\rho'} + \underline{u}') e^{-\frac{1}{2}u.\mathcal{G}^{-1}.u}$$

where

$$u \equiv \begin{pmatrix} \underline{u} \\ \underline{u}' \end{pmatrix} \qquad \text{and} \qquad \mathcal{G} \equiv \begin{pmatrix} \underline{\underline{G}}_\rho & \underline{\underline{G}}^\sigma_{\rho\rho'} \\ \underline{\underline{G}}^\sigma_{\rho'\rho} & \underline{\underline{G}}_{\rho'} \end{pmatrix}.$$

iii) We may write down similar, but more complicated expressions for all of the higher order cumulants similarly. For example

$\chi_3 =$ 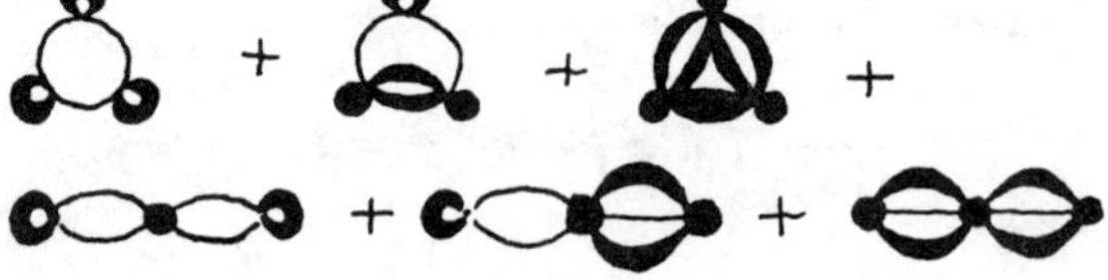

We may thus write down each cumulant in a closed form, which sums up each of its component 'tadpoles' and 'cocoons', and is furthermore valid at all temperatures (values of β) [Ranninger, 1965]. In particular as

a) T $\rightarrow 0$, $\beta \rightarrow \infty, G \sim \beta^{-1} \rightarrow 0$ (if we ignore quantum effects), and $\chi_k \sim (\beta^{-1})^n$, where $n = [\frac{3}{2}k]$, so the cumulant expansion is a power series in T, and as

b) $T \rightarrow \infty, \beta \rightarrow 0, G \sim \beta^{-1} \rightarrow \infty$, and $\chi_k \sim 1$, so the cumulant expansion is a power series in $\frac{1}{T}$ (at least for soft short range interactions).

In fact this high temperature region is one of giant motional fluctuations which has nothing to do with the solid state - it is the Mayer cluster expansion for an imperfect gas. So although each of the cumulants is well behaved over the full range of temperature variations, the sum will probably display a finite radius of convergence, corresponding to a phase transition. To take up this question it is necessary to sum up partial series of the cumulant expansion, which is what we will do next.

2.2 Summing the Rings

For a given number of vertices, one would expect that the most significant contributions arise when the vertices are most loosely connected (since we are interested in the region where G is (in some sense) small). So we separate the diagrams into classes according to the number of internal lines which must be cut in order to separate the diagrams into two disjoint pieces (this will be called 'reducing' the diagram). Thus, proper diagrams are diagrams which cannot be reduced by cutting one internal line. Improper diagrams with no external legs vanish for lattices in which every ion is a center of symmetry. So the most important set of diagrams which contribute to the cumulants are the 'ring' diagrams; these are all the diagrams which can be reduced by cutting two lines. They take the form

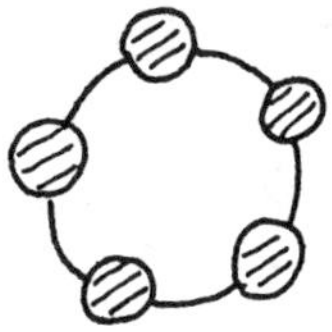

where  are self energy insertions. The remaining diagrams in the cumulants will be called 'irreducible'; in these diagrams at least three linking lines must emerge from each vertex. In particular, we define irreducible cumulants:

$\chi_1^{ir} = \chi_1 =$ 

$\chi_2^{ir} =$ 

$\chi_3^{ir} =$  $+$ 

We may obtain a closed expression for the sum of all the ring diagrams in a similar way to the Dyson summation:

$$\begin{aligned}
\chi^{(ring)} &= \sum_{r=2}^{\infty} \tfrac{1}{2}\beta^{-1}\tfrac{1}{r}\sum_{\underline{q}} tr \prod_{i=1}^{r}\{(\beta^{-1}\int_o^\beta d\tau_i)(-\beta\underline{\underline{\Sigma}}_{\underline{q}}(\tau_i)\underline{\underline{G}}_{\underline{q}}(\tau_i-\tau_{i+1}))\} \\
&= \sum_{r=2}^{\infty} \tfrac{1}{2}\beta^{-1}\tfrac{1}{r}\sum_{\underline{q}\nu} tr(-\beta\underline{\underline{\Sigma}}_{\underline{q}\nu}\underline{\underline{G}}_{\underline{q}\nu})^r \\
&= \tfrac{1}{2}\beta^{-1}\sum_{\underline{q}\nu} tr(-\beta\underline{\underline{\Sigma}}_{\underline{q}\nu}\underline{\underline{G}}_{\underline{q}\nu}) + \tfrac{1}{2}\beta^{-1}\sum_{\underline{q}\nu} tr\ln(1+\beta\underline{\underline{\Sigma}}_{\underline{q}\nu}\underline{\underline{G}}_{\underline{q}\nu}). \qquad (2.6)
\end{aligned}$$

Thus, using (1.30) and (1.50)

$$\begin{aligned}
F_o + \chi^{(ring)} + \tfrac{1}{2}\beta^{-1}\sum_{\underline{q}\nu} tr\beta\underline{\underline{\Sigma}}_{\underline{q}\nu}\underline{\underline{G}}_{\underline{q}\nu} &= \tfrac{1}{2}\beta^{-1}\sum_{\underline{q}\nu} tr\ln\beta^{-1}(\underline{\underline{G}}_{\underline{q}\nu}^{-1} + \beta\underline{\underline{\Sigma}}_{\underline{q}\nu}) + \text{const.} \\
&= -\tfrac{1}{2}\beta^{-1}\sum_{\underline{q}\nu} tr\ln\beta\mathbf{G}_{\underline{q}\nu} + \text{const.} \qquad (2.7)
\end{aligned}$$

Now naively one would perhaps expect the free energy (1.34) to be given by $F_o + \chi^{(ring)} + \mathcal{H}^{ir}$, where $\mathcal{H}^{ir}$ is the sum of the irreducible cumulants,

$$\mathcal{H}^{ir} \equiv -\beta^{-1}\sum_{m=1}^{\infty}\frac{(-\beta)^m}{m!}\chi_m^{ir}, \qquad (2.8)$$

evaluated using the full propagator **G**, i.e. we group the vacuum diagrams according to

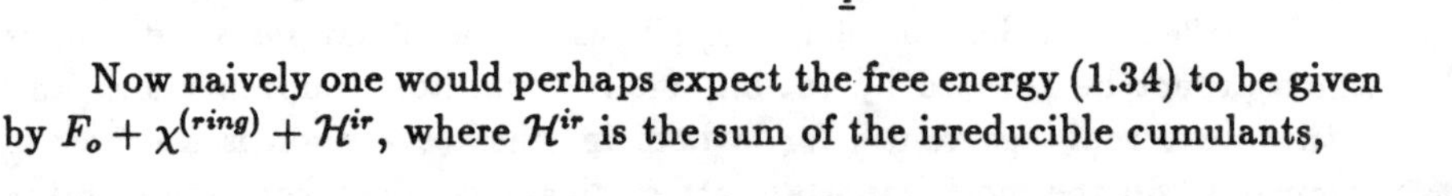

However althought his would be true if we were to consider resumming only the generic two-part function  $= \mathcal{D}_q$ (in which case we recover just the ordinary cumulant expansion), for general self energy insertions there will be some overcounting. For example, consider the diagram

which will be included both in the ring diagrams  (twice) and also in the second order irreducible cumulant  .

In fact the required result is [Luttinger and Ward, 1960, Lee and Yang, 1960]

$$\begin{aligned} F &= F_o + \chi^{(ring)} + \tfrac{1}{2}\beta^{-1} \sum_{\underline{q}\nu} tr \beta \underline{\underline{\Sigma}}_{\underline{q}\nu} (\underline{\underline{\mathbf{G}}}_{\underline{q}\nu} - \underline{\underline{G}}_{\underline{q}\nu}) + \mathcal{H}^{ir}[\mathbf{G}] \\ &= -\tfrac{1}{2}\beta^{-1} \sum_{\underline{q}\nu} tr \ln \beta \underline{\underline{\mathbf{G}}}_{\underline{q}\nu} - \tfrac{1}{2}\beta^{-1} \sum_{\underline{q}\nu} tr \beta \underline{\underline{\Sigma}}_{\underline{q}\nu} \underline{\underline{\mathbf{G}}}_{\underline{q}\nu} + \mathcal{H}^{ir}[\mathbf{G}], \end{aligned} \tag{2.9}$$

where the extra term is precisely that required to avoid the overcounting. Note that (2.9) now no longer involves the bare propagator explicitly, but is written entirely in terms of the full renormalized propagator $\underline{\underline{\mathbf{G}}}_{\underline{q}\nu}$, and the self energy $\underline{\underline{\Sigma}}_{\underline{q}\nu}$.

We now sketch an algebraic proof of (2.9):
From (1.30) and (1.34) we have

$$F[G] = -\tfrac{1}{2}\beta^{-1} \sum_{\underline{q}\nu} tr \ln \beta \underline{\underline{G}}_{\underline{q}\nu} + \mathcal{H}[G], \tag{2.10}$$

and we may define

$$F'[G,\mathbf{G}] \equiv -\tfrac{1}{2}\beta^{-1} \sum_{\underline{q}\nu} tr \ln \beta \underline{\underline{\mathbf{G}}}_{\underline{q}\nu} - \tfrac{1}{2}\beta^{-1} \sum_{\underline{q}\nu} tr (\underline{\underline{1}} - \underline{\underline{G}}_{\underline{q}\nu}^{-1} \underline{\underline{\mathbf{G}}}_{\underline{q}\nu}) + \mathcal{H}^{ir}[\mathbf{G}] \tag{2.11}$$

from (2.9) using the Dyson equation $-\beta \underline{\underline{\Sigma}}_{\underline{q}\nu} = \underline{\underline{G}}_{\underline{q}\nu}^{-1} - \underline{\underline{\mathbf{G}}}_{\underline{q}\nu}^{-1}$. We wish to show that $F = F'$ (at least up to a constant). Consider varying $\underline{\underline{G}}_{\underline{q}\nu}$, with $\underline{\underline{\mathbf{G}}}_{\underline{q}\nu}$ and $\underline{\underline{\Sigma}}_{\underline{q}\nu}$ considered as functionals of $\underline{\underline{G}}_{\underline{q}\nu}$.

When $\underline{\underline{G}}_{q\nu} = \underline{\underline{\mathbf{G}}}_{q\nu}$, $F = F'$ since $\mathcal{H} = \mathcal{H}^{ir}$ (they differ only through the ring diagrams and these vanish if $\underline{\underline{\Sigma}}_{q\nu} = 0$). Now

$$\begin{aligned}\frac{\delta F}{\frac{1}{2}\delta \underline{\underline{G}}_{q\nu}} &= \beta^{-1}\underline{\underline{G}}_{q\nu}^{-1} - \beta^{-1}\underline{\underline{G}}_{q\nu}^{-1}(\underline{\underline{\mathbf{G}}}_{q\nu} - \underline{\underline{G}}_{q\nu})\underline{\underline{G}}_{q\nu}^{-1} \\ &= -\beta^{-1}\underline{\underline{G}}_{q\nu}^{-1}\underline{\underline{\mathbf{G}}}_{q\nu}\underline{\underline{G}}_{q\nu}^{-1},\end{aligned} \tag{2.12}$$

since cutting a line in a diagram in $\mathcal{H}[G]$ produces a contribution (not necessarily proper) to $\mathbf{G}$, with the two external propagators amputated. It is easy to see how the numerical factors work out: in particular the $\frac{1}{n!}$ factor associated with n propagators joining the same two vertices becomes $\frac{1}{(n-1)!}$ when one of these propagators is cut, since this gives n identical contributions to $\mathbf{G}$. Now on the other hand

$$\frac{\delta F'}{\frac{1}{2}\delta \underline{\underline{G}}_{q\nu}} = \frac{\partial F'}{\frac{1}{2}\partial \underline{\underline{G}}_{q\nu}}\Big|_{\mathbf{G}} + \sum_{q'\nu'} \frac{\delta \underline{\underline{\mathbf{G}}}_{q'\nu'}}{\delta \underline{\underline{G}}_{q\nu}} : \frac{\partial F'}{\frac{1}{2}\partial \underline{\underline{\mathbf{G}}}_{q'\nu'}}\Big|_{G}, \tag{2.13}$$

and

$$\frac{\partial F'}{\frac{1}{2}\partial \underline{\underline{G}}_{q\nu}}\Big|_{\underline{\underline{\mathbf{G}}}} = -\beta^{-1}\underline{\underline{G}}_{q\nu}^{-1}\underline{\underline{\mathbf{G}}}_{q\nu}\underline{\underline{G}}_{q\nu}^{-1}, \tag{2.14}$$

which is the same as $\delta F/\delta G$. But cutting a line in any diagram in $\mathcal{H}^{ir}[\mathbf{G}]$ produces a contribution to the self energy, since $\underline{\underline{\Sigma}}_{q\nu}$ consists of all irreducible diagrams with two external lines evaluated using the full propagator $\mathbf{G}$ (note that there is no double counting problem for contributions to Σ). So

$$\frac{\delta \mathcal{H}^{ir}[\mathbf{G}]}{\frac{1}{2}\delta \underline{\underline{\mathbf{G}}}_{q\nu}} = \underline{\underline{\Sigma}}_{q\nu}, \tag{2.15}$$

and thus

$$\frac{\partial F'}{\frac{1}{2}\partial \mathbf{G}_{q\nu}}\Big|_{G} = -\beta^{-1}\underline{\underline{\mathbf{G}}}_{q\nu}^{-1} + \beta^{-1}\underline{\underline{G}}_{q\nu}^{-1} + \underline{\underline{\Sigma}}_{q\nu} = 0, \tag{2.16}$$

using the Dyson equation. Putting together (2.12), (2.13), (2.14) and (2.16) we have $\delta F/\delta G = \delta F'/\delta G$, whence $F = F'$ up to a numerical constant.

We note in passing that equations (2.10) and (2.11) constitute a generalised Legendre transformation [DeDominicis and Martin, 1964] from the variables $\underline{\underline{G}}_{q\nu}$ (or more precisely $\underline{\underline{\mathcal{D}}}_{q\nu}$) to new variables $\underline{\underline{\mathbf{G}}}_{q\nu}$: defining

$$W[\mathcal{D}] \equiv -\beta F[G] = \tfrac{1}{2}\sum_{q\nu} trln\beta\underline{\underline{G}}_{q\nu} - \beta\mathcal{H}[G],$$

where
$$\underline{\underline{G}}_{q\nu} = \beta^{-1}(\Omega_\nu^2 + \underline{\underline{\mathcal{D}}}_{q\nu})^{-1}, \tag{2.17}$$

$$\Gamma[\mathbf{G}] \equiv -\tfrac{1}{2}\sum_{q\nu} trln\beta\underline{\underline{\mathbf{G}}}_{q\nu} - \beta\mathcal{H}^{ir}[\mathbf{G}]$$

we have

$$\Gamma[\mathbf{G}] = W[\mathcal{D}] - \tfrac{1}{2}\beta\sum_{q\nu} tr\underline{\underline{\mathcal{D}}}_{q\nu}\underline{\underline{\mathbf{G}}}_{q\nu} \tag{2.18}$$

since $\sum_{q\nu}$ *tr* $\underline{\underline{G}}_{q\nu}$ vanishes. Then

$$\frac{\partial W}{\partial \mathbf{G}}\Big|_{\mathcal{D}} = 0, \qquad \frac{\partial \Gamma}{\partial \mathcal{D}}\Big|_{\mathbf{G}} = 0,$$

so that

$$\frac{\partial \Gamma}{\frac{1}{2}\partial \underline{\underline{\mathbf{G}}}_{q\nu}}\Big|_{\mathcal{D}} = -\underline{\underline{\mathcal{D}}}_{q\nu}, \qquad \frac{\partial W}{\frac{1}{2}\partial \underline{\underline{\mathcal{D}}}_{q\nu}}\Big|_{G} = \underline{\underline{\mathbf{G}}}_{q\nu}, \tag{2.19}$$

in accord with (2.12) and (2.14). Note that W consists of all connected diagrams, while Γ consists of all irreducible diagrams. We have a generalisation of the Legendre transformation (6) for the generation of *proper* graphs. The algebra in this section amounts to an algebraic (rather than diagrammatic) proof of this statement. This is the approach adopted by [Horner, 1967].

To summarise, in this section we have rearranged the original cumulant expansion into the form (2.9), an irreducible cumulant expansion which we hope will have improved convergence (we will take up this point again in § 3.3). However, we still do not have a procedure for actually evaluating this expansion, since we have no way of determining G (or equivalently Σ) except through the original perturbation series.

2.3 Self Consistency: the Dressed Phonon

From equations (1.23), (1.50), (2.9) and (2.15) we now have the following implicit expression for the free energy;

$$F = -\tfrac{1}{2}\beta^{-1}\sum_{\underline{q\nu}} tr \ln \beta \underline{\underline{\mathbf{G}}}_{q\nu} - \tfrac{1}{2}\beta^{-1}\sum_{q\nu} tr \beta \underline{\underline{\Sigma}}_{q\nu} \underline{\underline{\mathbf{G}}}_{q\nu} + \mathcal{H}^{ir}[\mathrm{G}]$$

where

$$\begin{aligned} \underline{\underline{\mathbf{G}}}_{q\nu} &= \beta^{-1}(\Omega_\nu^2 \underline{\underline{1}} + \underline{\underline{\mathcal{D}}}_{q\nu} + \underline{\underline{\Sigma}}_{q\nu})^{-1}, \\ \underline{\underline{\Sigma}}_{q\nu} &= \frac{\delta \mathcal{H}^{ir}[\mathrm{G}]}{\frac{1}{2}\delta \underline{\underline{\mathbf{G}}}_{q\nu}} . \end{aligned} \tag{2.20}$$

To solve these equations, we could start from the usual assumption A1) for $\mathcal{D}_\rho$ and v_ρ (equation (1.42)), use these to calculate $\underline{\underline{G}}_{q\nu}$, calculate $\underline{\underline{\Sigma}}_{q\nu}$ using (2.20), truncated to order k, use this value of $\underline{\underline{\Sigma}}_{q\nu}$ to calculate $\underline{\underline{\mathbf{G}}}_{q\nu}$, and then iterate. We know from (2.20) that if this procedure converges, we will get the correct result for F (to order k) by evaluating (2.9) using the self consistent Green's function $\underline{\underline{\mathbf{G}}}_{q\nu}$. However the iterations need not necessarily converge at all, and will not even make sense if $\underline{\underline{\mathcal{D}}}_\rho$ is not positive definite.

The crucial observation to be made now is that since (2.9) depends only on $\underline{\underline{\mathbf{G}}}_{q\nu}$, the initial value of $\underline{\underline{G}}_{q\nu}$ is irrelevant provided the iterations converge to the correct $\underline{\underline{\mathbf{G}}}_{q\nu}$. So we are now free to relax our remaining assumption A1). We consider a trial harmonic potential

$$W(\beta) = \sum_{\underline{q}} \beta^{-1} \int_0^\beta d\tau \int_0^\beta d\tau' \tfrac{1}{2} \underline{u}_q(\tau) . \underline{\mathcal{D}}_q(\tau - \tau') . \underline{u}_{-q}(\tau') , \tag{2.21}$$

where we have a trial force constant matrix $\underline{\underline{\mathcal{D}}}_{q\nu}$ and Green's function $\underline{\underline{G}}_{q\nu}$. As we expect $\underline{\underline{\mathbf{G}}}_{q\nu}$ to be positive definite, we will assume $\underline{\underline{\mathcal{D}}}_{q\nu}$ to be taken to be positive definite. Then instead of the traditional separation (1.26) and (1.43) we now let

$$V = W(\beta) + (V - W(\beta)) , \tag{2.22}$$

and consider $V - W(\beta)$ as the perturbation. $W(\beta)$ will lead to an extra (trivial) contribution to the self energy of $-\underline{\underline{\mathcal{D}}}_{q\nu}$, so equations (2.20) become

$$\begin{aligned} \underline{\underline{\mathbf{G}}}_{q\nu} &= \beta^{-1}(\Omega_\nu^2 \underline{\underline{1}} + \underline{\underline{\mathcal{D}}}_{q\nu} + \underline{\underline{\Sigma}}_{q\nu})^{-1} , \\ \underline{\underline{\Sigma}}_{q\nu} &= -\underline{\underline{\mathcal{D}}}_{q\nu} + \frac{\delta \mathcal{H}^{ir}[\mathbf{G}]}{\frac{1}{2}\delta \mathbf{G}_{\underline{q\nu}}} \end{aligned} \tag{2.23}$$

Now $\underline{\underline{\mathcal{D}}}_{q\nu}$ are totally arbitrary ((2.23) only defines $\mathcal{D} + \Sigma$), so we can choose them so that the self energy vanishes;

$$\underline{\underline{\Sigma}}_{q\nu} = 0 , \tag{2.24}$$

and thus

$$\begin{aligned} \underline{\underline{\mathbf{G}}}_{q\nu} &= \underline{\underline{G}}_{q\nu} = \beta^{-1}(\Omega_\nu^2 + \underline{\underline{\mathcal{D}}}_{q\nu})^{-1} , \\ \underline{\underline{\mathcal{D}}}_{q\nu} &= -\beta \frac{\delta \mathcal{H}^{ir}[G]}{\frac{1}{2}\delta G_{\underline{q\nu}}} . \end{aligned} \tag{2.25}$$

Because of (2.24), all ring diagrams will now be zero, and $\chi_k = \chi_k^{ir}$ for $k > 1$. For $k = 1$

$$\chi_1 = -\tfrac{1}{2} \sum_{q\nu} tr \underline{\underline{\mathcal{D}}}_{q\nu} \underline{\underline{G}}_{q\nu} + \chi_1^{ir} ,$$

where the first term is just $< -W(\beta) >$: note that this is compatible with (2.23). Then

$$F = -\frac{1}{2}\beta^{-1} \sum_{\underline{q\nu}} tr \ell n \beta \underline{\underline{G}}_{q\nu} - \tfrac{1}{2} \sum_{\underline{q\nu}} tr \underline{\underline{\mathcal{D}}}_{q\nu} \underline{\underline{G}}_{q\nu} + \mathcal{H}^{ir}[G] . \tag{2.26}$$

In fact the above argument is actually contained in the algebraic proof of the previous section, but now written in terms of $\mathcal{D}$ rather than G. It is now clear however that $\mathcal{D}$ plays the role of an effective force constant matrix (just as $\mathbf{G}$ was the effective (full) propagator), and that we can solve (2.9) and (2.20) by taking some initial form for $\mathcal{D}$ (perhaps that suggested by (1.42), but perhaps instead some other more inspired guess), and then iterating to self consistency, with a truncated cumulant expansion.

In principle, these general ideas can be extended to all the irreducible Green's functions [De Dominicis and Martin, 1964], although the integral equations which must then be solved (analogous to (2.25) for $G_{q\nu}$) become increasingly complicated. The simple example presented above however is a useful reminder that field theory and perturbation expansions are living, subtle things, depending on the full, dressed Green's functions and not on the bare ones evident in the classical Hamiltonian, or indeed the choice of bare fields. In this context reparameterisation invariance becomes self evident. It is important to remember this not only for quantum mechanical systems (for example the rare gases) but also in relativistic situations such as scalar field theory or low energy QCD, in which the structure of the classical theory bears little resemblance to that of the full quantum theory.

Variational Principles:

In the course of the proof in the previous section (§ 2.2), we derived a variational principle (equation (2.16)), namely [Luttinger and Ward, 1960, Sham, 1965]

$$\frac{\delta F}{\delta \mathbf{G}_{q\nu}} = 0\,. \qquad (2.27)$$

So F is stationary with respect to variations of $\mathbf{G}$ provided $\mathbf{G}$ satisfies the Dyson equation. If $\mathcal{D}$ is the trial force constant matrix and satisfies (2.25), then using (2.26) we may prove the related result

$$\frac{\delta F}{\frac{1}{2}\delta \underline{\underline{\mathcal{D}}}_{q\nu}} = \beta^{-1}\underline{\underline{\mathbf{G}}}_{q\nu} - \beta^{-1}\underline{\underline{\mathcal{D}}}_{q\nu} : \frac{\delta \underline{\underline{G}}_{q\nu}}{\delta \underline{\underline{\mathcal{D}}}_{q\nu}} - \beta^{-1}\underline{\underline{G}}_{q\nu} + \frac{\delta \underline{\underline{G}}_{q\nu}}{\delta \underline{\underline{\mathcal{D}}}_{q\nu}} : \frac{\delta \mathcal{H}^{ir}}{\delta \underline{\underline{G}}_{q\nu}} = 0 \qquad (2.28)$$

i.e. F is stationary with respect to variations of $\mathcal{D}$. Note that this result will be true order by order in $\mathcal{H}^{ir}$: it also holds for truncations of $\mathcal{H}^{ir}$. This variational principle is used by many authors [notably Koehler, 1966-1968, and Werthamer, 1970] to derive self consistent phonon equations. It carries with it the usual implication that even a poor approximation to $\underline{\underline{\mathcal{D}}}_{q\nu}$ will produce a reasonable value for F.

We may also show that, provided $\underline{\underline{\mathbf{G}}}_{q\nu}$ remains positive definite, the variational principle is a minimisation principle:

Consider

$$\frac{\delta^2 F}{\frac{1}{2}\delta \underline{\underline{G}}_{q\nu}\frac{1}{2}\delta \underline{\underline{G}}_{q'\nu'}} = 2\beta^{-1}\underline{\underline{G}}_{q\nu}^{-2}\delta_{q\nu,q'\nu'} + \frac{\delta^2 \mathcal{H}^{ir}}{\frac{1}{2}\delta \underline{\underline{G}}_{q\nu}\frac{1}{2}\delta \underline{\underline{G}}_{q'\nu'}}$$

Now $\delta^2\mathcal{H}^{ir}/\delta\underline{\underline{G}}_{\underline{q}\nu}\delta\underline{\underline{G}}_{\underline{q}'\nu'}$ is the sum of all connected four-point graphs with the internal legs removed, i.e.

$$\begin{aligned}\frac{\delta^2\mathcal{H}^{ir}}{\delta\underline{\underline{G}}_{\underline{q}\nu}\delta\underline{\underline{G}}_{\underline{q}'\nu'}} &= \beta^{-1}\underline{\underline{G}}^{-2}_{\underline{q}\nu} : \underline{\underline{G}}^{c}_{4\,\underline{q}\nu,-\underline{q}\nu,\underline{q}'\nu,-\underline{q}'\nu} : \underline{\underline{G}}^{-2}_{\underline{q}'\nu'} \\ &= \beta^{-1}G^{-2}_{\underline{q}\nu} :< \underline{u}_{\underline{q}\nu}\underline{u}_{-\underline{q}\nu}\underline{u}_{\underline{q}'\nu'}\underline{u}_{-\underline{q}'\nu'} >_c : \underline{G}^{-2}_{\underline{q}'\nu'} \\ &= -\beta^{-1}\underline{\underline{G}}^{-1}_{\underline{q}\nu}\underline{\underline{G}}^{-1}_{\underline{q}'\nu'} + \beta^{-1}\underline{\underline{G}}^{-2}_{\underline{q}\nu} :< \underline{u}_{\underline{q}\nu}\underline{u}_{-\underline{q}\nu}\underline{u}_{\underline{q}'\nu'}\underline{u}_{\underline{q}'\nu'} >: \underline{\underline{G}}^{-2}_{\underline{q}'\nu'},\end{aligned}$$

and $< \underline{u}_{\underline{q}\nu}\underline{u}_{-\underline{q}\nu}\underline{u}_{\underline{q}'\nu'}\underline{u}_{-\underline{q}'\nu'} >$ is positive semi-definite (in the space $(ij\underline{q}\nu)$) since the matrix $a_i a_j$ is positive definite, and the measure $[\mathcal{D}u]e^{-S}$ is strictly positive. Then letting $\underline{\underline{G}}^{-1}_{\underline{q}\nu} \equiv (b_i)$, the remaining terms are $2b_i^2\delta_{ij} - b_i b_j$ which is also positive definite, hence $\delta^2 F/\delta\underline{\underline{G}}_{\underline{q}\nu}\delta\underline{\underline{G}}_{\underline{q}'\nu'}$ is positive definite. Hence the result.

Note that the result will also hold order by order in k provided that $|< \underline{u}_{\underline{q}\nu}\underline{u}_{\underline{q}\nu}\underline{u}_{\underline{q}'\nu'}\underline{u}_{\underline{q}'\nu'} >|_{\text{order k}} < |\underline{\underline{G}}_{\underline{q}\nu}\underline{\underline{G}}_{\underline{q}'\nu'}|$. It is true trivially to first order (see also § 3.1 below).

3 Self Consistent Phonons

3.1 Free SCP

Self Consistent phonon (SCP) theory is most commonly used truncated to first order in the irreducible cumulant expansion, where it amounts to a summation of simple tadpole insertions. The ring diagrams are them of the form

where

$$\underline{\underline{\Sigma}}_{\underline{q}}^{(1)} \equiv$$

$$\begin{aligned} &= \frac{1}{M}\sum_{\rho} \eta_\rho(\underline{q})\eta_\rho(-\underline{q})\underline{\nabla}\,\underline{\nabla} < v_\rho > - \underline{\underline{\mathcal{D}}}_{\underline{q}}, \\ < v_\rho > &= exp(\tfrac{1}{2}\underline{\underline{G}}_{\rho} : \underline{\nabla}\,\underline{\nabla})v_\rho. \end{aligned} \tag{3.1}$$

They sum to give

$$F^{(1)} = -\tfrac{1}{2}\beta^{-1}\sum_{\underline{q}\nu} tr\ell n\beta \underline{\underline{G}}_{\underline{q}\nu} + \tfrac{1}{2}N\sum_{\rho}(1 - \tfrac{1}{2}\underline{\underline{G}}_{\rho} : \underline{\nabla}\,\underline{\nabla}) < v_\rho >,$$

so setting $\underline{\underline{\Sigma}}_{\underline{q}}^{(1)}$ to zero,

$$\underline{\underline{\mathcal{D}}}_{\underline{q}}^{(1)} = \frac{1}{M}\sum_{\rho} \eta_\rho(\underline{q})\eta_\rho(-\underline{q})\underline{\nabla}\,\underline{\nabla} < v_\rho >, \tag{3.2}$$

and

$$F^{(1)} = -\tfrac{1}{2}\beta^{-1}\sum_{\underline{q}\nu} tr\ell n\beta \underline{\underline{G}}_{\underline{q}\nu}^{(1)} - \tfrac{1}{4}\sum_{\underline{q}\nu} \underline{\underline{G}}_{\underline{q}\nu}^{(1)} : \mathcal{D}_{\underline{q}}^{(1)} + \tfrac{1}{2}N\sum_{\rho} < v_\rho >, \tag{3.3}$$

which, together with the appropriate definition of $\underline{\underline{G}}_{\underline{q}\nu}^{(1)}$ in terms of $\underline{\underline{\mathcal{D}}}_{\underline{q}\nu}^{(1)}$(1.23), and the relevant Fourier transform (1.24), are the basic first order SCP equations. Notice that(3.2) may be derived directly from(3.3) using the first order variational principle $\delta F^{(1)}/\delta \underline{\underline{\mathcal{D}}}_{\rho}^{(1)} = 0$.

If we diagonalise $\underline{\underline{\mathcal{D}}}_q^{(1)}$, and its eigenvalues are $\omega_q^2 = \underline{e}_q^\lambda . \underline{\underline{\mathcal{D}}}_q^{(1)} . \underline{e}_q^\lambda$, the first order free energy (3.3) may be written in the form

$$F^{(1)} = \sum_q (\beta^{-1} \ln 2 \sinh \tfrac{1}{2}\beta_q - \tfrac{1}{4}\omega_q \coth \tfrac{1}{2}\beta_q) + \tfrac{N}{2} \sum_\rho < v_\rho >,$$

where $\beta_q = \beta\hbar\omega_q$. This is only possible since $\underline{\underline{\mathcal{D}}}_q^{(1)}$ is independent of ν, which is due to $\underline{\underline{\Sigma}}_q^{(1)}$ being a *local* self energy insertion. This will no longer be true to higher order, as we shall see.

Since $\underline{\underline{\mathcal{D}}}_q^{(1)}$ is not the most general form for the harmonic ansatz, the minimisation principle (2.28) means that in fact $F^{(1)} > F$. To demonstrate this more directly one may use the inequality $< e^{-f} > \; > e^{<f>}$ [Feynman, 1973]. From (11),(14), and (1.25).

$$e^{-\beta F} = < e^{-(S-S_0)} >_{S_0} e^{-\beta F_0} > e^{-<S-S_0>-\beta F_0}.$$

Now $F_0 + \beta^{-1} < S - S_0 >$ is just $F^{(1)}$ as given by (3.3), so using the convexity of e^{-x}, $F < F^{(1)}$ as required.

Because the first order SCP equations (3.1)-(3.3), by including all distinct graphs of the form

(sometimes called "daisy graphs" or "cactus graphs") minimize $F^{(1)}$, this gives the best possible approximation to F given a ν independent ansatz for the dynamical matrix. At high temperatures only small values of ν give significant contributions to F, so here the first order results should be particularly accurate. Furthermore, a positive definite dynamical matrix is now guaranteed, whatever the bare force constants, because $F^{(1)}$ is bounded below.

The expression "self consistent phonon theory" is commonly used to describe only this first order truncation of the full SCP equations, which is basically a self consistent Hartree-Fock approximation. The phonons are self consistent, but non-interacting; there are no non-trivial self consistent phonon loops in scattering amplitudes. In this form, SCP theory has been used to calculate thermodynamic properties of solid rare gases (and in particular *bcc* He^3 and He^4, Ne, Ar, Kr and Xe [Horner, 1967, Gillis et al., 1968, Koehler, 1969, Goldman et al., 1970]. Although a useful leading approximation, many important dynamical effects will be missing, however, since in particular only even derivatives of the potential are involved, the phonons are undamped, and the crystal will not expand when heated (for example). To remedy these defects one must include higher orders in the self consistent phonon expansion.

3.2 Interacting SCP

The second order cummulant is (schematically)

The first of these terms was included in the first order ring summation (it is reducible), so we need only consider the second (the irreducible part). This gives rise to two self energy insertions, formed by cutting self-linked and linking lines respectively:

(2s) (2ℓ)

with

$$\underline{\underline{\Sigma}}^{(2s)}_{\underline{q}} = -\frac{1}{M}\tfrac{1}{2}\sum_{\rho\rho'\sigma} \tfrac{1}{2}\eta_\rho(\underline{q})\eta_\rho(-\underline{q}) \int_0^\beta d\tau \epsilon(\tfrac{1}{2}\underline{\underline{G}}^\sigma_{\rho\rho'}(\tau) : \underline{\nabla\nabla}' < \underline{\nabla\nabla} v_\rho >< v_{\rho'} >,$$

$$\underline{\underline{\Sigma}}^{(2\ell)}_{\underline{q}} = -\frac{1}{M}(\tfrac{1}{2})^2 \sum_{\rho\rho'\sigma} \eta_\rho(\underline{q})\eta_{\rho'}(-\underline{q}) e^{i\underline{q}\cdot\underline{R}_\sigma} \int_0^\beta d\tau e^{i\Omega_\nu \tau}$$

$$\epsilon_2(\tfrac{1}{2}G^\sigma_{\rho\rho'}(\tau) : \underline{\nabla\nabla}') < \underline{\nabla} v_\rho >< \underline{\nabla} v_{\rho'} > . \qquad (3.4)$$

Summing the second order ring diagrams yields the self consistent force constants: setting $\underline{\underline{\Sigma}}_{\underline{q}}^{(1)} + \underline{\underline{\Sigma}}_{\underline{q}}^{(2s)} + \underline{\underline{\Sigma}}_{\underline{q}\nu}^{(2\ell)}$ to zero, we get

$$\begin{aligned} \underline{\underline{\mathcal{D}}}_{\underline{q}}^{(2s)} = & \quad \frac{1}{2M}\sum_{\rho} \eta_{\rho}(\underline{q})\eta_{\rho}(-\underline{q})\underline{\underline{\mathcal{D}}}_{\rho}^{(2s)}, \\ \underline{\underline{\mathcal{D}}}_{\underline{q}\nu}^{(2\ell)} = & \quad \frac{1}{2M}\sum_{\rho\rho'\sigma} \eta_{\rho}(\underline{q})\eta_{\rho'}(-\underline{q})e^{i\underline{q}\cdot\underline{R}_{\sigma}}\underline{\underline{\mathcal{D}}}_{\rho\rho',\nu}^{(2\ell)\sigma} \end{aligned}$$

where

$$\underline{\underline{\mathcal{D}}}_{\rho}^{(2s)} = -\tfrac{1}{2}\sum_{\rho'\sigma}\int_0^{\beta} d\tau\, \epsilon_3(\tfrac{1}{2}\underline{\underline{G}}_{\rho\rho'}^{\sigma}(\tau) : \underline{\nabla}\,\underline{\nabla}') < \underline{\nabla}\,\underline{\nabla} v_{\rho} >< \nu_{\rho'} >, \qquad (3.5)$$

$$\underline{\underline{\mathcal{D}}}_{\rho\rho',\nu}^{(2\ell)\sigma} = -\tfrac{1}{2}\int_0^{\beta} d\tau\, e^{i\Omega_{\nu}\tau}\epsilon_2(\tfrac{1}{2}\underline{\underline{G}}_{\rho\rho'}^{\sigma}(\tau) : \underline{\nabla}\,\underline{\nabla}') < \underline{\nabla} v_{\rho} >< \underline{\nabla}' v_{\rho'} > .$$

The dynamical matrix $\underline{\underline{\mathcal{D}}}_{\underline{q}\nu}^{(2)} = \underline{\underline{\mathcal{D}}}_{\underline{q}}^{(1)} + \underline{\underline{\mathcal{D}}}_{\underline{q}}^{(2s)} + \underline{\underline{\mathcal{D}}}_{\underline{q}\nu}^{(2\ell)}$ has now acquired a ν dependent piece due to the non-local diagram (2ℓ). This leads to phonon damping.

The free energy to this order is

$$\begin{aligned} F^{(2)} = & -\tfrac{1}{2}\beta^{-1}\sum_{\underline{q}\nu} tr \ln \beta \underline{\underline{G}}_{\underline{q}\nu} - \tfrac{1}{4}\sum_{\underline{q}\nu}\underline{\underline{G}}_{\underline{q}\nu} : \underline{\underline{\mathcal{D}}}_{\underline{q}\nu} + \tfrac{N}{2}\sum_{\rho} < v_{\rho} >_2 \\ & - \tfrac{N}{4}\tfrac{1}{2}\sum_{\rho\rho'\sigma}\int_0^{\beta} d\tau\, \epsilon_3(\tfrac{1}{2}\underline{\underline{G}}_{\rho\rho'}^{\sigma}(\tau) : \underline{\nabla}\,\underline{\nabla}') < v_{\rho} >< v_{\rho'} >, \end{aligned} \qquad (3.6)$$

where the last term is the contribution of the irreducible part of χ_2, i.e.  . Note that (3.4) may be derived directly from (3.6) using (2.25) and observing that $\frac{d}{dx}\epsilon_n(x) = \epsilon_{n-1}(x)$.

Since the complete second order equations (3.5),(3.6) are quite involved, many suggestions for obtaining approximate results to second order have been proposed. The simplest of these is to estimate higher order corrections to F by calculating diagrams in the Peierls expansion, but using the new renormalised vertices obtained from $< v_{\rho} >$ in place of the bare ones [Götze,

1966, Beck and Meier, 1971]. The propagator is that obtained from the (first order) self consistent dynamical matrix, so all diagrams with tadpole insertions are omitted. To $0(\lambda^2)$ one thus includes

$\Delta F_{33} =$ 

and to $0(\lambda^4)$

$\Delta F_{44} =$  ,  ,  ,  .

a) b) c) d)

Such calculations are called "improved self consistent phonons" or the "renormalised harmonic approximation" [Goldman et al., 1968, Klein et al., 1970, 1974, Koehler et al., 1970, 1972, Glyde, 1071, Horner, 1972]. This renormalised Peierls expansion is expected to converge more rapidly than the original Peierls expansion since the characteristic length of the potential is increased and thus λ is reduced. At very low temperatures ΔF_{33} indeed gives a significant improvement over the first order results, but at higher temperatures the expansion suffers the same fate as Peierls' expansion: ΔF_{44} soon becomes equally important . This is essentially because multiphonon processes begin to make significant contributions; $\epsilon_n(x) \sim x^n$ only for very small x. So at high temperatures it makes more sense to group higher order terms according to the irreducible cumulant expansion. The first correction is then  . To improve on this by including multiple scatterings (for example diagram b) above) one might as well go all the way to the second order self consistent phonon scheme. On the other hand, diagram c) is a part of the third order irreducible cumulant, and d) is fourth order.

<u>Local and Non-local Force Constants</u>

Beyond first order it is convenient to separate $\mathcal{D}^{(s)}_{\underline{q}}$ into two parts $\mathcal{D}^{(s)}_{\underline{q}}$ and $\mathcal{D}^{(\ell)}_{\underline{q}\nu}$ due to short range (local) and long range (non-local) interactions respectively. Then

$$\mathcal{D}^{(\ell)}_{\underline{q}\nu} = \frac{1}{\beta}\int_0^\beta d\tau e^{i\hbar\Omega_\nu\tau}\mathcal{D}^{(\ell)}_{\underline{q}}(\tau),$$

and we define $\underline{\underline{\mathcal{D}}}^{(s)}_{\rho}$ and $\mathcal{D}^{(\ell)\sigma}_{\rho\rho'}(\tau)$ by

$$\underline{\underline{\mathcal{D}}}^{(s)}_{\underline{q}} = \frac{1}{2M}\sum_{\rho}\eta_\rho(\underline{q})\eta_\rho(-\underline{q})\underline{\underline{\mathcal{D}}}^{(s)}_{\rho},$$

$$\underline{\underline{\mathcal{D}}}^{(\ell)}_{\underline{q}}(\tau) = \frac{1}{2M}\sum_{\rho\rho'\sigma}\eta_\rho(\underline{q})\eta_{\rho'}(-\underline{q})e^{i\underline{q}\cdot\underline{R}_\sigma}\underline{\underline{\mathcal{D}}}^{(\ell)\sigma}_{\rho\rho'}(\tau).$$

Then (2.25) becomes

$$\begin{aligned}\underline{\underline{\mathcal{D}}}^{(s)}_{\rho} &= -\frac{\beta}{N}\left(\frac{\delta\mathcal{H}^{ir}}{\frac{1}{4}\delta\underline{\underline{G}}_\rho}\right)_{G^\sigma_{\rho\rho'}}, \\ \underline{\underline{\mathcal{D}}}^{(\ell)\sigma}_{\rho\rho'}(\tau) &= -\frac{\beta}{N}\left(\frac{\delta\mathcal{H}^{ir}}{\frac{1}{2}\delta G^\sigma_{\rho\rho'}(\tau)}\right)_{G_\rho},\end{aligned} \tag{3.7}$$

so that graphically contributions to $\underline{\underline{\mathcal{D}}}^{(s)}_{\rho}$ have the tadpole form

whereas contributions to $\underline{\underline{\mathcal{D}}}^{(\ell)\sigma}{}_{\rho\rho'}(\tau)$ have the cocoon form

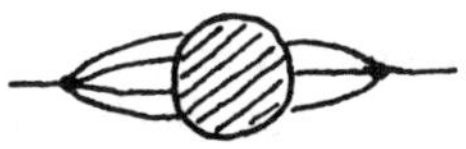

As we shall see in paragraph 4.3, the interesting features relating to phonon damping are all contained in the long range part $\mathcal{D}^{(\ell)}_q(\tau)$.

A useful numerical approximation in interacting SCP is to use only a subset of all possible self energy insertions to a given order. Usually those which are non-local are more important than the tadpoles which provide small additional frequency renormalisations (necessary to maintain the correct dispersion relations in real time, and thus unitarity). So at second order one drops diagram (2s) [Werthamer, 1970] and retains only (2ℓ):

At third order the diagram 

is probably the most significant self energy insertion.

3.3 Convergence and Stability

We already have several qualitative reasons for believing that the irreducible cumulant expansion should converge faster than the original perturbation expansion:

1. One expands in powers of renormalized vertices so one expects the effective λ parameter to be reduced (the renormalised potential is smoother than the bare one).
2. Multiphonon processes are included self consistently (phonons scatter off phonons which themselves scatter), so one expects all the dominant contributions to phonon scattering processes to be included.
3. Further summations of graphs give corrections which remain small over the whole temperature range, at least for soft interactions [Choquard, 1967].

We now examine the behaviour of each of the terms in the SCP expansion in various temperature ranges, and consider the stability of solutions to the self consistency equations and the possible implications for convergence. The arguements presented here will only be qualitative however; to the best of our knowledge no quantitative analysis exists.

i) Quantum region, $kT \ll \hbar\omega$. Then $G \sim \hbar/2\omega_{\underline{q}}$ and $\chi_k \sim \lambda^k \beta^{-k+1}$ with $\lambda \ll 1$. So the kth term in $\mathcal{H}^{ir}$ believes as λ^k and the series converges rapidly.

ii) Low Temperature region, $\hbar\omega \lesssim kT \ll kT_m$. Then we may assume that $G \sim \beta^{-1}$ and has no significant τ dependence. The effective potential will now be largely T independent, so the dominant contribution to each irreducible cumulant will be that with the least number of propagator factors:

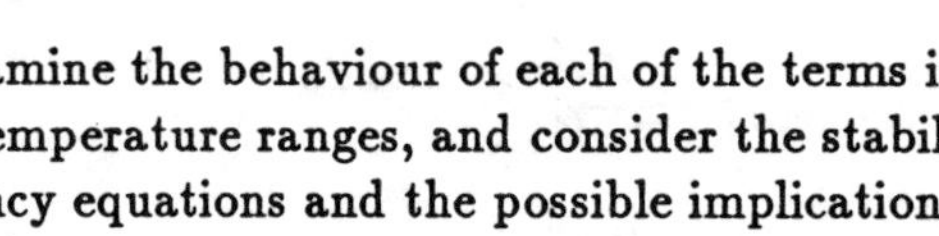

i.e. $\beta^{-(2k-1)}$ so that the kth term in $\mathcal{H}^{ir}$ believes as $\beta^{-k}\lambda^k$ and the series converges rapidly. Note that the ring diagrams behaves as β^{-k}, so contribute $\beta^{-1}\lambda^k$ at kth order, which is why it was necessary to sum them.

iii) High Temperature region: $T \lesssim T_m$. It is very difficult to give simple power counting arguments in this region, since the effective potential and irreducible cumulants will be complicated functions of temperature which will depend on the form of the bare potential. In addition, the propagator will acquire significant τ dependence, so we may no longer approximate $\beta^{-1}\int_0^\beta d\tau$ by 1.

iv) Very High Temperatures $T \gg T_m$. If we let $T \to \infty(\beta \to 0)$ then $G \to \infty$ and $\chi_k \sim 1$ (for soft core interations). So the kth contribution to $\mathcal{H}^{ir} \sim \beta^{k-1}$ and we have a Mayer type expansion for a perfect gas.

Stability

In order to gain more insight into the high temperature behaviour of the SCP expansion, we now consider the qualitative behaviour of the free energy as a functional of the propagator matrices $G_{q\nu}$ (or equivalently of the effective force constants $\mathcal{D}_{q\nu}$) at the same fixed temperature [Choquard, 1967]. Denote by λ the magnitude of the quantity $\underline{\underline{G}} : \underline{\nabla}\,\underline{\nabla} \sim \delta/R$. Then we know from (3.15) that $\partial F/\partial\lambda = 0$, and that a stationary point is a minimum if we calculate F to all orders, since $\partial^2 F/\partial\lambda^2 > 0$.

Now consider truncating the irreducible cumulant expansion for F. Then provided λ is small enough, since $\delta^2\mathcal{H}^{ir}/\delta\lambda^2 \sim const. + 0(\lambda), \delta^2 F/\delta\lambda^2 \sim \lambda^{-2} + const + 0(\lambda)$ will be positive. However for larger values of λ it is quite possible that a truncated expression for $\delta^2\mathcal{H}^{ir}/\delta\lambda^2$ will change sign, and for some values of $\lambda, \delta^2 F/\delta\lambda^2 < 0$. Consider for example the first order approximation. Then classically

$$\frac{1}{N}F = -\tfrac{1}{2}kT\ell n(kT\lambda) - \tfrac{1}{2}\phi(\lambda)$$

where $\phi(\lambda)$ is the total potential energy calculated using the effective potential, and will have the generic form

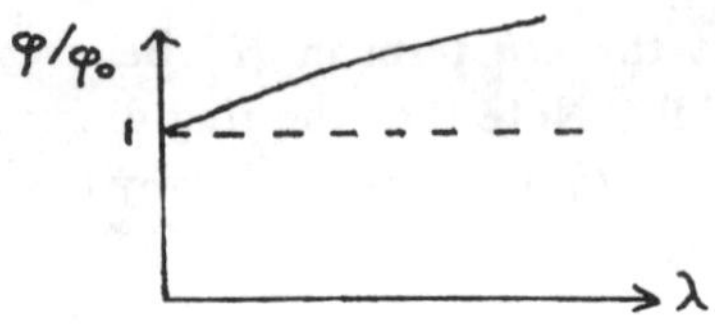

Thus F will have the form

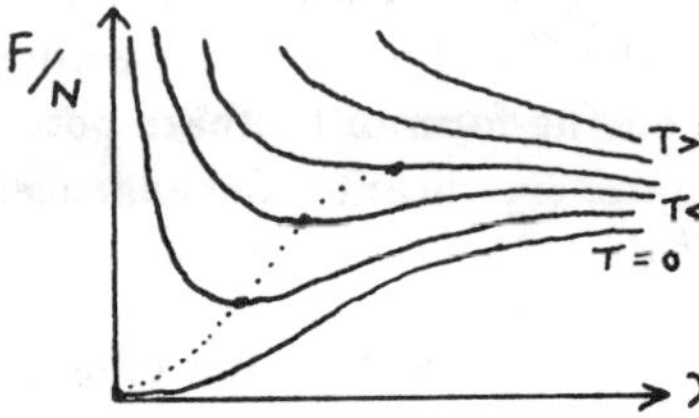

and we can see that as T increases, the effective value of λ increases (presumably approximately linearly) until at some $T = \bar{T}$, F no longer has a minimum (to this degree of approximation) and a stable value of λ no longer exists.

These arguments raise three basic questions which demand further investigation:

1) Is it possible to give λ (and hence G) a physical meaning away from the stationary point? More particularly, in nonequilibrium situations does it play the role of a generalised Boltzman H-function?

2) When F is truncated, can the stationary condition $\partial F/\partial\lambda = 0$ have more than one solution at a given temperature, and if so do the other solutions correspond to metastable configurations? For example, we know that the Hartree-Fock equations can indeed admit more than one solution. It is tempting to spectulate on the parallel between the evolution in time of a metastable system and the expansion of F: in both descriptions higher order terms involve more collisions, which lead ultimately to the decay of the metastable state. So the approximations made in treating metastable states as stable could be associated with truncations of the force energy functional that admit metastable states as solutions.

3) To what extent is one justified in considering truncations of the free energy functional? In particular does the truncation we have considered above have sensible properties, and how quickly does it converge to the exact free energy F?

If the above questions could be answered positively, self consistent phonons would give us a model of phase transitions in solids: schematically

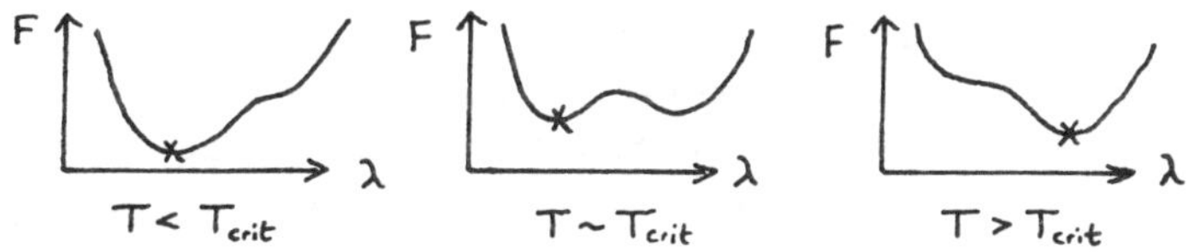

The sharpness of the transition should increase as we include more terms F. However in all the above discussion a single lattice structure is assumed throughout, which severely limits the usefulness of these ideas to phase transitions in which the lattice structure changes, and in particular to crystallization. Presumably if for $T > T_{crit}$ the stable value of $\lambda \gg 1$ i.e. $\delta \gg R$, the minimum will be an artifact of the model.

In a wider context, no such limitations need exist, indeed irreducible cumulant expansions have been used (in various guises) to examine the phases of $\lambda\phi^4$, dynamical symmetry breaking (and in particular chiral symmetry breaking) and even the confinement transition in QCD.

3.4 The Hard Core Problem

In the SCP scheme described above, the affects of soft multiple phonon scatterings are summed up into effective temperature dependent vertices and propagators which provide an optimal basis for the phonon scattering. However, since the trial Hamiltonian is built up of Gaussian distribution functions, nonvanishing (through small) penetrations of the atoms into the cores of their co-moving neighbours are permitted. This means that when calculating the effective potential $< v_\rho >$ by eqn. (2.4), one integrates over the core part of the bare potential function $v_\rho(\underline{R}_\rho)$. Now the soft long range parts of such potentials (including Coulomb potentials) are generally integrable, but if the potential has a hard core repulsive part the integration (2.4) may diverge.

This is in some ways analogous to the well known problem of ultraviolet divergencies, caused by short distance propagator singularities, which plague many quantum field theories (both polynomial and non-polynomial); as we shall see below it may be cured by a similar process of renormalization. Note however that when the Taylor expansion of the bare potential is truncated (approximation A3 of § 1.3) the hard core problem does not arise, since then multiple scatterings are omitted altogether. The hard core problem should thus be seen as an additional renormalization problem which may be present in the irreducible cumulant expansion of any nonpolynomial theory.

One obvious solution to the problem is to replace the Gaussian in eqn. (2.4) with a more exact expression for the two point correlation function

$g(r)$, which has the property that is everywhere equal to the appropriate Gaussian except at short distances (i.e. near a core) where it tends to zero very rapidly [Horner,1971]. It may be calculated by solving an appropriate scattering problem. This is equivalent to the Jastrow function approach [Jastrow, 1955, DeDominicis, 1962, Nosanow, 1966], where one considers a trial wave function

$$\psi(\underline{r}_1, \ldots \underline{r}_N) = \Pi_\rho f_\rho(\underline{r}_\rho) e^{-\frac{1}{4}\underline{u}_\rho \cdot \underline{\underline{\Gamma}}_\rho \cdot \underline{u}_\rho},$$

$$f_\rho(r) \to \begin{cases} 0 & r \to 0, \\ 1 & r \to \infty. \end{cases}$$

The Gaussian factor then gives the usual renormalized harmonic approximation and the Jastrow function f_ρ suppresses the hard cores. To find an appropriate form for the Jastrow function one usually assumes a trial form and then adopts a variational principle [Koehler and Werthamer, 1971, Werthamer, 1973]. These ideas may be developed further to an alternative perturbation procedure which takes into account both long and short range effects, called the 'hyper-netted chain expansion' [Ripka, 1979]. We will not pursue these ideas further here, as they are generally very involved, and only well understood at zero temperature.

An alternative approach would be to account for the multiple hard core scatterings by summing up additional classes of diagrams in our original perturbation expansion. In particular the Bruekner diagrams (or "ladder diagrams") [Bruekner and Sawada, 1957, Glyde and Khanna, 1971]

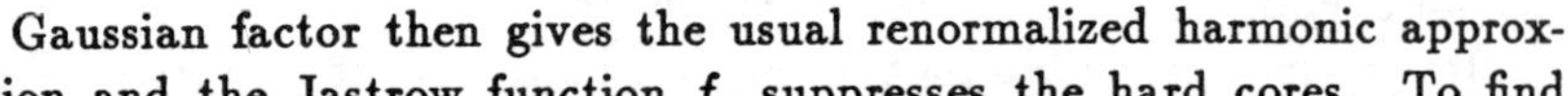

will include hard core effects by giving new effective vertices. One arrives in this way at an expansion similar to (2.4) but in which the integral converges. Formally the summation is achieved by solving the Bethe-Saltpeter equations

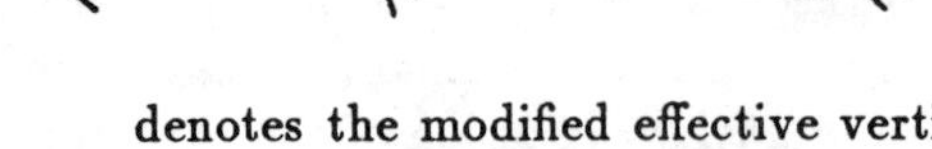

where denotes the modified effective vertices. However, the solution of these equations is notoriously difficult and will not be attempted here.

Instead, we shall consider the hard cores as only a minor modification of the SCP scheme. The potential $\phi_\rho(R_\rho)$ is separated into two parts which are then treated by different methods of approximation, namely a soft part $\phi_\rho(R_\rho)$ which gives rise to the self consistent phonons as described above, and a hard part $\phi_\rho^{(h)}(R_\rho)$ which we assume vanishes outside a typical hard core radius $d \ll$ the nearest neighbour distance a. This separation

$$\phi_\rho(R_\rho) = \phi_\rho^{(s)}(R_\rho) + \phi_\rho^{(h)}(R_\rho) \tag{3.8}$$

is certainly not unique however; we will return to this point later. The crystal potential $V = V_s + V_h$, with any external static potential included in V_s.

We may now formulate the hard-core problem as

$$Z = < e^{-\beta V_h} > Z_s$$

where

$$< e^{-\beta V_h} >_s = \int [\mathcal{D}u] e^{-S_s[u] - \beta V_h[u]},$$

and $S_s[u]$ is the action evaluated with potential V_s (cf. (11) and (12)). We must thus evaluate $< e^{-\beta V_h} >_s$ without having recourse to a cumulant type of expansion. We use instead the Mayer cluster expansion, originally developed to treat the imperfect gas (with hard short range interactions only).

The Mayer Cluster Expansion

Let μ denote a pair of indices (σ, σ'), where $\sigma' = \sigma + \rho > \sigma$. Then define

$$f_\mu \equiv 1 - e^{-\beta \phi_\mu^{(h)}(r_\mu)}, \quad r_\mu = | \underline{R}_\rho + \eta_\rho \underline{u}_s |,$$

(which is the Mayer function for the hard potential). So defined, f_μ varies from one to zero in a small region around $r = d$:

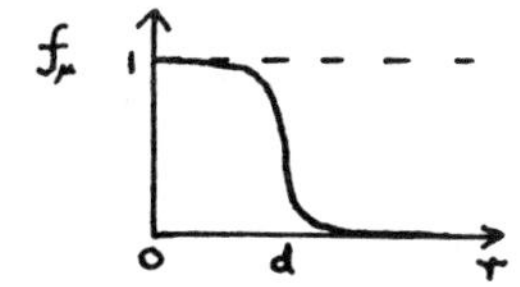

Then $e^{-\beta V_h} = \prod_\mu (1 - f_\mu)$.

Define

$$\begin{aligned} F_\mu &\equiv < f_\mu >_s , \\ F_{\mu\mu'} &\equiv < f_\mu f_{\mu'} >_s - < f_\mu >_s < f_{\mu'} >_s, \ldots, \end{aligned} \tag{3.9}$$

i.e. the cumulants $F_{\mu\mu'....}$ of f_μ, with the familiar property of depending only on relative indices, leading to an overall dependence on N as required. It is now easy to see that (3.9) may written as

$$\ln < e^{-\beta V_h} >_s = \sum_\mu \ln(1 - F_\mu) + \sum_{\mu<\mu'} \ln(1 + \frac{F_{\mu\mu'}}{(1-F_\mu)(1-F_{\mu'})}) + ...$$

In practice we can safely expand the logarithms to give

$$\begin{aligned} \ln < e^{-\beta V_h} >_s &= -\sum_\mu F_\mu - \tfrac{1}{2}\sum_\mu F_\mu^2 + \sum_{\mu<\mu'} F_{\mu\mu'} + ... \\ &= -\sum_\mu (F_\mu + \tfrac{1}{2}\tilde{F}_{\mu\mu}) + \tfrac{1}{2}\sum_{\mu\mu'} F_{\mu\mu'} + ..., \end{aligned} \tag{3.10}$$

where $\tilde{F}_{\mu\mu} =< f_\mu^2 >$. These terms constitute the beginning of a cumulant expansion applied to the evaluation of $< e^{-\beta V_h} >_s =< e^{\Sigma_\mu \ell n(1-f_\mu)} >$.

Now we expect all products and powers of mean values higher than the first to be small (or order $(d/a)^3$), so we can drop them to obtain the virial approximation to the hard core effects in solids:

$$\begin{aligned} \ln < e^{-\beta V_h} >_s &= -\sum_\mu F_\mu + \sum_{\mu<\mu'} F_{\mu\mu'}.... \\ &= < \prod_\mu (1 - f_\mu) - 1 >_{s,c} \end{aligned} \tag{3.11}$$

where the suffix C denotes connected graphs only.

<u>Pure Core Effects</u>

We now consider the ring diagram summation procedure which leads to the renormalized harmonic approximation; taking only the first term of (3.11) leads us directly to the result

$$\begin{aligned} F &= -\tfrac{1}{2}\beta^{-1}\sum_q tr\ell n\beta G_q - \tfrac{1}{2}\sum_q tr\mathcal{D}_q G_q \\ &+ \tfrac{1}{2}N\sum_\rho < \phi_\rho^{(s)} > + \tfrac{1}{2}N\beta^{-1}\sum_\rho < f_\rho >, \end{aligned} \tag{3.12}$$

where

$$\begin{aligned} \underline{\underline{\mathcal{D}_q}} &= N\frac{\delta}{\delta \underline{\underline{G_q}}}\sum_\rho \{ < \phi_\rho^{(s)} > + \beta^{-1} < f_\rho > \} \\ &= \frac{1}{M}\sum_\rho (1 - cos\underline{q}.\underline{R}_\rho)\underline{\nabla}\,\underline{\nabla} < \phi_\rho^{(s)} + \beta^{-1} f_\rho > \end{aligned} \tag{3.13}$$

We immediately notice that, in this extended renormalized harmonic approximation, the bare potential $\phi_\rho(\underline{R}_\rho)$ is modified to a bare pseudopotential

$$\phi^*(R) = \phi^{(s)}(R) + \beta^{-1}f(r) = \phi^{(s)}(R) + \beta^{-1}(1 - e^{-\beta\phi^{(h)}(R)}), \qquad (3.14)$$

smeared by the Greens function $\underline{\underline{G}}_q$ to become $< \phi^{(s)} > +\beta^{-1} < f >$. Thus $\beta^{-1} < f >$ acts as a repulsive potential modifying through its tail the effective force constants (by only small amounts under normal conditions). However, the non-analytic character of this effect must be kept in mind.

We may now offer an answer to the non-uniqueness question posed by the separation eqn.(3.8). Assuming a particular form for $\phi^{(h)}(R)$, with $|\phi^{(h)}(R)| \ll |\phi^{(s)}(R)|, R > d$, and $|\phi^{(h)}(R) \gg |\phi^{(s)}(R)|, R < d$, let $\xi(R)$ be a function which parameterises the separation: $\xi(R) \sim 0$ for $R > d$, and

$$\phi(R) = (1 - \xi(R))\phi^{(s)}(R) + (\phi^{(h)}(R) + \xi(R)\phi^{(s)}(R)),$$

so that

$$\phi^*(R) = (1 - \xi(R))\phi^{(s)}(R) + \beta^{-1}(1 - e^{-\beta(\phi^{(h)}(R)+\xi(R)\phi^{(s)}(R))}).$$

Then

$$\frac{\delta\phi^*(R)}{\delta\xi(R)} = \phi^{(s)}(R)(1 - e^{-\beta(\phi^{(h)}(R)+\xi(R)\phi^{(s)}(R))})$$

which is much smaller than $\phi^*(R)$ over the whole range of R so that $\phi^*(R)$ is indeed approximately independent of the separation (3.8). However the form of ξ adopted may affect the convergence rate of the virial expansion (3.11).

Schematically the three potentials ϕ, ϕ^* and $< \phi^* >$ will take the form

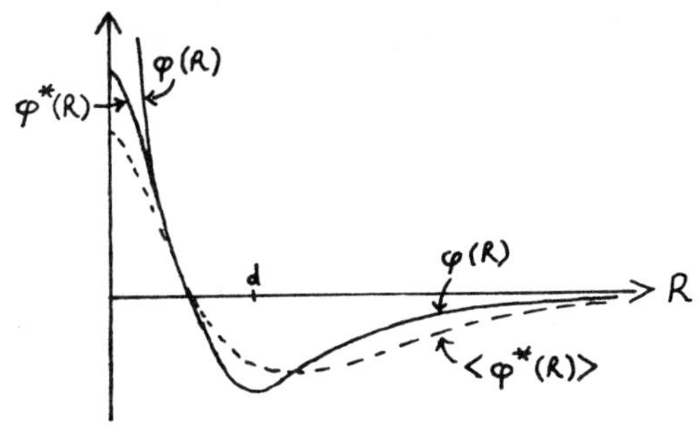

Higher Order Effects

It is now straightforward to calculate higher order corrections (3.12)-(3.13). Firstly, we have to take into account the remaining terms in the virial expansion (3.11), to give in the renormalized harmonic approximation

$$-\beta\mathcal{H}_h^{ir} = -\sum_\mu < f_\mu > + < \prod_\mu(1 - f_\mu) - 1 >_{s,ir,c}$$

where the second term includes all irreducible connected graphs with vertices f_ρ and propagators $\underline{\underline{G}}_q$. For example the second order term is

$$\tfrac{1}{2}N\sum_{\rho\rho'\sigma}\epsilon_3(\underline{\underline{G}}^\sigma_{\rho\rho'} : \underline{\nabla}\,\underline{\nabla}') < f_\rho >< f_{\rho'} >\equiv$$  .

Secondly, we consider the effect due to interference between the hard part and the soft part of the potential. To do this we must go beyond the renormalized harmonic approximation. Since

$$F_{\mu_1\ldots\mu_n} =< f_{\mu_1\ldots.f_{\mu_n}} >_{W,c} + < f_{\mu_1}\ldots f_{\mu_n}(e^{-\beta V_s} - 1) >_{W,ir,c},$$

and noting that the first terms belong to $\mathcal{H}_h^{ir}$, we may define the remainder as

$$\begin{aligned}\mathcal{H}_{sh}^{ir} &= < (\prod_\mu(1 - f_\mu) - 1)(e^{-\beta V_s} - 1) >_{W,ir,c} \\ &= \sum_{n=1}^{\infty}\frac{(-\beta)^n}{n!} < \prod_\mu(1 - f_\mu) - 1)V^n >_{W,ir,c}\,.\end{aligned}$$

So finally we obtain the result

$$F = -\tfrac{1}{2}\beta^{-1}\sum_q tr\ell n\beta G_q - \tfrac{1}{2}\sum_q tr\mathcal{D}_q G_q + \mathcal{H}^{ir}\,,$$

where

$$\mathcal{H}^{ir} = \mathcal{H}_s^{ir} + \mathcal{H}_h^{ir} + \mathcal{H}_{sh}^{ir},$$

and

$$\underline{\underline{\mathcal{D}}}_q = \frac{\delta\mathcal{H}}{\frac{1}{2}\delta\underline{\underline{G}}_q}\,.$$

$\mathcal{H}_h^{ir} + \mathcal{H}_{sh}^{ir}$ may be thought of as counterterms to the originally divergent expression $\mathcal{H}^{ir}$. For example, in second order, the irreducible cumulant is now given by

$$\underbrace{\ominus}_{(\mathcal{H}_s^{ir})_2} + \underbrace{\ominus\!\!* + *\!\!\ominus}_{(\mathcal{H}_{sh}^{ir})_2} + \underbrace{*\!\!\ominus\!\!*}_{(\mathcal{H}_h^{ir})_2}$$

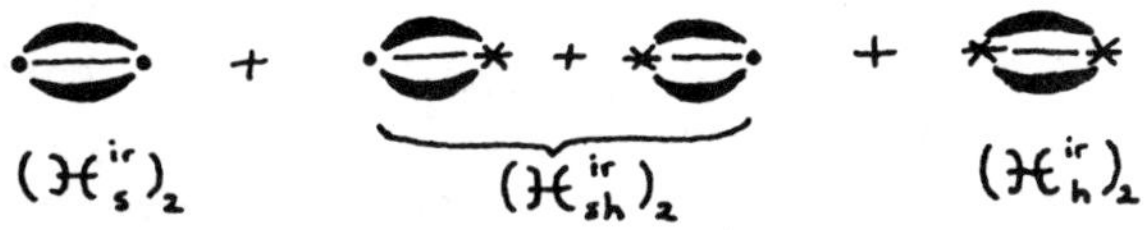

where the middle terms remove the divergencies in each vertex, while the last one removes the overall divergence.

4 Physics from Phonons

4.1 Thermodynamics of Crystals

Once we have evaluated the free energy F of the crystal, other quantities of interest - the entropy, internal energy, specific heats, equation of state, stress tensor, etc. may be obtained (at least in principle) by functional differentiation. In particular, we have to first order in derivatives the entropy,

$$S = -\left(\frac{\partial F}{\partial T}\right)_V , \tag{4.1}$$

the internal energy,

$$U = F + TS, \tag{4.2}$$

the thermal pressure,

$$P = -\left(\frac{\partial F}{\partial V}\right)_T , \tag{4.3}$$

which when written in terms of U and T gives an equation of state for the crystal, and the stress energy tensor

$$\sigma_{ij} = \frac{1}{V}\left(\frac{\partial F}{\partial u_{ij}}\right)_T , \tag{4.4}$$

where u_{ij} are homogeneous linear strain totation parameters; under such a strain

$$\underline{R}_\ell \to \underline{R}'_\ell = (\underline{\underline{1}} + \underline{\underline{u}}).\underline{R}_\ell .$$

In the harmonic approximation, we have (1.14))

$$F = \beta^{-1} \sum_q ln(2 \sinh \tfrac{1}{2}\beta_q), \tag{4.5}$$

so that

$$S \;=\; k \sum_q \left(\tfrac{1}{2}\beta_q \coth \tfrac{1}{2}\beta_q - ln(2 \sinh \tfrac{1}{2}\beta_q)\right), \tag{4.6}$$

$$U = \beta^{-1} \sum_q \tfrac{1}{2}\beta_q \coth \tfrac{1}{2}\beta_q, \tag{4.7}$$

and

$$P = V^{-1}\beta^{-1} \sum_q \left(\tfrac{1}{2}\beta_q \coth \tfrac{1}{2}\beta_q\right) \gamma_q, \tag{4.8}$$

where

$$\gamma_q \equiv \left(\frac{\partial ln\omega_q}{\partial lnV}\right)_T, \tag{4.9}$$

are the mode Gruneisen parameters. Defining the Gruneisen constant

$$\gamma \equiv \frac{\sum_q \frac{1}{2}\beta_q \coth \frac{1}{2}\beta_q \gamma_q}{\sum_q \frac{1}{2}\beta_q \coth \frac{1}{2}\beta_q}, \tag{4.10}$$

the equation of state (4.8) takes on the familiar Mie-Gruneisen form

$$P = \frac{U}{V}\gamma. \tag{4.11}$$

This approximation, in which anharmonic corrections to the phonon spectrum are ignored, but the strain dependence of the frequencies is retained and used to calculate elastic properties, is called the "quasiharmonic approximation" [Born and Huang, 1954].

For anisotropic strains, (4.8)-(4.11) may be generalized to

$$\sigma_{ij} = V^{-1}\beta^{-1} \sum_q \left(\tfrac{1}{2} \coth \tfrac{1}{2}\beta_q\right) (\gamma_q)_{ij} = \frac{U\gamma_{ij}}{V},$$

where

$$(\gamma_q)_{ij} = \left(\frac{\partial ln\omega_q}{\partial u_{ij}}\right)_T. \tag{4.12}$$

In the classical, or high temperature, limit ($T \to \infty, \beta \to 0$), we have

$$F \simeq \sum_q \hbar\omega_q \simeq 0, \qquad S \simeq 3Nk, \qquad U \simeq 3NkT,$$

and

$$P = \gamma U/V = 3\gamma NkT/V \tag{4.13}$$

where now the Gruneisen constant $\gamma = \frac{1}{3N}\sum_q \gamma_q$; each mode carries equal weight.

Quantities which depend on several orders of partial derivatives of the free energy are also of interest. Differentiating the pressure and internal energy with respect to V and T, we have the isothermal compressibility

$$\frac{1}{\chi_T} = -V\left(\frac{\partial P}{\partial V}\right)_T, \tag{4.14}$$

the specific heat at constant volume

$$C_V = T\left(\frac{\partial S}{\partial T}\right)_V = \left(\frac{\partial U}{\partial T}\right)_V, \tag{4.15}$$

and the latent heat of expansion

$$L_V = T\left(\frac{\partial P}{\partial T}\right)_V = P + \left(\frac{\partial U}{\partial V}\right)_T, \tag{4.16}$$

which may be taken the three fundamental quantities. Using the reciprocity theorem

$$\left(\frac{\partial V}{\partial P}\right)_T \left(\frac{\partial P}{\partial T}\right)_V \left(\frac{\partial T}{\partial V}\right)_P = -1$$

we can relate these quantities to other well known second order quantities, for example the coefficient of thermal expansion (or volume expansivity)

$$\alpha_p = \frac{1}{V}\left(\frac{\partial V}{\partial T}\right)_P = \frac{1}{T}\chi_T L_V, \tag{4.17}$$

the specific heat at constant pressure

$$C_P = T\left(\frac{\partial S}{\partial T}\right)_P = c_V + \frac{VT\alpha_p^2}{\chi_T} = c_V + V\alpha_p L_V = c_V + \frac{V}{T}\chi_T L_V^2, \tag{4.18}$$

and the adiabatic compressibility

$$\chi_S = \frac{C_V}{C_p}\chi_T. \tag{4.19}$$

In the harmonic approximation, we may use the expressions (4.6)-(4.8) to give

$$
\begin{aligned}
\frac{1}{\chi_T} &= V^{-1}\beta^{-1}\sum_q \left(\tfrac{1}{2}\beta_q \coth \tfrac{1}{2}\beta_q\right)\left(\gamma_q - \gamma_q'\right), \\
C_V &= k\sum_q \left(\tfrac{1}{2}\beta_q \operatorname{cosech}\tfrac{1}{2}\beta_q\right)^2, \\
L_V &= V^{-1}\beta^{-1}\sum_q \left(\tfrac{1}{2}\beta_q \operatorname{cosech}\tfrac{1}{2}\beta_q\right)^2 \gamma_q,
\end{aligned}
\tag{4.20}
$$

where $\gamma_q' \equiv \left(\frac{\partial \gamma_q}{\partial \ln V}\right)_T$.

In the classical limit these reduce to

$$
\begin{aligned}
\frac{1}{\chi_T} &\simeq (\gamma - \gamma')\frac{3NkT}{V}, \\
C_V &\simeq 3Nk\,(\mathit{the\ Dulong - Petit\ law}), \\
L_V &\simeq \left(\frac{3NkT}{V}\right) \qquad \alpha_p \simeq \frac{1}{T}\frac{\gamma}{\gamma - \gamma'}, \\
C_p &\simeq 3Nk\left(1 + \frac{\gamma^2}{\gamma - \gamma'}\right), \quad \frac{1}{\chi_S} \simeq (\gamma + \gamma^2 - \gamma')\frac{3NkT}{V}.
\end{aligned}
\tag{4.21}
$$

The low temperature limit is more subtle since it depends on the details of the spectrum; in the Debye approximation $C_V \sim C_p \sim T^3, L_V \sim T^4$, and $\alpha_p \sim T^3$ as $T \to 0$.

4.2 Self Consistent Thermodynamics

We now consider the calculation of the first and second derivatives of the free energy with respect to external macroscopic parameters, for interacting phonons, and in particular for the interacting self consistent phonons described in section II. This is of more theoretical than practical value - it allows us to make contact with (in certain approximations) various conjectured forms for the equation of state, for example. In practice it is probably most convenient to evaluate all derivatives of the free energy numerically.

Consider the free energy F as a thermodynamic function depending upon T,V (or alternatively the inhomogeneous strain parameters u_{ij}) and upon the set of propagator functions $G_{q\nu}$. The latter are regarded here as additional, though intermediate, state variables which are ultimately eliminated by means of the self consistency equations (2.25). This point of view is suggested by the fact that the pressure, internal potential energy and dynamical matrices are given in terms of F and its first order partial derivatives at fixed $(T, G_{q\nu})$; $(V, G_{q\nu})$ and (V, T) respectively. A given set of the matrices $\{G_{q\nu}\}$ will characterize a state of motional fluctuations of the system, and this state will be labelled by $\lambda \equiv \{\frac{1}{M} G_{q\nu}\}$ (we divide by the mass M for reasons of dimensionality).

It will prove useful to introduce a representation for $\mathcal{H}^{ir}$ as a parametric integral. Consider switching on the propagator functions $G_{q\nu}$ from zero to their actual value and thus replacing $G_{q\nu}$ by $\xi G_{q\nu}, 0 \leq \xi \leq 1$. Then we integrate (2.25) by writing (ignoring static energies)

$$\begin{aligned} \mathcal{H}^{ir}(T, U, \lambda) &= \int_0^1 d\xi \frac{d\mathcal{H}^{ir}}{d\xi}(T, V, \xi\lambda) \\ &= \tfrac{1}{2} \sum_{q\nu} tr \underline{\underline{G}}_{q\nu} \int_0^1 d\xi \; \underline{\underline{\mathcal{D}}}^{\lambda'}_{q\nu}(T, V), \end{aligned} \tag{4.22}$$

where $\lambda' \equiv \xi\lambda$, and $D^{\lambda'}_{q\nu}(T, V)$ is calculated by replacing $G_{q\nu}$ by $\xi G_{q\nu}$ wherever they occur in the evaluation of the appropriate diagrams. Note in this context that

$$\frac{d}{dx}\varepsilon_n(x) = \varepsilon_{n-1}(x), \qquad n \geq 1. \tag{4.23}$$

<u>First Order Derivatives</u>

Using the representation (4.22) , we may write the pressure (4.3) in the form

$$P = -\left(\frac{\partial F}{\partial V}\right)_{T,\lambda} = -\left(\frac{\partial \mathcal{H}^{ir}}{\partial V}\right)_{T,\lambda} = \frac{1}{V} \sum_{q\nu} tr \underline{\underline{G}}_{q\nu} \int_0^1 d\xi (-\tfrac{1}{2}) \left(\frac{\delta \underline{\underline{\mathcal{D}}}^{\lambda'}_{q\nu}}{\delta ln V}\right)_{T,\lambda'} \tag{4.24}$$

So it is useful to define "mode Gruneisen parameters"

$$\underline{\underline{\gamma}}^{(T\lambda)}_{q\nu} \equiv -\tfrac{1}{2} (\underline{\underline{\mathcal{D}}}_{q\nu})^{-1} \left(\frac{\delta \underline{\underline{\mathcal{D}}}_{q\nu}}{\delta ln V}\right)_{T,\lambda} \tag{4.25}$$

or, writing

$$\underline{\underline{\gamma}}_{q\nu}^{(T\lambda)} = \frac{1}{\beta}\int_0^\beta d\tau e^{i\Omega_\nu\tau}\underline{\underline{\gamma}}_{q}^{(T\lambda)}(\tau), \tag{4.26}$$

$$-\tfrac{1}{2}\left(\frac{\delta\underline{\underline{\mathcal{D}}}_{q\nu}}{\delta lnV}\right)_{T,\lambda} = \frac{1}{\beta}\int_0^\beta d\tau'\underline{\underline{\mathcal{D}}}_{q}^{\lambda}(\tau-\tau').\underline{\underline{\gamma}}_{q}^{(T\lambda)}(\tau'). \tag{4.27}$$

Then the pressure (4.24) becomes

$$P = \frac{1}{V}\sum_{\underline{q}\nu} tr\int_0^1 d\xi\underline{\underline{G}}_{q\nu}\underline{\underline{\mathcal{D}}}_{q\nu}^{\lambda'}\underline{\underline{\gamma}}_{q\nu}^{(T\lambda')}. \tag{4.28}$$

Now if we define an averaging operation

$$\overline{X_{q\nu}^{\lambda}} \equiv \left(\mathcal{D}_{q\nu}^{\lambda}\right)^{-1}\int_0^1 d\xi\mathcal{D}_{q\nu}^{\lambda'}X_{q\nu}^{\lambda'}, \tag{4.29}$$

then (4.28) becomes

$$P = \frac{1}{V}\sum_{\underline{q}\nu} tr\underline{\underline{G}}_{q\nu}\underline{\underline{\mathcal{D}}}_{q\nu}\bar{\underline{\underline{\gamma}}}_{q\nu}^{(T\lambda)}. \tag{4.30}$$

In the classical theory

$$\tfrac{1}{2}E = \tfrac{1}{2}tr\sum_{\underline{q}\nu}\underline{\underline{G}}_{q\nu}\underline{\underline{\mathcal{D}}}_{q\nu} = \tfrac{3}{2}NkT \tag{4.31}$$

which is just the mean kinetic energy, so 4.30) constitutes a weighted average of the classical contributions to the kinetic energy. We thus define a suitable generalization of the classical Gruneisen parameter γ by

$$\gamma_1 \equiv tr\sum_{\underline{q}\nu}\underline{\underline{G}}_{q\nu}\underline{\underline{\mathcal{D}}}_{q\nu}\bar{\underline{\underline{\gamma}}}_{q\nu}^{(T\lambda)}\Big/ tr\sum_{\underline{q}\nu}\underline{\underline{G}}_{q\nu}\underline{\underline{\mathcal{D}}}_{q\nu}, \tag{4.32}$$

so the equation of state reads

$$P = \frac{E}{V}\gamma_1, \tag{4.33}$$

which goes over into the familiar Mie-Gruneisen equation of state (4.11) in the harmonic approximation (where $\gamma_1 = \gamma$).

For an anisotropic situation, the same arguments give the stress energy tensor

$$\sigma_{ij} = \frac{1}{V}\sum_{q\nu} tr\underline{\underline{G}}_{q\nu}\underline{\underline{\mathcal{D}}}_{q\nu}(\bar{\underline{\underline{\gamma}}}^{(T\lambda)}_{q\nu})_{ij},$$

where

$$(\underline{\underline{\gamma}}^{(T\lambda)}_{q\nu})_{ij} = -\tfrac{1}{2}(\underline{\underline{\mathcal{D}}}_{q\nu})^{-1}\left(\frac{\delta\underline{\underline{\mathcal{D}}}_{q\nu}}{\delta u_{ij}}\right)_{T,\lambda}. \tag{4.34}$$

For the internal energy

$$U = \left(\frac{\partial}{\partial\beta}(\beta F)\right)_V = \left(\frac{\partial}{\partial\beta}(\beta F)\right)_{V,\beta^2\mathcal{D}},$$

since

$$-\beta F = \tfrac{1}{2}\sum_{q\nu} tr(\beta^2\underline{\underline{\mathcal{D}}}_{q\nu})(\beta^{-1}\underline{\underline{G}}_{q\nu}) - \beta\mathcal{H}^{ir}(\beta, V, G) + \tfrac{1}{2}\sum_{q\nu} tr ln\beta^{-1}\underline{\underline{G}}_{q\nu},$$

so

$$\begin{aligned} U &= \left(\frac{\partial}{\partial\beta}(\beta F)\right)_{V,\beta^{-1}G} = \left(\frac{\partial}{\partial\beta}(\beta\mathcal{H}^{ir})\right)_{V,\beta^{-1}G} \\ &= \left(\frac{\partial}{\partial\beta}(\beta\mathcal{H}^{ir})\right)_{V,\lambda} + \sum_{q\nu} tr\left(\frac{\partial(\beta\mathcal{H}^{ir})}{\partial\underline{\underline{G}}_{q\nu}}\right)_{V,\beta}\left(\frac{\partial(\beta^{-1}\underline{\underline{G}}_{q\nu})}{\partial\beta}\right)_{V,\underline{\underline{G}}_{q\nu}} \\ &= \mathcal{H}^{ir} - T\left(\frac{\partial\mathcal{H}^{ir}}{\partial T}\right)_{V,\lambda} + \tfrac{1}{2}\sum_{q\nu} tr\underline{\underline{\mathcal{D}}}_{q\nu}\underline{\underline{G}}_{q\nu}. \end{aligned} \tag{4.35}$$

Introducing the isochoric fluctuation matrices

$$\underline{\underline{\gamma}}^{(V\lambda)}_{q\nu} = -\tfrac{1}{2}(\underline{\underline{\mathcal{D}}}_{q\nu})^{-1}\left(\frac{\delta\underline{\underline{\mathcal{D}}}_{q\nu}}{\delta lnT}\right)_{V,\lambda}, \tag{4.36}$$

we have, using (4.22)

$$\begin{aligned} U &= \tfrac{1}{2}\sum_{\underline{q\nu}} tr\underline{\underline{G}}_{q\nu}\underline{\underline{\mathcal{D}}}_{q\nu} + \tfrac{1}{2}\sum_{\underline{q\nu}} tr\underline{\underline{G}}_{q\nu}\int_0^1 d\xi \underline{\underline{\mathcal{D}}}^{\lambda'}_{q\nu}(1+2\underline{\underline{\gamma}}^{(V\lambda')}_{q\nu}) \\ &= \tfrac{1}{2}\sum_{\underline{q\nu}} tr\underline{\underline{G}}_{q\nu}\underline{\underline{\mathcal{D}}}_{q\nu}\{\underline{\underline{1}} + (\underline{\underline{\bar{1}}} + 2\underline{\underline{\bar{\gamma}}}^{(V\lambda)}_{q\nu})\} \\ &= E(1+\gamma_2), \end{aligned} \tag{4.37}$$

where

$$\gamma_2 = \tfrac{1}{2}(1-\bar{1}) + \sum_{\underline{q\nu}} tr\underline{\underline{G}}_{q\nu}\underline{\underline{\mathcal{D}}}_{q\nu}\underline{\underline{\bar{\gamma}}}^{(V\lambda)}_{q\nu} \Big/ \sum_{\underline{q\nu}} tr\underline{\underline{G}}_{q\nu}\underline{\underline{\mathcal{D}}}_{q\nu}, \tag{4.38}$$

and $\bar{1} = \mathcal{D}^{-1}_{\underline{q\nu}} \int_0^1 d\xi \mathcal{D}^{\lambda'}_{\underline{q\nu}} \neq 1$ in general.

In the harmonic approximation $\underline{\underline{\gamma}}^{(V\lambda)}_{q\nu} = 0, \bar{1} = 1$, and $\gamma_2 = 0$, which explains why such a parameter does not arise in the usual harmonic treatment.

The two sets of parameters $\underline{\underline{\gamma}}^{(T\lambda)}_{q\nu}$ and $\underline{\underline{\gamma}}^{(V\lambda)}_{q\nu}$ which characterize the first order variations of the free energy may be calculated directly by diagramatic methods; one adds extra external vertices $\cdots\otimes$ to the vacuum diagrams, whose contribution can be determined by varying the classical action with respect to lnV and lnT respectively. To second order, these diagrams take the form

where

which is the only contribution in the quasiharmonic approximation.

However, it is probably easier from a purely practical point of view to calculate $\underline{\underline{\gamma}}^{(T\lambda)}_{q\nu}$ and $\underline{\underline{\gamma}}^{(V\lambda)}_{q\nu}$ directly from their definitions (4.25) and (4.36) by numerical differentiation (making small changes in the external parameters V and T, holding the correlation functions λ fixed) than to attempt to evaluate a new set of diagrams.

Second Order Derivatives

Explicit evaluation of the second order partial derivaties of F, and thus compressibilities, specific heats, and expansion coefficients, is rather elaborate; we will only sketch the outlines here, and refer the interested reader to [Choquard, 1967] for further details.

We first consider the classical limit ($\beta \to 0$) for simplicity. Then using the definitions (4.26) and (4.36) for $\underline{\underline{\gamma}}_{\underline{q}}^{(T\lambda)}$ and $\underline{\underline{\gamma}}_{\underline{q}}^{(V\lambda)}$ (which have no ν dependence in the classical limit), we define

$$\underline{\underline{\gamma}}_{\underline{q}}^{(T\lambda)\prime} = \left(\frac{\partial \underline{\underline{\gamma}}_{\underline{q}}^{(T\lambda)}}{\partial \ln V}\right)_{T,\lambda},$$

$$\underline{\underline{\dot{\gamma}}}_{\underline{q}}^{(T\lambda)} = \left(\frac{\partial \underline{\underline{\gamma}}_{\underline{q}}^{(T\lambda)}}{\partial \ln T}\right)_{V,\lambda}, \qquad (4.39)$$

and the matrix

$$\tfrac{1}{2} C_{\underline{q}\,\underline{q}'} = \underline{\underline{G}}_{\underline{q}} \cdot \frac{\delta^2 H^{ir}(T,V,\lambda)}{\delta \underline{\underline{G}}_{\underline{q}} \, \delta \underline{\underline{G}}_{\underline{q}'}} \cdot \underline{\underline{G}}_{\underline{q}'}^{\dagger}. \qquad (4.40)$$

Then the isothermal Gruneisen matrix

$$\underline{\underline{\gamma}}_{\underline{q}}^{(T)} \equiv -\tfrac{1}{2}(\underline{\underline{D}}_{\underline{q}})^{-1} \cdot \left(\frac{\partial \underline{\underline{D}}_{\underline{q}}}{\partial \ln V}\right)_{T}, \qquad (4.41)$$

and isochoric Gruneisen matrix

$$\underline{\underline{\gamma}}_{\underline{q}}^{(V)} = -\tfrac{1}{2}(\underline{\underline{D}})_{\underline{q}}^{-1} \cdot \left(\frac{\partial \underline{\underline{D}}_{\underline{q}}}{\partial \ln T}\right)_{V}, \qquad (4.42)$$

satisfy the integral equations

$$\gamma_q^{(T)} = \gamma_q^{(T\lambda)} - \sum_{q'} \beta^{-1} C_{qq'} : \gamma_{q'}^{(T)},$$

$$\gamma_q^{(V)} = \gamma_q^{(V\lambda)} - \sum_{q'} \beta^{-1} C_{qq'} : \gamma_{q'}^{(V)}, \qquad (4.43)$$

whence

$$\gamma_q^{(T)} = \sum_{q'} (1 + \beta^{-1} C)_{qq'}^{-1} : \gamma_{q'}^{(T\lambda)}$$

$$\gamma_q^{(V)} = \sum_{q'} (1 + \beta^{-1} C)_{qq'}^{-1} : \gamma_{q'}^{(V\lambda)}. \qquad (4.44)$$

Then, using the definitions (4.14)-(4.16) and the expressions (4.30) and (4.37) for P and U, one can show after a considerable amount of algebra that

$$\begin{aligned} \frac{1}{\chi_T} &= \frac{3N\kappa T}{V}\gamma_{11}, \\ C_V &= 3N\kappa\gamma_{22}, \\ L_V &= \frac{3N\kappa T}{V}\gamma_{12}, \end{aligned} \tag{4.45}$$

with the three new Gruneisen parameters γ_{11}, γ_{22} and γ_{12} defined by

$$\begin{aligned} \gamma_{11} &= \frac{1}{3N}\sum_q tr(\overline{\gamma_q^{(T\lambda)}} + 2\overline{\gamma_q^{(T\lambda)^2}} - \overline{\gamma_q^{(T\lambda)\prime}}) \\ &- \frac{2}{3N}\sum_{qq'}\gamma_q^{(T\lambda)\dagger} : (1+\beta^{-1}C)^{-1}_{qq'} : \gamma_{q'}^{(T\lambda)} \\ \gamma_{22} &= \frac{1}{2} - \frac{1}{3Nk}T\Big(\frac{\partial^2 H^{ir}}{\partial T^2}(T,V,\lambda)\Big)_{V,\lambda} \\ &+ \frac{1}{3N}\frac{1}{2}\sum_{qq'}(1+2\gamma_q^{(V\lambda)})^\dagger : (1+\beta^{-1}C)^{-1}_{qq'} : (1+2\gamma_{q'}^{(V\lambda)}), \\ \gamma_{12} &= \frac{1}{3N}\sum_q tr(-2\overline{\gamma_q^{(V\lambda)}\gamma_q^{(T\lambda)}} + \overline{\dot{\gamma}_q^{(T\lambda)}}) \\ &+ \frac{1}{3N}\sum_{qq'}\gamma_q^{(T\lambda)\dagger} : (1+\beta^{-1}C)^{-1}_{qq'} : (1+2\gamma_{q'}^{(V\lambda)}). \end{aligned} \tag{4.46}$$

It is useful to consider the diagrammatic representation of these expressions. In fact to first order (ie in the renormalized quasiharmonic approximation) they represent the sum of an infinite sequence of diagrams of the generic form [Götz and Michel, 1968, Werthamer, 1970]

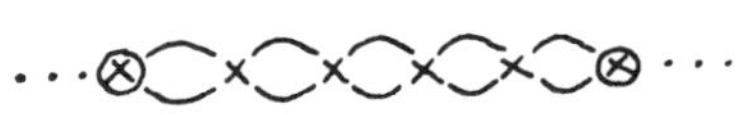

where 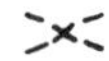 represents the vertex $\partial^2\mathcal{H}/\partial G\partial G$. Indeed, the Dyson equations (4.43) which generate these diagrams can be written

This complication arises essentially because the propagators themselves acquire a (V,T) dependence through the self consistency condition, and this must be included in the calculation of second order derivatives. Numerical calculations show that even at quite low temperatures, $\gamma^{(V)}$ and $\gamma^{(V\lambda)}$ can indeed be significantly different [Klein et al., 1970].

A useful extension of the Mie-Gruneisen scheme (summarized by (4.21)) may be obtained by working in the renormalized harmonic approximation, so that D_q depends on V and λ, but not T explicitly, and the isochoric isofluctuational derivatives $\gamma_q^{(V\lambda)}$ are zero, as are $\dot{\gamma}_q^{(T\lambda)}$. We may also make an isotropy assumption on the vertex functions C; we write in place of (4.40)

$$\frac{\delta^2 \mathcal{H}^{ir}}{\delta \underline{\underline{G}}_q \delta \underline{\underline{G}}_{q'}} \simeq \underline{\underline{D}}_q \frac{1}{N} Y(V,\lambda) \underline{\underline{D}}_{q'}, \tag{4.47}$$

where the scalar function $Y(V,\lambda)$ is the trace of C. In fact $Y \sim < \nabla \nabla \nabla \nabla \varphi > / < \nabla \nabla \varphi >^2$ and measures the anharmonicity of the renormalized potential. With these assumptions, the results (4.46) becomes

$$\begin{aligned}
\gamma_{11} &= \overline{\gamma^{(T\lambda)}} - \overline{\gamma^{(T\lambda)'}} + 2(\overline{\gamma^{(T\lambda)^2}} - \gamma^{(T\lambda)^2}) \\
&+ 2kTY(1+kTY)^{-1}\gamma^{(T\lambda)^2}, \\
\gamma_{22} &= (1+\tfrac{1}{2}kTY)/(1+kTY), \\
\gamma_{12} &= (1+kTY)^{-1}\gamma^{(T\lambda)} = \gamma^{(T)},
\end{aligned} \tag{4.48}$$

where $\gamma^{(T\lambda)} \equiv \frac{1}{3N}\sum_q \gamma_q^{(T\lambda)}$.

In the harmonic approximation $Y = 0$, so $\gamma_{11} = \gamma - \gamma', \gamma_{22} = 1, \gamma_{12} = \gamma$, as required. Note that the usual relation $VL_V/TC_V = \gamma$ is no longer true away from the harmonic approximation, justifying our adoption of L_V as a fundamental physical quantity.

In the quantum case, the situation is considerably more involved; here we will only give the results in the renormalized harmonic approximation and isotropic approximation (4.47) described above. We will also assume that the Gruneisen parameters are independent of q. Then instead of (4.45) and (4.48) we now have

$$
\begin{aligned}
\frac{1}{\chi_T} &= \frac{E^h}{V}(\gamma_{11}^a - \gamma_{11}^b) - \frac{C^h T \gamma_{11}^b}{V}, \\
C_V &= C^h \gamma_{22}, \\
L_V &= \frac{C^b T \gamma_{12}}{V},
\end{aligned}
\tag{4.49}
$$

where, if Ω_q^2 are the eigenvalues of $\underline{\underline{D}}_q$, the internal energy and specific heat in the renormalized harmonic approximation are (cf. (4.7), (4.20))

$$
\begin{aligned}
E^h &= \sum_q \tfrac{1}{2}\hbar\Omega_q \coth \tfrac{1}{2}\beta\hbar\Omega_q, \\
C^h &= k\sum_q (\tfrac{1}{2}\beta\hbar\omega_q cosech \tfrac{1}{2}\beta\hbar\Omega_q)^2,
\end{aligned}
\tag{4.50}
$$

and the Gruneisen parameters are given by

$$
\begin{aligned}
\gamma_{11}^a &= \overline{\gamma^{(T\lambda)}} + 2\overline{\gamma^{(T\lambda)^2}} - \overline{\gamma^{(T\lambda)'}}, \\
\gamma_{11}^b &= \gamma^{(T\lambda)}(1 + \tfrac{1}{2}(\varepsilon + \varepsilon')Y)^{-1}\gamma^{(T\lambda)} = \gamma^{(T\lambda)}\gamma^{(T)}, \\
\gamma_{22} &= (1 + \tfrac{1}{2}\varepsilon Y)/(1 + \tfrac{1}{2}(\varepsilon + \varepsilon')Y), \\
\gamma_{12} &= (1 + \tfrac{1}{2}(\varepsilon + \varepsilon')Y)^{-1}\gamma^{(T\lambda)} = \gamma^{(T)}, \\
\varepsilon &= \frac{1}{N}E^h, \quad \varepsilon' = \frac{1}{N}C^h T.
\end{aligned}
\tag{4.51}
$$

In the classical limit $\varepsilon = \varepsilon' = kT$ and the classical results (4.45) and (4.48) are recovered.

Obviously one could generalize this framework by dropping the assumptions of q-independence or isotropy, or allowing $\underline{\underline{D}}_q$ some explicit temperature dependence. However, in numerical work it is probably simplest to evaluate χ_T, C_V and L_V explicitly by numerical differentiation of the free energy. The above analytic results (4.45)-(4.51) could then be fitted to the resulting curves to investigate the accuracy of the approximations involved.

4.3 Phonon Damping

When we go beyond first order in the self consistent phonon scheme, the self consistent dynamical matrix $D_{q\nu}$ becomes dependent on ν due to non-local (or long range) self energy insertions (see §3.2). We will show in this section that the phonons are then damped out; they propagate with a finite lifetime.

To make this phonon damping explicit, it is necessary to make the analytic continuation to real time (so that $\tau \to -it, \nu \to i\omega$) since the energies and lifetimes of the physical states depend on the real time Green's functions (or "response functions" or "correlation functions") $\underline{\underline{G}}(\underline{x}, t; \underline{x}', t')$. These quantities will also be useful if we wish to consider non-equilibrium properties, for example transport properties (eg thermal conductivity), collective modes (eg second sound) or scattering (eg thermal neutron scattering [Maradudin and Fein, 1962, Werthamer, 1970, 1972]). In principle, they should be determined using the full real time formalism for finite temperature field theory [Martin and Schwinger, 1959, Schwinger, 1961, Keldysh, 1964]; in practice we shall see that for our purposes analytic continuation from the imaginary time formalism will be sufficient [Luttinger and Ward, 1960, Baym and Mermin, 1961, Abrikosov, et al., 1963].

To analytically continue the Green's function $G_{q\nu}$ defined at the discrete set of frequencies Ω_ν to Green's functions $G_q(\omega)$ defined on the whole of the real line, we introduce the spectral representation (or Lehman representation)

$$\beta G_{q\nu} = \int_0^\infty \frac{\omega' d\omega'}{\pi} \frac{\rho_q(\omega')}{\Omega_\nu^2 + \omega'^2}, \tag{4.52}$$

where $\rho_q(\omega)$ is the spectral density function. Setting $\nu = 0, \rho_q(\omega) \sim \omega^2$ as $\omega \to 0$. Letting $\nu \to \infty$ in (4.52), we obtain the sum rule

$$\int_0^\infty \frac{d\omega^2}{2\pi} \rho_q(\omega) = 1. \tag{4.53}$$

Furthermore if the physical Hilbert space has a positive norm metric,

$$\rho_q(\omega) \geq 0 \qquad \text{for} \qquad \omega > 0, \tag{4.54}$$

and the phonon-phonon scattering matrix will be unitary.

Extending $\rho_q(\omega)$ to the negative real axis according to $\rho_q(-\omega) = -\rho_q(\omega)$ (note that $\rho_q(0) = 0$) we may write the representation (4.52) in the equivalent form

$$\beta G_{q\nu} \;=\; \int_{-\infty}^\infty \frac{d\omega'}{2\pi} \frac{\rho_q(\omega'}{\omega' - i\Omega_\nu}. \tag{4.55}$$

Now

$$\beta G_q(\tau) \equiv \sum_\nu e^{i\Omega_\nu \tau\hbar} \beta G_{q\nu}, \quad 0 \leq \tau \leq \beta,$$

so using the spectral representation (4.55) for $G_{q\nu}$, and extending the range of τ to $-\beta \leq \tau \leq \beta$, we have

$$\begin{aligned}\beta G_q(\tau) &= \frac{1}{2\pi}\int_o^\infty d\omega \frac{\cosh(\frac{1}{2}\beta + |\tau|)}{\sinh\frac{1}{2}\beta\omega}\rho_q(\omega) \\ &= \frac{1}{2\pi}\int_o^\infty d\omega(\coth\tfrac{1}{2}\beta\omega\cosh\omega\tau - \sinh\omega|\tau|)\rho_q(\omega). \quad (4.56)\end{aligned}$$

For local propagation, $\mathcal{D}_{q\nu} = \mathcal{D}_q = \omega_q^2$, the spectral density takes the form

$$\rho_q(\omega) = \frac{\pi}{\omega}(\delta(\omega - \omega_q) + \delta(\omega + \omega_q)) = 2\pi\delta(\omega^2 - \omega_q^2), \quad (4.57)$$

so that $G_{q\nu}$ and $G_q(\tau)$ are given by (1.16) and (1.20). More generally, as we shall see below, these singular peaks in $\rho_q(\omega)$ will be broadened out, so that the phonons become damped quasi-particles.

Now consider the function

$$\mathcal{G}_q(z) \equiv \int_o^\infty \frac{\omega' d\omega'}{\pi}\frac{\rho_q(\omega')}{\omega' - z^2} = \int_{-\infty}^\infty \frac{d\omega'}{2\pi}\frac{\rho_q(\omega')}{\omega' - z}, \quad (4.58)$$

defined in such a way that $\beta G_{q\nu} = \mathcal{G}_q(i\Omega_\nu)$. However $\mathcal{G}_q(z)$ is well defined for all z away from the real axis. In fact as z crosses the real axis, $\mathcal{G}_q(z)$ picks up a pole contribution $i\rho_q(z)$; $\mathcal{G}_q(z)$ has a branch cut on the real axis.

We may now define the retarded and advanced Green's functions

$$\begin{aligned}G_q^R(\omega) &\equiv \mathcal{G}_q(\omega + i\varepsilon) \\ G_q^A(\omega) &\equiv \mathcal{G}_q(\omega - i\varepsilon) \quad (4.59)\end{aligned}$$

or, using the spectral representations (4.58)

$$G_q^R(\omega) = \int_0^\infty \frac{\omega' d\omega'}{2\pi}\frac{\rho_q(\omega')}{\omega'^2 - (\omega + i\varepsilon)^2} = \int_{-\infty}^\infty \frac{d\omega'}{2\pi}\frac{\rho_q(\omega')}{\omega' - \omega - i\varepsilon}, \textit{etc.} \quad (4.60)$$

$G_q^R(\omega)$ is analytic in the upper half-plane; $G_q^A(\omega)$ is analytic in the lower half-plane. As $\omega \to \infty$, using the sum rule (4.53), $G_q^R(\omega) \sim -1/\omega^2$.

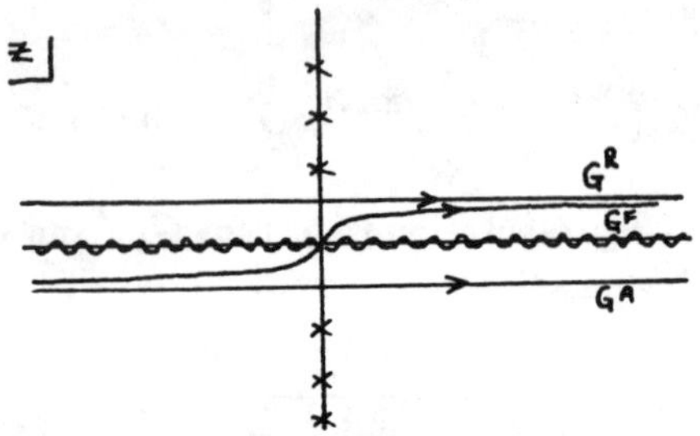

Since ω and $\rho_q(\omega)$ are both real,

$$G_q^A(\omega) = G_q^R(\omega)^* = G_q^R(-\omega). \tag{4.61}$$

In fact using the result

$$\frac{1}{x + i\varepsilon} = P\frac{1}{x} - i\pi\delta(x),$$

we may derive the dispersion relations

$$ReG_q^R(\omega) = \frac{1}{\pi}P\int_{-\infty}^{\infty}\frac{\mathcal{I}_m G_q^R(\omega')}{\omega' - \omega}d\omega', \tag{4.62}$$

since

$$\mathcal{I}_m G_q^R(\omega) = -\mathcal{I}_m G_q^A(\omega) = \tfrac{1}{2}\rho_q(\omega). \tag{4.63}$$

The real-time retarded and advanced Green's functions are obtained by Fourier transforming (4.58); this yields

$$G_q^R(t) = i\int_{-\infty}^{\infty}\frac{d\omega}{2\pi}e^{-i\omega t}G_q^R(\omega) = \Theta(t)\int_{-\infty}^{\infty}\frac{d\omega}{2\pi}e^{-i\omega t}\rho_q(\omega), \tag{4.64}$$

and $G_q^A(t) = G_q^R(-t)$. The vanishing of $G_q^R(t)$ for $t < 0$ follows directly from analyticity of $G_q^R(\omega)$ in the upper half-plane. Indeed the retarded Green's functions were constructed with precisely this property in mind. However, the Green's function which is used to calculate physical scattering amplitudes must be symmetric in t (canonically it is the expectation value of the time-ordered product of two phonon fields); the phonon is its own antiparticle. This Green's function is known as the causal Green's function, or Feynman Green's function, $G_q^F(t)$. It is obtained most easily by a Wick rotation of the imaginary time Green's function (4.56) on the interval $-\beta < \tau < \beta$, and then using analytic continuation; $G_q^F(t) = G_q(-i\tau)$, for $-\beta < t < \beta$, so that

$$\begin{aligned} G_q^F(t) &= \frac{1}{2\pi}\int_o^\infty d\omega \frac{\cosh\left(\frac{1}{2}\beta - i|t|\right)\omega}{\sinh\frac{1}{2}\beta\omega}\rho_q(\omega) \\ &= \frac{1}{2\pi}\int_o^\infty d\omega(\coth\tfrac{1}{2}\beta\omega\cos\omega t + i sin\,\omega|t|)\rho_q(\omega), \end{aligned} \tag{4.65}$$

for $-\infty < t < \infty$.

In terms of the phonon and antiphonon Green's functions

$$G_q^\pm(t) \equiv \int_{-\infty}^\infty \frac{d\omega}{2\pi}\frac{\rho_q(\omega)}{1 - e^{\mp\beta\omega}}e^{i\omega t},$$

we have $G_q^F(t) = \Theta(t)G_q^+(t) + \Theta(-t)G_q^-(t)$. Comparing with (4.64), the fourier transform of the Feynman Green's function is thus a linear combination of the retarded and advanced Green's functions, weighted in frequency space according to the Bose-Einstein distribution:

$$G_q^F = \frac{1}{1 - e^{-\beta\omega}}G_q^R(\omega) + \frac{1}{1 - e^{\beta\omega}}G_q^A(\omega) \tag{4.66}$$

$$= \int_{-\infty}^\infty \frac{d\omega'}{2\pi}\rho_q(\omega')\{P\frac{1}{\omega' - \omega} + i\pi\coth\tfrac{1}{2}\beta\omega\delta(\omega' - \omega)\}. \tag{4.67}$$

As $T \to 0$, we have

$$\begin{aligned} G_q^F(\omega)\,|_{T=0} &= \Theta(\omega)G_q^R(\omega) + \Theta(-\omega)G_q^A(\omega) \\ &= \int_{-\infty}^\infty \frac{d\omega'^2}{4\pi}\frac{\rho_q(\omega')}{\omega'^2 - \omega^2 - i\varepsilon}, \end{aligned} \tag{4.68}$$

which is the familiar "$i\epsilon$-prescription". Positive frequency modes propagate forwards in time, whereas negative frequency modes propagate backwards.

The Feynman Green's function is analytic in neither the upper nor the lower half plane; indeed from (4.67) one deduces the modified dispersion relation

$$ReG_q^F(\omega) = \frac{1}{\pi} P \int_{-\infty}^{\infty} d\omega' \frac{\mathcal{I}mG_q^F(\omega')}{\omega' - \omega} \tan \tfrac{1}{2}\beta\omega' \tag{4.69}$$

with $\mathcal{I}mG_q^F(\omega) = \frac{1}{2} \coth \frac{1}{2}\beta\omega\rho_q(\omega)$. (4.70)

Notice that $ReG_q^F(\omega) = ReG_q^R(\omega)$, whereas $\mathcal{I}mG_q^F(\omega) = \coth \frac{1}{2}\beta\omega\mathcal{I}mG_q^R(\omega) \geq 0$ for all ω.

To summarize, we have obtained real time Green's functions $G_q^R(t), G_q^A(t), G_q^F(t)$ (with Fourier transforms $G_q^R(\omega), G_q^A(\omega), G_q^F(\omega)$) by analytic continuation from $G_q(\tau)$ (with transform G_q). Since the consistent dynamical matrix $\mathcal{D}_q(\tau)$ (or $\mathcal{D}_{q\nu}$) is nothing but a self energy insertion, it too may be analytically continued to real time dynamical matrices $\mathcal{D}_q^R(t), \mathcal{D}_q^A(t)\mathcal{D}_q^F(t)$ (or $\mathcal{D}_q^R(\omega), \mathcal{D}_q^A(\omega), \mathcal{D}_q^F(\omega)$, respectively) satisfying the same analyticity relations as the corresponding Green's functions (viz. eqns. (4.61), (4.62), (4.69)). These are related via the appropriate continuation of the Dyson equations

$$\begin{aligned} G_q^R(\omega) &= (\mathcal{D}_q^R(\omega) - (\omega + i\epsilon)^2)^{-1}, \\ G_q^A(\omega) &= (\mathcal{D}_q^A(\omega) - (\omega - i\epsilon)^2)^{-1}. \end{aligned} \tag{4.71}$$

The spectral density function $\rho_q(\omega)$ is thus given by

$$\rho_q(\omega) = 2Im(\mathcal{D}_q^R - (\omega + i\epsilon)^2)^{-1}. \tag{4.72}$$

Separating $\underline{\underline{\mathcal{D}}}_q^R(\omega), \underline{\underline{\mathcal{D}}}_q^A(\omega)$ into real and imaginary parts

$$\begin{aligned} \underline{\underline{\mathcal{D}}}_q^R(\omega) &\equiv (\underline{\underline{\omega}}_q(\omega) - i\underline{\underline{\Gamma}}_q(\omega))^2, \qquad \underline{\underline{\mathcal{D}}}_q^A(\omega) \equiv (\underline{\underline{\omega}}_q(\omega) + i\underline{\underline{\Gamma}}_q(\omega))^2, \\ \underline{\underline{\mathcal{D}}}_q^F(\omega) &= \underline{\underline{\omega}}_q(\omega)^2 - \underline{\underline{\Gamma}}_q(\omega)^2 - i\coth \tfrac{1}{2}\beta\omega\{\underline{\underline{\omega}}_q(\omega), \underline{\underline{\Gamma}}_q(\omega)\}, \end{aligned} \tag{4.73}$$

where $\underline{\underline{\omega}}_q(\omega)$ and $\underline{\underline{\Gamma}}_q(\omega)$ are both real. For values of ω such that the eigenvalues of $\underline{\underline{\Gamma}}_q(\omega)$ are nonvanishing, we then have

$$\underline{\underline{\rho}}_q(\omega) = [(\underline{\underline{\omega}}_q(\omega)^2 - \underline{\underline{\Gamma}}_q(\omega)^2 - \omega^2)^2 + \{\underline{\underline{\omega}}_q(\omega), \underline{\underline{\Gamma}}_q(\omega)\}^2]^{-1} 2\{\underline{\underline{\omega}}_q(\omega), \underline{\underline{\Gamma}}_q(\omega)\} \tag{4.74}$$

For values of ω where $\underline{\underline{\Gamma}}_q(\omega)$ has a vanishing eigenvaue, the appropriate part of $\underline{\underline{\rho}}_q(\omega)$ will also vanish, unless ω is equal to the appropriate part of $\underline{\underline{\omega}}_q(\omega)$ in which case we will have a δ-function singularity like equation (4.57). The positivity of $\underline{\underline{\rho}}_q(\omega)$ (eqn. (4.54)) may now be used to infer that

$$Im\underline{\underline{\mathcal{D}}}^F_q(\omega) \leq 0 \tag{4.75}$$

for consistency with unitarity.

It remains to write down the analytic continuation of equations (3.7), ie of the self energy insertions. The short range contributions are trivial, since they have no frequency dependence. For the long range contribution, we first consider the simplest non-local graph

We may evaluate this as in §1.4, but now using the full propagator in the Lehmann representation (4.52), and the renormalized vertices $< v_\rho >$. Performing the frequency sums by contour integration given in place of (1.52)

$$\tilde{\Sigma}^b_{q\nu} = \tfrac{1}{2}\sum_{q_1 q_2} \mid \tilde{V}(q,q_1,q_2) \mid^2 \int d\omega_1 \int d\omega_2 \rho_{q_1}(\omega_1)\rho_{q_2}(\omega_2)\left(\frac{\bar{n}_1 + \bar{n}_2 + 1}{\omega_1 + \omega_2 + i\Omega_\nu}\right), \tag{4.76}$$

which is in the Lehmann form as expected. The analytic continuation is now performed as before by letting $i\Omega_\nu \to \omega \pm i\epsilon$, to obtain the retarded and advanced self energies $\tilde{\Sigma}^{bR}_q(\omega)$, $\tilde{\Sigma}^{bA}_q(\omega)$, from which one has $\tilde{\Sigma}^{bF}_q(\omega)$. Note that all the frequency sums are evaluated *before* the continuation is performed [Balian and DeDominicis, 1960, Dzyaloshinski, 1962, Baym and Sessler, 1963].

Such methods are also available for the higher order graphs, but with ever increasing complexity; what is required is a continuation of the more general expressions (3.9) (given explicitly (to second order) by (3.7)). The secret here is to work in position space rather than momentum space; for example to lowest order, instead of (4.76) we have (cf (1.51))

$$(\underline{\underline{\tilde{\Sigma}}}^b)^\sigma_{\rho\rho'}(\tau) = \tfrac{1}{2}(\tfrac{1}{2}G^\sigma_{\rho\rho'}(\tau) : \underline{\nabla}\,\underline{\nabla}')^2 < \underline{\nabla}\phi_\rho >< \underline{\nabla}'\phi_{\rho'} >, \tag{4.77}$$

which may be analytically continued by Wick rotation $\tau \to -it$ to give the Feynman self energy $(\underline{\underline{\tilde{\Sigma}}}^b)^{\sigma F}_{\rho\rho'}(\tau)$, from which the retarded and advanced self energies may be deduced by Fourier transformation. In particular, the *complete* non-local second order contribution (3.5) to the dynamical matrix may also be continued in this way; it becomes

$$\underline{\underline{\mathcal{D}}}^{(2\ell)\sigma}_{\rho\rho' F}(\omega) = -\tfrac{1}{2}\int_0^\infty dt\ 2cos\omega t\ \epsilon_2(\tfrac{1}{2}\underline{\underline{G}}^{\sigma}_{\rho\rho'}{}^{F}(t)\ :\ \underline{\nabla}\,\underline{\nabla}') < \underline{\nabla} v_\rho >< \underline{\nabla}' v_{\rho'} > , \tag{4.78}$$

whence $\underline{\underline{\mathcal{D}}}^F_q(\omega), \underline{\underline{\mathcal{D}}}^R_q(\omega), \underline{\underline{\mathcal{D}}}^A_q(\omega)$ and in particular $\underline{\underline{\rho}}_q(\omega)$ (use (4.73) and (4.74)). These in turn give $\underline{\underline{G}}^R_q(\omega), \underline{\underline{G}}^A_q(\omega)$ by (4.71), whence $\underline{\underline{G}}^F_q(\omega), \underline{\underline{G}}^F_q(t)$ by (4.67); alternatively one could use (4.65) directly, once $\underline{\underline{\rho}}_q(\omega)$ is known. The important point is that one cannot work with the retarded/advanced functions above; the continuation (4.78) is possible only for Feynman functions.

Note that the contribution (4.78) will always in general generate an imaginary part for $\underline{\underline{\mathcal{D}}}^F_q(\omega)$, even if the original ansatz were real, since $\underline{\underline{G}}^F_q(t)$ is always complex (save possibly for isolated values of t). Thus we expect the spectral function $\underline{\underline{\rho}}_q(\omega)$ to be nonzero for all values of ω (save possibly for isolated points) when it is determined self consistently to second order or higher. The positivity and normalization of $\rho_q(\omega)$ (equations (4.53), (4.54)) should also be preserved if the phonon scattering is to be unitary; it may indeed be possible to turn this into a (non-perturbative) proof of unitarity.

Continuations similar to (4.78) may be made for the higher order contributions, with the proviso tht all of the *internal* frequency sums (or integrals over τ), be performed first; in practice one has to be careful over the ordering of the τ integrations and the final continuation. Local self energy insertions and vacuum graphs may be evaluated without continuation, ie using the propagator (4.52) or (4.56). Finally, we repeat the warning given at the beginning of this section that although the continuation offered here may give a useful description of phonon damping, for non-equilibrium properties it is probably better to use the more involved Green's functions obtained from the real-time formalism.

Lorentzian Line Shape

If the phonon damping is sufficiently weak, the above formulae may be somewhat simplified by adopting a Lorentzian form for the spectral function $\rho_q(\omega)$. This amounts to the assumption that

1. $\|\underline{\underline{\Gamma}}_q(\omega)\| \ll \|\underline{\underline{\omega}}_q(\omega)\|$.

2. $\underline{\underline{\omega}}_q(\omega)$ and $\underline{\underline{\Gamma}}_q(\omega)$ are each independent of ω, and

3. $\underline{\underline{\omega}}_q$ and $\underline{\underline{\Gamma}}_q$ are diagonal in the same basis.

This means in practice that we adopt a basis in which $\underline{\underline{\omega}}_q$ is diagonal, and thus drop the off-diagonal parts of $\underline{\underline{\Gamma}}_q$:

$$\underline{\underline{\omega}}_q . \underline{e}^\lambda(\underline{q}) = \omega_{\underline{q}\lambda} \underline{e}^\lambda(\underline{q}), \quad \Gamma_{\underline{q}\lambda} \equiv \underline{e}^\lambda(\underline{q}).\underline{\underline{\Gamma}}_q . \underline{e}^\lambda(\underline{q}) . \tag{4.79}$$

Then in this basis (4.71) and (4.74) become

$$G_q^R(\omega) = ((\omega_q + i\Gamma_q)^2 - \omega^2)^{-1}, \quad G_q^A(\omega) = ((\omega_q - i\Gamma_q)^2 - \omega^2)^{-1}, \tag{4.80}$$

$$\begin{aligned} \rho_q(\omega) &= sgn(\omega)\frac{4\omega_q\Gamma_q}{(\omega_q^2 + \Gamma_q^2 - \omega^2)^2 + 4\omega_q^2\Gamma_q^2} \\ &= \frac{2\Gamma_q}{\omega}\{\frac{1}{(\omega - \omega_q)^2 + \Gamma_q^2} - \frac{1}{(\omega + \omega_q)^2 + \Gamma_q^2}\} . \end{aligned} \tag{4.81}$$

To ensure unitarity and the correct behaviour for small ω^2, we must add to $\rho_q(\omega)$ a smooth background

$$\rho_q^B(\omega) = -sgn(\omega)\frac{4\omega_q\Gamma_q}{(\omega_q^2 + \Gamma_q^2)^2} f(\omega, \omega_q, \Gamma_q) ,$$

where $f(0, \omega_q, \Gamma_q) = 1$, and

$$\int_0^\infty dw^2 f(\omega, \omega_q, \Gamma_q) = \frac{(\omega_q^2 + \Gamma_q^2)^2}{4\omega_q\Gamma_q}\{2tan^{-1}\frac{\omega_q^2 - \Gamma_q^2}{2\omega_q\Gamma_q} - \pi\} \simeq \omega_q^2 ,$$

when $\Gamma_q \ll \omega_q$; a suitable choice of f would then be $e^{-\omega^2/\omega_q^2}$.

The spectral density thus consists essentially of a peak at $\omega = \omega_q$, of width Γ_q ;

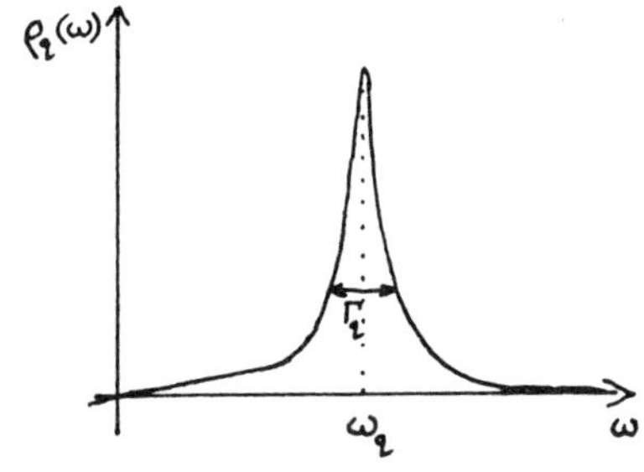

To show that this indeed corresponds to a damped quasiparticle (i.e. phonon), one evaluates the real-time Green's functions; for $\frac{1}{\omega_q} \ll t < \frac{1}{\Gamma_q}$, we have by contour integration

$$G_q^R(t) \simeq \frac{1}{2\omega_q} e^{-\Gamma_q t} e^{i\omega_q t}, \tag{4.82}$$

$$G_q^F(t) \simeq \frac{1}{2\omega_q} e^{-\Gamma_q t}(cos\omega_q t \coth \tfrac{1}{2}\beta\omega_q - isin\omega_q t). \tag{4.83}$$

Thus the phonons have a finite lifetime of order Γ_q.

The expression (4.83) for $G_q^F(t)$ may be used in (4.78) to approximate the second order nonlocal correction to the dynamical matrix; then the new values of ω_q and Γ_q are given by

$$\begin{aligned} \omega_q^2 &= \underline{e}^\lambda(\underline{q}).Re\underline{\underline{\mathcal{D}}}_{\underline{q}}^R(\omega_q).\underline{e}^\lambda(\underline{q}), \\ \Gamma_q &= \frac{1}{2\omega_q}\underline{e}^\lambda(\underline{q}).Im\underline{\underline{\mathcal{D}}}_{\underline{q}}^R(\omega_q).\underline{e}^\lambda(\underline{q}, \end{aligned} \tag{4.84}$$

and may be used to check on the consistency of the Lorentzian approximation.

The imaginary time Green's function $G_{q\nu}$ are evaluated directly from the Lehmann representation (4.52); at high temperatures one can show that

$$G_{q\nu} \underset{\beta\to 0}{\sim} \frac{1}{\omega_q^2 + \Omega_\nu^2}[1 - \kappa\frac{\Gamma_q}{\omega_q} + 0(\frac{\Omega_\nu^2\Gamma_q}{\omega_q^3})], \quad \Omega_\nu^2 \ll \omega_q^2$$

where the constant κ depends on the choice of the background ρ^B only.

Note that there is in general no reason to assume a priori that the self consistent phonon spectrum will be of Lorentzian form; this will only be so when the renormalised potential is almost harmonic. As the renormalized harmonicity increases, the consistent spectral function will be less and less sharply peaked, and thus the 'physical' phonons will become less and less like 'particles'.

5 Real Crystals

5.1 Cells, Crystallites and Shell Models

So far we have only considered a Bravais lattice system with a two body interionic potential. To extend this treatment to non-Bravais lattices is not difficult; all that is required is a generalisation of the Fourier transforms defined by eqns. (3)-(7). Instead of variables $(i,\ell) \leftrightarrow (\lambda,\underline{q})$ we now have $(i,\kappa,\ell) \leftrightarrow (\lambda,\kappa,\underline{q})$, where κ labels the ions in a unit cell. Then

$$\begin{aligned} \underline{u}_{q\kappa} &= \frac{1}{\sqrt{N}}\sum_{\ell} e^{i\underline{q}.\underline{R}_{\ell\kappa}}\underline{u}_{\ell\kappa}, \\ \underline{u}_{\ell\kappa} &= \frac{1}{\sqrt{N}}\sum_{\underline{q}} e^{i\underline{q}.\underline{R}_{\ell\kappa}}\underline{u}_{q\kappa}, \end{aligned} \tag{5.1}$$

so

$$\begin{aligned} \underline{u}_{\ell'\kappa'} - \underline{u}_{\ell\kappa} &= \frac{1}{\sqrt{N}}\sum_{\underline{q}} e^{i\underline{q}.\underline{R}_{\ell'\kappa'}}\underline{u}_{q\kappa'} - e^{i\underline{q}.\underline{R}_{\ell\kappa}}\underline{u}_{q\kappa} \\ &= \frac{1}{\sqrt{N}}\sum_{\underline{q}} (e^{i\underline{q}.\underline{R}_{\rho}}\underline{u}_{q\kappa'} - \underline{u}_{q\kappa})e^{i\underline{q}.\underline{R}_{\ell\kappa}}, \end{aligned} \tag{5.2}$$

where now $\rho \equiv (\ell\ell'\kappa\kappa')$ so $\underline{R}_{\rho} \equiv \underline{R}_{\ell'\kappa'} - \underline{R}_{\ell\kappa} \equiv \underline{S}_{\kappa'\kappa}$. Thus with

$$\begin{aligned} \underline{u}_{q\kappa} &= \sum_{\lambda} \frac{e^{\lambda}_{\kappa}(\underline{q})}{M_{\kappa}^{\frac{1}{2}}}\underline{u}_{q\lambda}, \\ \underline{u}_{\ell'\kappa'} - \underline{u}_{\ell\kappa} &= \frac{1}{\sqrt{N}}\sum_{\underline{q}\lambda} [e^{i\underline{q}.\underline{R}_{\rho}}\frac{\underline{e}^{\lambda}_{\kappa'}(\underline{q})}{M_{\kappa'}^{1/2}} - \frac{\underline{e}^{\lambda}_{\kappa}(\underline{q})}{M_{\kappa}^{1/2}}]u_{\underline{q}\lambda}e^{i\underline{q}.\underline{R}_{\ell\kappa}} \\ &\equiv \frac{1}{\sqrt{N}}\sum_{\underline{q}\lambda} \underline{E}^{\lambda}_{\rho}(\underline{q})u_{\underline{q}\lambda}e^{i\underline{q}.\underline{R}_{\ell\kappa}}, \end{aligned} \tag{5.3}$$

where

$$\underline{E}^{\lambda}_{\rho}(\underline{q}) \equiv e^{i\underline{q}.\underline{R}_{\rho}}\frac{\underline{e}^{\lambda}_{\kappa'}(\underline{q})}{M_{\kappa'}^{1/2}} - \frac{\underline{e}^{\lambda}_{\kappa}(\underline{q})}{M_{\kappa}^{1/2}}, \tag{5.4}$$

and $\{\underline{e}^{\lambda}_{\kappa}(\underline{q})\}$ are sets of orthonormal basis vectors. This is the appropriate generalisation of $\eta_{\rho}(\underline{q})\underline{e}^{\lambda}(\underline{q})$ to the non-Bravais case (here there is no factorisation).

To show how this works in practice, consider the generalisation of (1.24) for the correlation matrices. We now have

$$\begin{aligned}
\underline{\underline{G}}_\rho &\equiv \underline{\underline{G}}_{\kappa\kappa'\ell\ell'} = < (\underline{u}_{\ell\kappa} - \underline{u}_{\ell'\kappa'})(\underline{u}_{\ell\kappa} - \underline{u}_{\ell'\kappa'})^\dagger > \\
&= \frac{1}{N}\sum_{\underline{q}} < (\underline{u}_{q\kappa} - e^{i\underline{q}.\underline{R}_\rho}\underline{u}_{q\kappa'})(\underline{u}_{q\kappa} - e^{i\underline{q}.\underline{R}_\rho}\underline{u}_{q\kappa'})^\dagger > \\
&= \frac{1}{N}\sum_{\underline{q}}\{\underline{\underline{G}}_{\kappa\kappa}(\underline{q}) + \underline{\underline{G}}_{\kappa'\kappa'}(\underline{q}) - 2\cos \underline{q}.\underline{R}_\rho Re\underline{\underline{G}}_{\kappa\kappa'}(\underline{q}) + 2sin\underline{q}.\underline{R}_\rho Im\underline{\underline{G}}_{\kappa\kappa'}(\underline{q})\},
\end{aligned} \tag{5.5}$$

where $Re\underline{\underline{G}}_{\kappa\kappa'} \equiv \frac{1}{2}(\underline{\underline{G}}_{\kappa\kappa'} + \underline{\underline{G}}_{\kappa'\kappa})$, $iIm\underline{\underline{G}}_{\kappa\kappa'} \equiv \frac{1}{2}(\underline{\underline{G}}_{\kappa\kappa'} - \underline{\underline{G}}_{\kappa'\kappa})$, and $\underline{\underline{G}}_{\kappa\kappa'}(\underline{q}) = < \underline{u}_{q\kappa}\underline{u}^\dagger_{q\kappa'} >$. Then choosing a basis set in which $\underline{\underline{G}}_{\kappa\kappa'}(\underline{q})$ is diagonal,

$$\begin{aligned}
\underline{\underline{G}}_\rho &= \frac{1}{N}\sum_{\underline{q}\lambda}\{\frac{\underline{e}^\lambda_\kappa\underline{e}^{\lambda\dagger}_\kappa}{M_\kappa} + \frac{\underline{e}^\lambda_{\kappa'}\underline{e}^{\lambda\dagger}_{\kappa'}}{M_{\kappa'}} - 2\cos \underline{q}.\underline{R}_\rho Re(\frac{\underline{e}^\lambda_\kappa\underline{e}^\lambda_{\kappa'}}{\sqrt{M_\kappa M_{\kappa'}}}) \\
&\qquad + 2sin\underline{q}.\underline{R}_\rho Im(\frac{{}_\kappa{}^\lambda e_{\kappa'}}{\sqrt{M_\kappa M_{\kappa'}}})\}G^\lambda(\underline{q}) \\
&= \frac{1}{N}\sum_{\underline{q}\lambda}\underline{E}^\lambda_\rho(\underline{q})\underline{E}^\lambda_\rho(\underline{q})^\dagger G^\lambda(\underline{q}).
\end{aligned} \tag{5.6}$$

Similarly one can show that

$$\underline{\underline{G}}^\sigma_{\rho\rho'}(\tau) = \frac{1}{N}\sum_{\underline{q}\lambda\nu} e^{i\Omega_\nu\tau\hbar}e^{i\underline{q}.\underline{R}_\sigma}\underline{E}^\lambda_{\rho,\nu}(\underline{q})\underline{E}^\lambda_{\rho',\nu}(\underline{q})^\dagger G^\lambda_\nu(\underline{q}), \tag{5.7}$$

where

$$\rho = (\ell_1, \ell_2, \kappa_1, \kappa_2), \quad \rho' = (\ell'_1, \ell'_2, \kappa'_1, \kappa'_2), \quad \text{and } \underline{R}_\sigma = \underline{S}_{\kappa_1,\kappa'_1}.$$

Of course the inverse transformations (e.g. for eqns. (3.4),(3.5),(3.7)) may be generalized in a precisely similar fashion:

$$\underline{\underline{\mathcal{D}}}^{(s)}_{\kappa\kappa'}(\underline{q}) = \sum_S \sum_{\kappa''} [\delta_{\kappa\kappa'} - \delta_{\kappa'\kappa''} e^{-i\underline{q}.\underline{S}_{\kappa\kappa''}}] \underline{\underline{\mathcal{D}}}^{(s)}(\underline{S}_{\kappa\kappa''}),$$

$$\underline{\underline{\mathcal{D}}}^{(\ell)}_{\kappa\kappa',\nu}(\underline{q}) = \sum_{SS'S''} \sum_{\kappa_1\kappa_1'} \sum_{\kappa_2\kappa_2'} e^{-i\underline{q}.\underline{S}''_{\kappa_1\kappa_2}} [\delta_{\kappa\kappa_1} - \delta_{\kappa\kappa_1'} e^{i\underline{q}.\underline{S}_{\kappa_1\kappa_1'}}]$$

$$[\delta_{\kappa'\kappa_2} - \delta_{\kappa'\kappa_2'} e^{i\underline{q}.\underline{S}'_{\kappa_2\kappa_2'}}] \underline{\underline{\mathcal{D}}}^{(\ell)}_{\nu}(\underline{S}_{\kappa_1\kappa_1'}, \underline{S}'_{\kappa_2\kappa_2'}, \underline{S}''_{\kappa_1\kappa_2}), \tag{5.8}$$

where $\underline{\underline{\mathcal{D}}}^{(s)}(\underline{S}_{\kappa\kappa'}) \equiv \underline{\underline{\mathcal{D}}}^{(s)}_{\rho}$, and $\underline{\underline{\mathcal{D}}}^{(\ell)}_{\nu}(\underline{S}_{\kappa_1\kappa_1'}, \underline{S}'_{\kappa_2\kappa_2'}, \underline{S}''_{\kappa_1\kappa_2}) \equiv \underline{\underline{\mathcal{D}}}^{(\ell)\sigma}_{\rho\rho',\nu}$. Then since $\{e^{\lambda}_{\kappa}(\underline{q})\}$ diagonalises $\underline{\underline{\mathcal{D}}}_{\kappa\kappa',\nu}(\underline{q}) = \underline{\underline{\mathcal{D}}}^{(s)}_{\kappa\kappa'}(\underline{q}) + \underline{\underline{\mathcal{D}}}^{(\ell)}_{\kappa\kappa',\nu}(\underline{q})$,

$$\underline{\underline{\mathcal{D}}}_{\kappa\kappa',\nu}(\underline{q}) = \sum_{\lambda} \underline{e}^{\lambda}_{\kappa}(\underline{q}) \underline{e}^{\lambda}_{\kappa'}(\underline{q}) \omega^2_{\underline{q}\lambda\nu}$$

so that

$$\omega^2_{q\nu} = \sum_{\sigma\rho\rho'} e^{-i\underline{q}.\underline{R}_{\sigma}} \underline{E}^{\lambda}_{\rho,\nu}(\underline{q})^{\dagger} . \underline{\underline{\mathcal{D}}}^{\sigma}_{\rho\rho',\nu} . \underline{E}^{\lambda}_{\rho',\nu}(\underline{q}),$$

as naively expected.

For a molecular system or 'crystallite', which may be considered as one large unit cell, we simply drop the suffices ℓ and $\underline{q}$; the Fourier transforms are no longer relevant.

Shell Models

The description of self consistent phonon dynamics presented above is limited to lattices of atoms and ions interacting via a two-body potential $V[\underline{u}_{\rho}] = \sum_{\ell,\rho} \phi_{\rho}[\underline{u}_{\rho}]$. This potential is supposed to include all effective interactions between the nuclei, including the effects of the electron clouds. The normal approach is to treat these adiabatically [Born and Oppenheimer, 1927]; one first solves an equation for the electron motion assuming that the nuclei are fixed, and then uses these solutions to supply the potential $V[\underline{u}_{\rho}]$ for the nuclei [Baym, 1961, Meissner 1970]. This adiabatic approximation is good because the electron mass is so small [Moshinsky and Kittel, 1968].

Now instead of solving the electronic motion explicitly (which is very difficult) it has been found that a good estimate of $V[\underline{u}_{\rho}]$ for ionic crystals, at least in the harmonic approximation, may be obtained by treating the outermost electrons as massless shells, with empirical potentials between them.

The shells and cores are then bound together by a harmonic "spring" potential, and all interact via long range Coulomb forces. We refer the reader to [Cochran 1959, 1971, Dick and Overhauser, 1958 and Sinha, 1969] for a more detailed description of the harmonic shell model and its derivation from the full theory. Here we wish to describe the generalisation of the harmonic shell model to an anharmonic shell model, which should give a good description of ionic crystals even at high temperatures.

The action is now a functional of both nuclear (core, u) and electronic (shell, v) coordinates. However only the former are genuine dynamical degrees of freedom, so

$$S[u] = S[u, v[u]] = \frac{1}{\hbar}\int_0^{\beta\hbar} d\tau(\tfrac{1}{2}M\dot{u}(\tau)^2 + V[u, v[u]]) \tag{5.9}$$

The functional v[u] is determined by the equation of motion of the electrons, which amounts to a constraint on the shell coordinates v, namely $\delta V[u,v]/\delta v = 0$. In the harmonic theory this constraint is linear in v and may be solved explicity; for an anharmonic potential it must be solved perturbatively. It may be shown [Ball, 1986] that this leads to Feynman rules very similar to those for the rigid ion models. The propagators $G^{\alpha\alpha'}$ are now two by two matrices, with indices $\alpha = (u, v)$ denoting cores and shells respectively. Defining bare force constants

$$F^{uu} = \frac{\delta^2 V}{\delta u \delta u}, \quad F^{uv} = \frac{\delta^2 V}{\delta u \delta v}, \quad F^{vv} = \frac{\delta^2 V}{\delta v \delta v}, \tag{5.10}$$

and the dynamical matrix

$$\mathcal{D} = F^{uu} - F^{uv}.F^{vv-1}.F^{vu}, \tag{5.11}$$

the propagators are

$$G_\alpha^{\alpha'} = \begin{pmatrix} G^{uu} & G^{uv} \\ G^{vu} & G^{vv} \end{pmatrix} = \begin{pmatrix} G & G.C^T \\ C.G & C.G.C^T \end{pmatrix} = \begin{pmatrix} 1 & C \\ C^T & 1 \end{pmatrix}\begin{pmatrix} G & 0 \\ 0 & 0 \end{pmatrix}\begin{pmatrix} 1 & C^T \\ C & 1 \end{pmatrix} \tag{5.12}$$

where $G_\nu = \beta^{-1}(M\Omega_\nu{}^2 + \mathcal{D})^{-1}$ as usual. Note that (5.12) demonstrates explicitly the trivial propagation of the shell-type modes. Then to each line of a given diagram we assign factors $\frac{1}{2}\underline{\underline{G}}^{\alpha\alpha'}(\tau-\tau') : \underline{\nabla}^{\alpha}\underline{\nabla}^{\alpha'}$, where $\underline{\nabla}^{\alpha}$ indicates that the potential V at the vertex at the end of the propagator is to be differentiated with respect to u or v. Furthermore, there are additional propagators in the diagrams, represented by and associated with a factor $-(F^{vv})^{-1}$; these never form closed loops, and in fact are there to enforce the adiabatic condition. The diagram resummation and imposition of self consistency may now be performed as before [Ball, 1986]. In particular the force constants $F^{\alpha\alpha'}$ are determined self consistently by summing irreducible self energy insertions;

$F^{\alpha\alpha'} =$ 

where the leading term is the renormalized bare contribution (5.10). To second order we thus have (schematically)

$$\begin{aligned}
&\underline{\underline{F}}^{\alpha\alpha'} = <\underline{\nabla}^{\alpha}\underline{\nabla}^{\alpha'}V> \\
&-\tfrac{1}{2}\int_0^{\beta} d\tau^{-i\Omega_\nu\tau\hbar}\epsilon_2\Big(\tfrac{1}{2}\sum_{\beta\beta'} G^{\beta\beta'}(\tau) : \underline{\nabla}^{\beta}\underline{\nabla}^{\beta'}\Big) <\underline{\nabla}^{\alpha}V><\underline{\nabla}^{\alpha'}V> \\
&+\tfrac{1}{2}\int_0^{\beta} d\tau e^{-i\Omega_\nu\tau\hbar}\epsilon_1\Big(\tfrac{1}{2}\sum_{\beta\beta'} G^{\beta\beta'}(\tau) : \underline{\nabla}^{\beta}\underline{\nabla}^{\beta'}\Big)F^{vv-1}(\tau) <\underline{\nabla}^{v}\underline{\nabla}^{\alpha}V><\underline{\nabla}^{v}\underline{\nabla}^{\alpha'}V>,
\end{aligned} \tag{5.13}$$

where $$<V>= exp\Big(\tfrac{1}{2}\sum_{\alpha\alpha'} G^{\alpha\alpha'} : \underline{\nabla}^{\alpha}\underline{\nabla}^{\alpha'}\Big)V. \tag{5.14}$$

To write out these equations more explicitly, we exploit the separability of the potential, $V = N\sum_\rho v_\rho$ where $\rho \equiv (\ell_1, \ell_2, \kappa_1, \kappa_2, \alpha_1, \alpha_2)$. Then (5.13) and (5.14) become, separating short-range from long-range pieces

$$\begin{aligned}
\underline{\underline{F}}^{(s)}_{\rho} &= <\underline{\nabla}\,\underline{\nabla}v_\rho> + \ldots, \\
\underline{\underline{F}}^{\sigma(\ell)}_{\rho\rho',\nu} &= -\tfrac{1}{2}\int_0^{\beta} d\tau e^{-i\Omega_\nu\tau\hbar}\big[\epsilon_2\big(\tfrac{1}{2}\underline{\underline{G}}^{\sigma}_{\rho\rho'}(\tau) : \underline{\nabla}\,\underline{\nabla}'\big) <\underline{\nabla}v_\rho><\underline{\nabla}v_{\rho'}> \\
&- \epsilon_1\big(\tfrac{1}{2}\underline{\underline{G}}^{\sigma}_{\rho\rho'} : \underline{\nabla}\,\underline{\nabla}'\big)\underline{\underline{\tilde{G}}}^{\sigma}_{\rho\rho'}(\tau) :<\underline{\nabla}\,\underline{\nabla}v_\rho><\underline{\nabla}'\,\underline{\nabla}'v_{\rho'}>\big],
\end{aligned} \tag{5.15}$$

$$v_\rho \equiv v^{\alpha_1\alpha_2}_{\kappa_1\kappa_2}(\underline{S}^{\alpha_1\alpha_2}_{\kappa_1\kappa_2}) \quad , \quad \underline{S}^{\alpha_1\alpha_2}_{\kappa_1\kappa_2} \equiv \underline{R}_\rho \equiv \underline{R}^{\alpha_2}_{\ell_2\kappa_2} - \underline{R}^{\alpha_1}_{\ell_1\kappa_1},$$

and

$$\begin{aligned} < v_\rho(\underline{R}_\rho) > &= exp(\tfrac{1}{2}\underline{G}_\rho : \underline{\nabla}\,\underline{\nabla}) v_\rho(\underline{R}_\rho) \\ &= \frac{1}{[(2\pi)^3 det \underline{G}_\rho]^{1/2}} \int d^3\underline{u}\ \ v_\rho(\underline{R}_\rho + \underline{u}) exp(-\tfrac{1}{2}\underline{u}.\underline{\underline{G}}_\rho^{-1}.\underline{u}). \end{aligned} \tag{5.16}$$

These force constants are tranformed to momentum space by a generalization of (5.8).

$$\begin{aligned} &\underline{\underline{F}}^{\alpha\alpha'}_{\kappa\kappa'}(\underline{q})^{(s)} = \sum_S \sum_{\kappa''\alpha''} [\delta^{\alpha\alpha'}_{\kappa\kappa'} - \delta^{\alpha'\alpha''}_{\kappa'\kappa''} e^{-i\underline{q}.\underline{S}^{\alpha\alpha''}_{\kappa\kappa''}}] \underline{\underline{F}}^{(s)}(\underline{S}^{\alpha\alpha''}_{\kappa\kappa''}), \\ &\underline{\underline{F}}^{\alpha\alpha'}_{\kappa\kappa',\nu}(\underline{q})^{(\ell)} = \sum_{S,S',S''} \sum_{\substack{\kappa_1\kappa_1' \\ \alpha_1\alpha_1'}} \sum_{\substack{\kappa_2\kappa_2' \\ \alpha_2\alpha_2'}} e^{-i\underline{q}.\underline{S}''^{\alpha_1\alpha_2}_{\kappa_1\kappa_2}} [\delta^{\alpha\alpha_1}_{\kappa\kappa_1} - \delta^{\alpha\alpha_1'}_{\kappa\kappa_1'} e^{i\underline{q}.\underline{S}^{\alpha_1\alpha_1'}_{\kappa_1\kappa_1'}}] \\ &[\delta^{\alpha'\alpha_2}_{\kappa'\kappa_2} - \delta^{\alpha'\alpha_2'}_{\kappa'\kappa_2'} e^{i\underline{q}.\underline{S}'^{\alpha_2\alpha_2'}_{\kappa_2\kappa_2'}}] \underline{\underline{F}}_\nu(\underline{S}'^{\alpha_1\alpha_1'}_{\kappa_1\kappa_1'}, \underline{S}'^{\alpha_1\alpha_1'}_{\kappa_2\kappa_2'}, \underline{S}''^{\alpha_1\alpha_2}_{\kappa_1\kappa_2}), \end{aligned} \tag{5.17}$$

where

$$\underline{\underline{F}}(\underline{S}^{\alpha\alpha''}_{\kappa\kappa''}) \equiv \underline{\underline{F}}_\nu(\underline{S}^{\alpha_1\alpha_1'}_{\kappa_1\kappa_1'}, \underline{S}'^{\alpha_2\alpha_2'}_{\kappa_2\kappa_2'}, \underline{S}''^{\alpha_1\alpha_2}_{\kappa_1\kappa_2}) \equiv \underline{\underline{F}}^{(\ell)\sigma}_{\rho\rho',\nu}.$$

Then $\underline{\underline{F}}^{\alpha\alpha'}_{\kappa\kappa',\nu}(\underline{q}) = \underline{\underline{F}}^{\alpha\alpha'}_{\kappa\kappa'}(\underline{q})^{(s)} + \underline{\underline{F}}^{\alpha\alpha'}_{\kappa\kappa'}(\underline{q})^{(\ell)}$, the dynamical matrix is

$$\begin{aligned} \underline{\underline{\mathcal{D}}}_{\kappa\kappa',\nu}(\underline{q}) &\equiv \underline{\underline{F}}^{uu}_{\kappa\kappa',\nu}(\underline{q}) - \sum_{\kappa_1\kappa_2} \underline{\underline{F}}^{uv}_{\kappa\kappa_1,\nu}(\underline{q}).(\underline{\underline{F}}^{vv}_\nu(\underline{q}))^{-1}_{\kappa_1\kappa_2}.\underline{\underline{F}}^{vu}_{\kappa_2\kappa',\nu}(\underline{q}) \\ &\equiv \sum_\lambda \underline{e}^{\lambda,\nu}_{\kappa,u}(\underline{q}) e^{\lambda,\nu}_{\kappa',u}(\underline{q})^\dagger \omega^2_{\underline{q}\lambda\nu}, \end{aligned} \tag{5.18}$$

and we define

$$e^{\lambda,\nu}_{\kappa,v}(\underline{q}) \equiv -\sum_{\kappa_1\kappa_2} (\underline{\underline{F}}^{vv}_\nu(\underline{q}))^{-1}_{\kappa\kappa_1}.\underline{\underline{F}}^{vu}_{\kappa_1\kappa_2,\nu}(\underline{q}).\underline{e}^{\lambda\nu}_{\kappa_2,u}(\underline{q}), \tag{5.19}$$

$$\underline{\underline{F}}^{vv}_{\kappa\kappa',\nu}(\underline{q}) \equiv \sum_\lambda \underline{\tilde{e}}^{\lambda,\nu}_\kappa(\underline{q})(\underline{\tilde{e}}^{\lambda,\nu}_{\kappa'}(\underline{q}))^\dagger \tilde{\omega}^2_{\underline{q}\lambda\nu}, \tag{5.20}$$

so that the propagators are

$$G_{\underline{q}\lambda\nu} = \beta^{-1}(\Omega_\nu^2 + \omega^2_{\underline{q}\lambda\nu})^{-1},$$

so that the propagators are

$$
\begin{aligned}
G_{\underline{q}\lambda\nu} &= \beta^{-1}(\Omega_\nu^2 + \omega_{\underline{q}\lambda\nu}^2)^{-1}, \\
\underline{\underline{G}}_\rho &= \frac{1}{N}\sum_{\underline{q}\lambda\nu} \underline{E}_\rho^{\lambda\nu}(\underline{q})\underline{E}_\rho^{\lambda\nu}(\underline{q})^\dagger G_{\underline{q}\lambda\nu}, \\
\underline{\underline{G}}_{\rho\rho'}^\sigma(q) &= \frac{1}{N}\sum_{\underline{q}\lambda\nu} e^{i\Omega_\nu\tau\hbar} e^{i\underline{q}\cdot\underline{R}_\sigma} \underline{E}_\rho^{\lambda\nu}(\underline{q})\underline{E}_{\rho'}^{\lambda\nu}(\underline{q})^\dagger G_{\underline{q}\lambda\nu}, \\
\underline{\underline{\tilde{G}}}_{\rho\rho'}^\sigma(\tau) &= \frac{1}{N}\sum_{\underline{q}\lambda\nu} e^{i\Omega_\nu\tau\hbar} e^{i\underline{q}\cdot\underline{R}_\sigma} \underline{\tilde{E}}_\rho^{\lambda\nu}(\underline{q})\underline{\tilde{E}}_{\rho'}^{\lambda\nu}(q)^\dagger (\tilde{\omega}_{\underline{q}\lambda\nu}^2)^{-1},
\end{aligned}
\tag{5.21}
$$

where

$$
\begin{aligned}
\underline{E}_\rho^{\lambda\nu}(\underline{q}) &= e^{i\underline{q}\cdot\underline{R}_\rho} \underline{e}_{\kappa'\alpha'}^{\lambda\nu}(\underline{q})/M_{\kappa'}^{1/2} - \underline{e}_{\kappa\alpha}^{\lambda\nu}(\underline{q})/M_\kappa^{1/2}, \\
\underline{\tilde{E}}_\rho^{\lambda\nu}(\underline{q}) &= e^{i\underline{q}\cdot\underline{R}_\rho} \delta_{\alpha' v} \tilde{e}_{\kappa'}^{\lambda\nu}(\underline{q}) - \delta_{\alpha v} \underline{\tilde{e}}_\kappa^{\lambda\nu}(\underline{q}).
\end{aligned}
\tag{5.22}
$$

Finally, after iterating (5.15)-(5.22) to self-consistency, the free energy is given by

$$
\begin{aligned}
F &= \tfrac{1}{2}\beta^{-1}\sum_{\underline{q}\lambda\nu} \ell n\beta G_{\underline{q}\lambda\nu} - \tfrac{1}{2}\sum_{\underline{q}\lambda\nu} \omega_{\underline{q}\lambda\nu}^2 G_{q\lambda\nu} + \tfrac{N}{2}\sum_\rho < v_\rho > \\
&- \frac{N}{4}\tfrac{1}{2}\sum_{\rho\rho'\sigma}\int_0^\beta d\tau[\epsilon_3(\tfrac{1}{2}\underline{\underline{G}}_{\rho\rho'}^\sigma(\tau) : \underline{\nabla}\,\underline{\nabla}') < \underline{\nabla} v_\rho >< \underline{\nabla}' v_{\rho'} > \\
&- \epsilon_2(\tfrac{1}{2}G_{\rho\rho'}^\sigma(\tau) : \underline{\nabla}\,\underline{\nabla}')\tilde{G}_{\rho\rho'}^\sigma(\tau) :< \nabla v_\rho >< \underline{\nabla} v_{\rho'} >] + \ldots
\end{aligned}
\tag{5.23}
$$

The spectral density function $\rho_{\underline{q}\lambda\nu}(\omega)$ may be obtained from (5.15) as described in §4.3.

5.2 Numerical Evaluation of SCP Equations

We can summarize the SCP procedure by the following steps (to be specific we will do this for the second order shell model; the simpler procedures for the rigid ion model may be obtained by ommitting all reference to the shell coordinates and setting $\tilde{G} = 0$);

(ii) Evaluate the force constant matrices $F^{\alpha\alpha'}$ (eqns. (5.15),(5.16)).

(iii) Perform the Fourier summations (5.17) for a lattice of points $\underline{q}$ in the first Briouillin zone.

(iv) For each value of $\underline{q}$, calculate the dynamical matrix $\mathcal{D}$ (eqn. (5.18)).

(v) Diagonalise $\mathcal{D}$, to give ω^2, e_u, and e_v by eqn. (5.19). Also diagonalise F^{vv} to give $\tilde{w}^2, \tilde{e}$ (5.20).

(vi) Calculate G_ρ, $G^\sigma_{\rho\rho'}$,$\tilde{G}^\sigma_{\rho\rho'}$, as given by (5.21), (5.22).

(vii) Return to step ii) with these new Green's functions, and iterate until self consistency is achieved.

(viii) Calculate F from (5.23). Any required thermodynamic or thermoelectric information may now be obtained by performing calculations for various values of the external parameters (the lattice parameter a, the temperature T, and possibly an external electric field $\underline{E}$) to evaluate derivatives of F numerically, or else by the more sophisticated methods of § 4.2.

This procedure is summarized in the flow diagram opposite.

Although rather elaborate, the only steps in this procedure which could prove troublesome in practice are the evaluation of the force constants (step ii)), and in particular the choice of values of ρ over which it is necessary to evaluate the Green's functions $G, \tilde{G}$. We shall show in the next section that the dynamical corrections to the force constants are all short range, so that only small values of ρ need be considered.

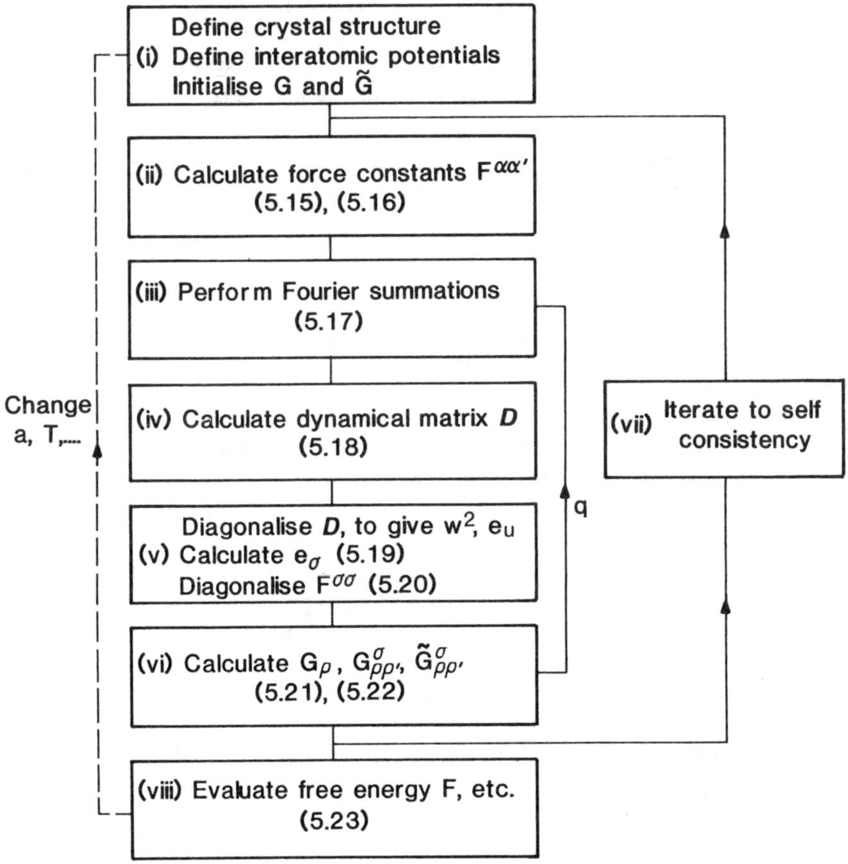
Define crystal structure
(i) Define interatomic potentials
Initialise G and G̃
(ii) Calculate force constants $F^{\alpha\alpha'}$
(5.15), (5.16)
(iii) Perform Fourier summations
(5.17)
(iv) Calculate dynamical matrix D
(5.18)
Diagonalise D, to give w^2, e_u
(v) Calculate e_σ (5.19)
Diagonalise $F^{\sigma\sigma}$ (5.20)
(vi) Calculate G_ρ, $G^\sigma_{\rho\rho'}$, $\tilde{G}^\sigma_{\rho\rho'}$
(5.21), (5.22)
(vii) Iterate to self consistency
(viii) Evaluate free energy F, etc.
(5.23)
q
Change
a, T,....

5.3 Numerical Calculations: Practicalities

1) Evaluation of Effective Potentials and Force Constants
We wish to consider the evaluation of the integrals

$$< \phi(\underline{R}) >= \frac{1}{[(2\pi)^3 det\underline{\underline{G}}]^{1/2}} \int d^3\underline{u}\, \phi(\underline{R}+\underline{u}) e^{-\frac{1}{2}\underline{u}.\underline{\underline{G}}^{-1}.\underline{u}}, \tag{5.24}$$

and the similar integral associated with the effective force constant matrix:

$$< \underline{\nabla}\,\underline{\nabla}\phi(R) >= \frac{1}{[(2\pi)^3 det\underline{\underline{G}}]^{1/2}} \int d^3\underline{u}\, \underline{\nabla}\,\underline{\nabla}\phi(\underline{R}+\underline{u}) e^{-\frac{1}{2}\underline{u}.\underline{\underline{G}}^{-1}.\underline{u}}. \tag{5.25}$$

For now we assume that ϕ is the soft part of the potential so that (5.14) and (5.25) are well defined integrals: in practice this may mean that we integrate only over $| \underline{R}+\underline{u} |> \epsilon$ for some $\epsilon \ll a$. Indeed, for $R \gg a$, the u integration can be usefully restricted to one unit cell only, or even to a sphere of radius $\sim 3\lambda$ (where $\lambda \equiv \frac{1}{3}$ tr $\underline{\underline{G}}$), with only a small error.

a) Isotropic fluctuations: as an initial approximation, or in the specific case of a cubic crystal, we may take $\underline{\underline{G}}$ to be isotropic;

$$\underline{\underline{G}} \simeq (\tfrac{1}{3} tr \underline{\underline{G}})\underline{\underline{1}} = \lambda \underline{\underline{1}}.$$

Then if $\phi(\underline{r}) = \phi(r)$ (the potential is spherically symmetric), we have

$$\begin{aligned} < \phi(R) > &= \frac{1}{\sqrt{2\pi\lambda^3}} \int \frac{d^3\underline{u}}{2\pi} \phi(| \underline{R}+\underline{u} |) e^{-\frac{1}{2}u^2/\lambda} \\ &= \frac{1}{\sqrt{2\pi\lambda}} \int_0^\infty dr\, \frac{2\lambda r}{R} \phi(r) g_-(r,R,\lambda)\,, \end{aligned} \tag{5.26}$$

where

$$g_\pm(r,R,\lambda) \equiv \tfrac{1}{2}\{exp(-\frac{(r-R)^2}{2\lambda}) \pm exp(-\frac{(r+R)^2}{2\lambda})\}. \tag{5.27}$$

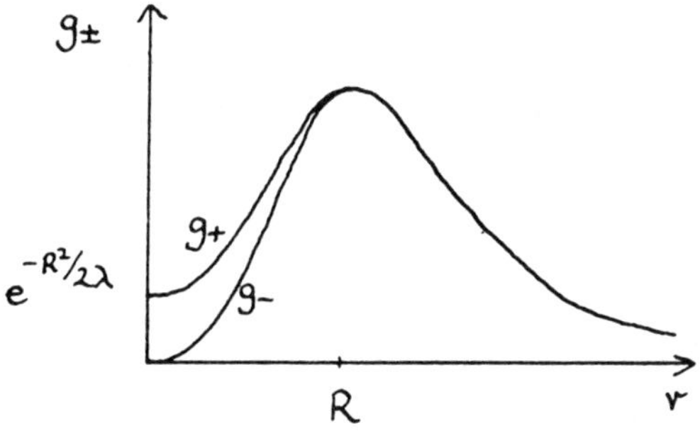

Similarly $<\underline{\nabla}\,\underline{\nabla}\phi>=(\underline{\underline{1}}-\underline{\hat{R}}\,\underline{\hat{R}})<\nabla\nabla\phi>_\perp+\underline{\hat{R}}\underline{\hat{R}}<\nabla\nabla\phi>_{||}$, where $\underline{\hat{R}}=\underline{R}/\mid\underline{R}\mid$, and

$$\begin{aligned}<\nabla\nabla\phi>_\perp &\equiv \frac{1}{\sqrt{2\pi\lambda}}\int_0^\infty dr\,2\{\frac{\lambda}{R^2}(\phi''-\frac{\phi'}{r})(g_+-\frac{\lambda}{Rr}g_-)+\frac{\phi'}{R}g_-\},\\ <\nabla\nabla\phi>_{||} &\equiv \frac{1}{\sqrt{2\pi\lambda}}\int_0^\infty dr 2\{\frac{r}{R}(\phi''-\frac{\phi'}{r})\Big((1+2(\frac{\lambda}{Rr})^2)g_-\\ &- 2(\frac{\lambda}{Rr})g_+\Big)+\frac{\phi'}{R}g_-\}\,. \qquad (5.28)\end{aligned}$$

In practice we can usually take the r integration in the range (ϵ,∞), or indeed $(R-3\lambda,R+3\lambda)$.

b) Coulomb terms: In the case of the Coulomb potential $\phi(R)\sim 1/\mid\underline{R}\mid$, we can obtain several useful analytic results. Firstly, in the limit as $\mid\underline{R}\mid\rightarrow\infty$

$$\begin{aligned}<1/R> &= \frac{1}{[(2\pi)^3 det\underline{\underline{G}}]^{1/2}}\int d^3\underline{u}\frac{1}{\mid\underline{R}+\underline{u}\mid}e^{-\frac{1}{2}\underline{u}.\underline{\underline{G}}^{-1}.\underline{u}}\\ &\sim \frac{1}{R}\{1+\tfrac{1}{2}(3\underline{\hat{R}}^T.\underline{\underline{G}}.\underline{\hat{R}}-tr\underline{\underline{G}})\frac{1}{R^2}+O(\frac{1}{R^4})\} \quad (5.29)\end{aligned}$$

whence

$$\begin{aligned} < \underline{\nabla}\,\underline{\nabla}(1/R) > \quad - \quad & \underline{\nabla}\,\underline{\nabla}(1/R) \sim \{\tfrac{15}{2}\underline{\hat{R}}.\underline{\underline{G}}.\underline{\hat{R}}(7\underline{\hat{R}\hat{R}} - \underline{\underline{1}}) - 30\underline{\hat{R}}^T\underline{\underline{G}}.\underline{R} \\ + \quad & 3\underline{\underline{G}} - \tfrac{3}{2}(5\underline{\hat{R}\hat{R}} - \underline{\underline{1}})tr\underline{\underline{G}}\}\frac{1}{R^5} + O(\frac{1}{R^7}) \qquad (5.30) \end{aligned}$$

Now $\lambda \sim 1/R$ as $R \to \infty$, so

$$\frac{< \nabla\nabla(1/R) > - \nabla\nabla(1/R)}{\nabla\nabla(1/R)} \sim O(\frac{\lambda}{R^2}) \sim O(\frac{1}{R^3}).$$

Note that if $\underline{\underline{G}}$ is isotropic, the first correction term in (5.29) vanishes, as does (5.30). In fact for isotropic $\underline{\underline{G}}$ we can evaluate $< 1/R >$ explicitly:

$$\begin{aligned} < 1/R > &= \int d^3\underline{u}\frac{1}{\mid \underline{R} + \underline{u} \mid}h(u) \\ &= \int_0^\infty du \int_{-1}^{1} d(\cos\theta)\, 2\pi u^2 \frac{h(u)}{(R^2 + u^2 + 2uR\cos\theta)^{1/2}} \\ &= \int_0^\infty 2\pi du\,(\frac{u}{R})(\mid u + R \mid - \mid u - R \mid)h(u) \\ &= \frac{1}{R} - \sqrt{\frac{2}{\pi\lambda}}\int_0^\infty \frac{d\epsilon}{\lambda}\frac{\epsilon}{R}(R + \epsilon)e^{-(R+\epsilon)^2/2\lambda} \\ &\sim \frac{1}{R} + \sqrt{\frac{2\lambda}{\pi}}e^{-R^2/2\lambda}\{\frac{1}{R^2} + O(\frac{1}{R^4})\} \qquad (5.31) \end{aligned}$$

as $R \to \infty$. So in the isotropic case, the corrections are exponentially damped, with screening length $\sqrt{\lambda}$.

If we remove the tails of the Gaussian, and approximate it by a distribution

$$h(u) \simeq \frac{15}{8\pi(2\lambda)^{3/2}}(1 - \tfrac{1}{2}\frac{u^2}{\lambda})$$

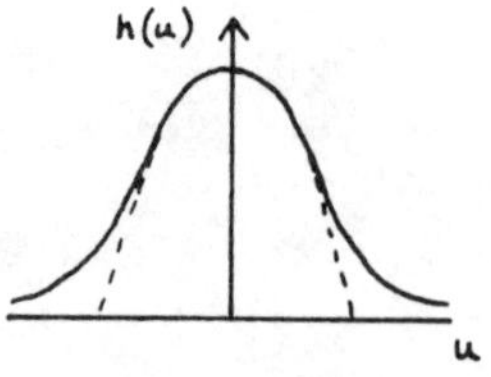

it is an easy exercise to show that

$$< 1/R >= \begin{cases} 1/R & R > \frac{1}{\sqrt{2\lambda}} \\ \frac{15}{4}\frac{1}{(2\lambda)^{3/2}}(\lambda - \frac{1}{3}R^2 + \frac{1}{20}\frac{R^4}{\lambda}) & R < \frac{1}{\sqrt{2\lambda}}, \end{cases}$$

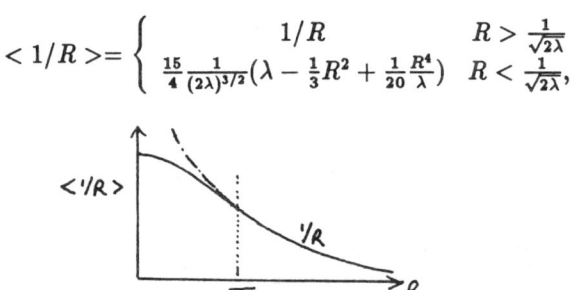

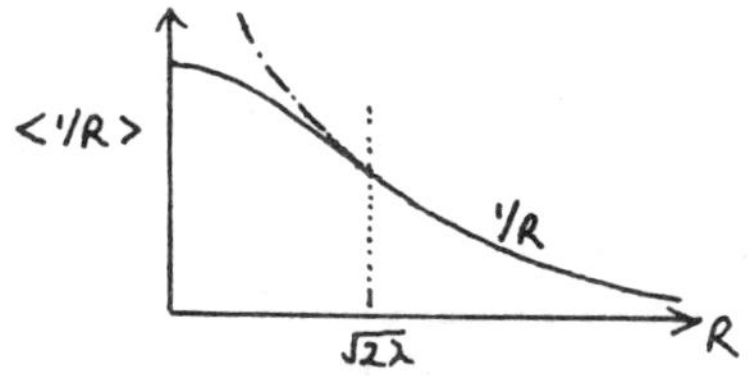

confirming (5.31), and showing how the averaging rounds off the Coulomb potential near the origin. Note that Coulomb potentials are soft; they need no hard core corrections.

c) Hard core corrections: as shown at the end of § 3.3, these have the form

$$< f(R) >= \frac{1}{(2\pi)^{3/2}(det\underline{\underline{G}})^{1/2}} \int d^3\underline{u}\, f(\underline{R} + \underline{u})e^{-\frac{1}{2}\underline{u}\cdot\underline{\underline{G}}^{-1}\cdot\underline{u}} \quad (5.32)$$

To approximate this, we set

$$f(R) = \begin{cases} 1 & 0 \le R \le d \\ 0 & R > d, \end{cases}$$

corresponding to hard spheres of radius d;

$$\phi^h(R) = \begin{cases} \infty & 0 \le R \le d \\ 0 & R > d. \end{cases}$$

Then in the isotropic approximation, we find that

$$< f(R) >= \tfrac{1}{2}(erf(\frac{R+d}{\sqrt{2\lambda}}) - erf(\frac{R-d}{\sqrt{2\lambda}})) - \sqrt{\frac{2\lambda}{\pi R^2}}g_-(d, R, \lambda),$$

where $erf x \equiv \frac{2}{\sqrt{\pi}} \int_0^x dte^{-t^2}$.

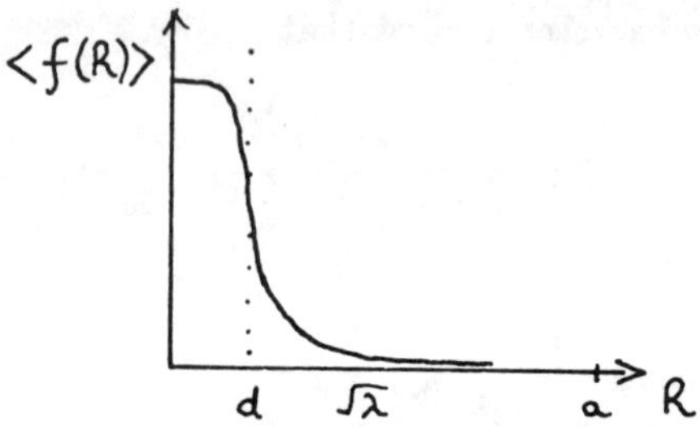

So for fixed λ, the step function $f(R)$ becomes smeared out and $< f(R) >$ exhibits a small tail extending toward values $R > d$ as expected. But as $R \to \infty, < f(R) > \sim e^{-R^2/2\lambda}$, so the hard core corrections are still only significant at short range; they have the same screening length as the corrections to the Coulomb potential. It is easy to see that with $\lambda, d \ll a, < f > \ll 1$ for $R \geq a$, justifying the virial expansion.

Note also that for fixed $R \geq a, < f(R) >_\lambda$ is very flat around $\lambda = 0$, goes through a maximum as λ increases (at $\lambda \sim R^2$, where $< f > \sim (d/R)^3$), and then decreases to zero as $\frac{2}{3}\frac{1}{\sqrt{\pi}}\frac{d^3}{\lambda^{3/2}}$.

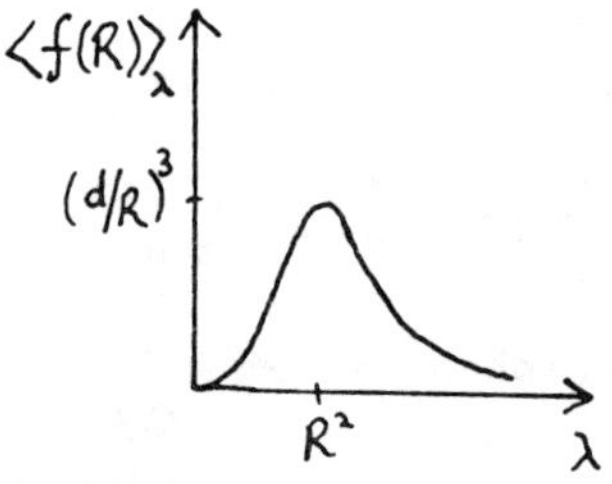

To summarise:

i) In an isotropic situation, all corrections to both short range repulsive and long range Coulomb potentials are of very short range (at most, say, two lattice spacings) and can be calculated explicitly using (5.26) - (5.28). The appropriate asymptotic expansions may then be used to estimate the size of the corrections for larger R. Note that this means that the correlation matrices $\underline{\underline{G}}_\rho$ need only be evaluated for small separations.

ii) In an anisotropic situation, not only must we calculate the corrections to the short range repulsions using the full three dimensional integrals (5.24) and (5.25), but we must also estimate the $O(1/R^6)$ corrections to the Coulomb forces in some outer region, using the asymptotic result (5.29), and possibly a suitable asymptotic expansion for $\underline{\underline{G}}_\rho - \frac{1}{3}(tr\underline{\underline{G}}_\rho)\underline{\underline{1}}$ at large values of ρ.

2) Calculation of Higher Order SCP terms

We will consider here for simplicity the second order corrections: third and higher order calculations involve similar considerations. For simplicity we consider a rigid ion model contribution . From (3.5)

$$\begin{aligned}
\underline{\underline{D}}^{(2)}_{q\nu} &= \frac{1}{2M}\sum_\rho \eta_\rho(\underline{q})\eta_\rho(-\underline{q}) < \nabla\underline{\nabla} v_\rho > \\
&- \frac{1}{2M}\tfrac{1}{2}\sum_\rho \eta_\rho(\underline{q})\eta_\rho(-\underline{q})\sum_{\rho'\sigma}\int_0^\beta d\tau\,\epsilon_3(\tfrac{1}{2}\underline{\underline{G}}^\sigma_{\rho\rho'}(\tau) : \underline{\nabla}\,\underline{\nabla}') < \underline{\nabla}\,\underline{\nabla} v_\rho > < v_{\rho'} > \\
&- \frac{1}{2M}\tfrac{1}{2}\sum_{\rho\rho'\sigma} \eta_\rho(\underline{q})\eta_{\rho'}(-\underline{q})e^{-i\underline{q}\cdot\underline{R}_\sigma}\int_0^\beta d\tau\, e^{i\hbar\Omega_\nu\tau} \\
&\qquad\qquad \epsilon_2(\tfrac{1}{2}\underline{\underline{G}}^\sigma_{\rho\rho'}(\tau) : \underline{\nabla}\,\underline{\nabla}') < \underline{\nabla} v_\rho > < \nabla' v_{\rho'} >,
\end{aligned}$$

and from (3.6).

$$F^{(2)} = F^{(1)} - \frac{N}{4}\tfrac{1}{2}\sum_{\rho\rho'\sigma}\int_0^\beta d\tau\,\epsilon_3(\tfrac{1}{2}G_{\rho\rho'}(\tau) : \underline{\nabla}\,\underline{\nabla}') < v_\rho > < v_{\rho'} > \ .$$

As far as the Fourier transforms are concerned, this is the most convenient form for numerical computation. The sums over ρ, ρ', σ will not be difficult to perform since we will only get significant contributions locally (for essentially the same reasons as in the effective potential calculations above). So the two points to consider are a) the evaluation of $\varepsilon_n(\underline{\underline{G}} : \nabla\nabla') < v > < v' >$, and b) the proper-time integrations.

a) The terms of the form $(\underline{\underline{G}} : \underline{\nabla\nabla'})^m < v > < v' >$ are evaluated trivially; to second order we must calculate $< \nabla^n v >$ for $n = 0, 1, 2, 3, 4$. We are then left with $\exp(\underline{\underline{G}} : \underline{\nabla}\,\underline{\nabla}') < v > < v' >$; taking Fourier transforms as for the effective potential, we have (cf (2.5) of § (2.1)).

$$exp(\tfrac{1}{2}\underline{\underline{G}}_{12}(\tau) : \underline{\nabla}_1\underline{\nabla}_2) < v_1(\underline{R}_1) >< v_2(\underline{R}_2) >$$

$$= \frac{1}{((2\pi)^6 det\mathcal{G})^{\frac{1}{2}}} \int d^3\underline{u}_1 \int d^3\underline{u}_2 v_1(\underline{R}_1 + \underline{u}_1) v_2(\underline{R}_2 + \underline{u}_2) e^{\frac{1}{2}u.\mathcal{G}^{-1}.u},$$

where

$$u \equiv \begin{pmatrix} \underline{u}_1 \\ \underline{u}_2 \end{pmatrix} \text{ and } \mathcal{G} \equiv \begin{pmatrix} \underline{\underline{G}}_{11} & \underline{\underline{G}}_{12} \\ \underline{\underline{G}}_{21} & \underline{\underline{G}}_{22} \end{pmatrix} = \begin{pmatrix} \underline{\underline{G}}_{\rho} & \underline{\underline{G}}^{\sigma}_{\rho\rho'} \\ \underline{\underline{G}}^{\sigma}_{\rho'\rho} & \underline{\underline{G}}_{\rho'} \end{pmatrix},$$

and the determinant and inverse are with respect to the six dimensional space. The explicit evaluation of these integrals would be time consuming (particular for a large number of values of $(\rho, \rho', \sigma, \tau)$). There is some simplification in the isotropic case ($\underline{\underline{G}}_{\rho}$ and $\underline{\underline{G}}^{\sigma}_{\rho\rho'}$ both isotropic); after performing the angular integrations (which are rather involved) a two dimensional integral remains to be performed numerically.

b) In the high temperature limit we need only consider a small number of values of ν, and correspondingly the τ integration in (3.5) and (3.6) is smooth and can be evaluated without difficulty (let $\tau \to 1/\tau$); for large ν the $e^{i\hbar\Omega_{\nu}\tau}$ factor in the non-local part of (3.5) makes the integrand oscillate rapidly, making the integral very small as $\beta \to 0$. Also notice that it is only this piece which need be evaluated at different values of ν; the local pieces are ν independent.

Acknowledgements

I would like to thank Marshall Stoneham, John Harding, and other members of the Theoretical Physics Division at AERE Harwell for their encouragement and hospitality during the period 1980-1986 during which this article was written, St. John's College, Cambridge, DAMTP, Silver St., Cambridge, the Theoretical Physics Department of Imperial College, London and the SERC for their support and toleration, and the secretaries of ITP, SUNY at Stony Brook for typing the manuscript.

References

Abers, E. and Lee, B. (1973) *Phys. Rep.* **9**, 1.

Abrikosov, A., Gorkov, L., and Dzyaloshinski, I. (1963) "Methods of Quantum Field Theory in Statistical Physics" (Prentice Hall).

Balian, R. and DeDominicis, C. (1960), *Nucl. Phys.* **16**, 502.

Ball, R. (1986) *Journ. Phys.* **C19**, 1293.

Bardeen, W., and Moshe, M. (1983) *Phys. Rev.* **D28**, 1372.

Barron, T. and Klein, M. (1974), in "Dynamical Properties of Solids," Vol I (Amsterdam, North Holland).

Baym, G. (1961), *Ann. Phys.* N.Y. **14**, 1.

Baym, G. and Mermin, N. (1961) *Jour. Math. Phys.* **2**, 232.

Baym, G. and Sessler, A. (1963), *Phys. Rev.* **131**, 2345.

Beck, M. and Meier, P. (1971) *Zeit Phys.* **247**, 189.

Boccarma, N. and Sarma, G. (1965) *Physics* **1**, 219.

Born, M. and Huang, K. (1954) "Dynamical Theory of Crystal Lattices" (London, OUP).

Born, M. and Oppenheimer, J. (1927), *Ann. Phys.* N.Y. **84**, 457.

Boyer, L. (1981) *Phys. Rev.* **B23**, 3673.

Bruekner, K. and Sawada, K. (1957) *Phys. Rev.* **106**, 1117.

Catlow, C., Corish, J., Jacobs, P. and Lidiard, A. (1981) *J. Phys.* **C14**, 1121.

Chattopadhyay, P., Chakrabarti, B. and Sinha, S. (1977) *J. Phys.* **C10**, 161.

Choquard, P. (1967) "The Anharmonic Crystal" (New York, Benjamin)

Cochran, W. (1959) *Proc. Roy. Soc.* **A253**, 260.

(1971) *Crit. Rev. Solid State Sci.*, **2**, 1.

Cornwall, J. and Norton, R. (1973) *Phys. Rev.* **D8**, 3338.

DeDominicis, C. (1962) *Jour. Math. Phys.* **3**, 983.

DeDominicis, C. and Martin, P. (1964) *Jour. Math. Phys.* **5**, 14, 31.

Dick, B., and Overhauser, A. (1958), *Phys. Rev.*, **112**, 90.

Dzyaloshinski, I. (1962) *Sov. Phys. JETP* **15**, 778.

Fetter, A. and Walecka, J. (1971) "Quantum Theory of Many Particle Systems", (New York, McGraw-Hill).

Feynman, R. (1984) *Rev. Mod. Phys.* **20**, 367

(1953) *Phys. Rev.* **81**, 1291

(1973) "Statistical Mechanics" (New York, Benjamin).

Feynman, R. and Hibbs, A. (1965) "Quantum Mechanics and Path Integrals" (New York, McGraw-Hill).

Fredkin, D. and Werthamer, N. (1965) *Phys. Rev.*, **138**, A1527.

Gillis, N., Werthamer, N. and Koehler, T. (1968) *Phys. Rev.* **165**, 95.

Gillis, N. and Werthamer, N. (1968) *Phys. Rev.* **167**, 607.

Gillis, N. Koehler, T. and Werthamer, T. (1968), *Phys. Rev.* **175**, 1110.

Glyde, H. (1971) *Can. Jour. Phys.* **49**, 761.

Glyde, H. and Khanna, R. (1971) *Can. Jour. Phys.* **49**, 2997.

Glyde, H. and Klein, M. (1971) *Crit. Rev. Solid State Sci.* **2**, 181.

Goldman, V., Horton, G. and Klein, M.(1968) *Phys. Rev. Lett.* **21**, 1527.

(1970) *Phys. Rev. Lett.* **24**, 1424.

Götze, W. (1966) *Phys. Rev.* **156**, 951.

Götze, W. and Michel, K. (1968) *Zeit für Phys.* **217**, 170.

Hooton, D. (1958), *Phil. Mag.* **3**, 49.

Horner, H. (1967) *Zeit für Phys.* **205**, 72.

(1970), *Phys. Rev.* **A1**, 1712, 1722.

(1971), *Zeit. für Phys.* **242**, 432.

(1972), *Jour. Low Temp. Phys.* **8**, 511.

(1974) in "Dynamical Properties of Solids", Vol I. (Amsterdam, North Holland).

Honerkamp, J. and Meetz, K. (1971) *Phys. Rev.* **D3**, 1976.

Honerkamp, J. and Ecker, G. (1972) *Nucl. Phys.* **B36**, 130.

Honerkamp, J., Krause, F. and Scheunert, M. (1973), *Nucl. Phys.* **B69**, 168.

Huckkaby, D. and Salzburg, Z. (1970), *Jour. Chem. Phys.* **53**, 2304.

Ingermanson, R., (1986) Ph.D. Thesis, Berkeley (unpublished).

Jackiw, R. and Johnson, K. (1973) *Phys, Rev.* **D8**, 2386.

Jackiw, R. Cornwall, J. and Tomboulis, E. (1974) *Phys. Rev.* **D10**, 2428.

Jastrow, R. (1955) *Phys. Rev.* **98**, 1479.
Kadanoff, L. and Baym, G. (1962) "Quantum Statistical Mechanics" (London, Benjamin/Cummings).
Keldysh, L. (1964) *Sov. Phys. JETP* **20**, 1018.
Kerr, W. and Sjölander, A. (1970) *Phys. Rev.* **B1**, 2723.
Klein, M. Goldman, V. and Horton, G. (1969) *Jour. Phys.* **C2**, 1542.
(1970) *Journ. Phys. Chem. Sol.* **31**, 2441.
Klein, M, Chell, G., Goldman, V. and Horton, G. (1970) *Jour. Phys.* **C3**,806.
Klein, M., Koehler, T. and Gray, R. (1973) *Phys. Rev.* **B7**, 1571.
Koehler, T., (1966) *Phys. Rev. Lett.* **17**, 89.
(1967) *Phys. Rev. Lett.* **18**, 516.
(1968) *Phys. Rev.* **165**, 942.
(1969) *Phys. Rev. Lett.* **22**, 777.
(1970) *Phys. Lett.* **33A**, , 359.
Koehler, T. and Werthamer, N. (1971) *Phys. Rev.* **A3**, 2074.
(1972) *Phys. Rev.* **A5**, 2230.
Lee, T., and Yang, C. (1969) *Phys. Rev.* **117**, 22, 897.
Lennard-Jones, J. and Devonshire, A. (1938) *Proc. Roy. Soc.* **53**, A167.
Lima, R. and Tsallis, C. (1980) *Phys. Rev.* **B21**, 458.
Luttinger, J. and Ward, J. (1960) *Phys. Rev.* **118**, 1417.
MacDonald, R. and Mountain, R. (1979) *Phys. Rev.* **B20**, 4012.
MacDonald, R. and MacDonald, W. (1981) *Phys. Rev.* **B24**, 1715.
Maradudin, A. and Fein, A. (1962) *Phys. Rev.* **128**, 2589.
Martin, P. and Schwinger, J. (1959) *Phys. Rev.* **115**, 1342.
Meissner, G. (1970) *Zeit. Phys.* **235**, 85.
Mills, R. (1969) "Propagators for Many-Particle Systems", (New York, Gordon and Breach).
Morita, T. and Hiroike, K. (1961) *Prog. Theor. Phys.* **25**, 537.
Moshinsky, M. and Kittel, C. (1968) *Proc. Nat. Acad. Sci.* USA, **60** ,4.
Niemi, A. and Semenoff, G. (1984) *Ann. Phys.* **152**, 105.
Nosanow, L. (1966) *Phys. Rev.* **146**, 120.
Ranninger, J. (1965) *Phys. Rev.* **140**, A2031.
Ree, F. and Holt, A. (1970) *Phys. Rev.* **B8**, 826.

Ripka, G. (1979) *Phys. Rep.* **56**, 1.
Salam, A. (1971) Proc. of Amsterdam Int. Conf. El. Part.,
Proc. of Rochester Meeting of APS/DDPF.
Salam, A., Isham, C. and Strathdee, J. (1971) *Phys. Rev.* **D3**, 1805.
(1972) *Phys. Rev.* **D5**, 2548.
Schwinger, J.(1961) *Jour. Math. Phys.* **2**, 407.
Sham, L. (1965) *Phys. Rev.* **139**, A1189.
Shukla, R. and Cowley, E. (1971) *Phys. Rev.* **B3**, 4055.
Shukla, R. (1980) *Int. Jour. of Thermophys.* **1**, 73.
Shukla, R. and MacDonald, R. (1980) *High Temp. High Press*, **12**, 291.
Shukla, R., Paskin, A., Welch, D. and Dienes, G. (1981) *Phys. Rev.* **B24**, 724.
Sinha, S. (1969) *Phys. Rev.* **177**, 1257.
Stevenson, P. (1984) *Zeit. Phys.* **C24**, 87.
(1984) *Phys. Rev.* **D30**, 1712.
(1985) *Phys. Rev.* **D32**, 1389.
Tripathi, R. and Pathak, K. (1974) *Nuov. Cim.* **21B**, 289.
Van Doren, V. (1973) *Solid State Com.* **12**, 943.
Welch, D. Dienes, G. and Paskin, A. (1978) *Jour. Phys. Chem. Sol.* **39**, 589.
Werthamer, N. (1970) *Phys. Rev.* **B1**, 572.
(1970) *Phys. Rev.* **A2**, 2050.
(1972) *Phys. Rev.* **B5**, 285.
(1973) *Phys. Rev.* **A7**, 254.
Witschel, W. and Bohmann, J. (1980) *Journ. Phys.* **A13**, 2735.
Yukalov, V. (1979) *Ann. Phys.* **36**, 419.
(1981) *Ann. Phys.* **38**, 31.
Zubov, V. and Terletsky, V. (1970) *Ann. Phys.* **24**, 97.
Zubov, V. (1974) *Ann. Phys.* **32**, 33, 93; **33**, 103
(1975) *Phys. Stat. Sol.* **72**, 483.
(1978) *Phys. Stat. Sol.* **87**, 385; **88**, 43.
(1981) *Phys. Stat. Sol.* **104**, 383.

Chapter 2

DEFECTS AND TRANSPORT IN IONIC SOLIDS

J.H. Harding

Theoretical Physics Division, Harwell Laboratory,
Didcot, Oxon OX11 0RA, U.K.

The phenomenon of transport, whether of mass (diffusion) or charge (conductivity) is a property of crystals that is of fundamental importance. Although it is true that transport will take place at any finite temperature, in practice, bulk transport of matter is important only when the crystal is heated to a substantial fraction (at least half) of its melting temperature. In ionic crystals, matter transport frequently implies charge transport and this also therefore becomes rapidly more important with rising temperature. In most cases, transport is associated with point defects. These may be created by deliberate (or accidental!) doping or by random thermal fluctuations, In this chapter we shall consider how progress made in calculating defect energies and entropies may be used in obtaining thermodynamic and transport properties of ionic crystals. Problems related to the surface will be briefly discussed, but the main concern here is with point defects in the bulk crystal. In the past, the calculations that have been done have normally been of energies. These have then been compared with the data obtained by analysing experimental data; that is to say, with enthalpies obtained by some multi-parameter fit to the data, usually diffusion or conductivity results. This raises two major issues.

There is first the question of whether such a procedure can produce unique values of the defect parameters. (For a discussion of the problems involved in fitting experimental data see Chadwick and Corish 1987). It would obviously be desirable to omit this step, but if we are to do so, it is not enough (though it is much!) to calculate energies of the various defect processes. We must calculate the pre-factors as well. These factors contain a number of terms, but only two present any difficulties of principle. These are the defect entropies and the defect activities. For any purposes, it is adequate to calculate defect activities by the Debye-Hückel theory (Lidiard 1957). We discuss the problems associated with defect activities in Section 2.3 where we shall be particularly concerned with situations beyond the dilute limit. The defect entropies present a more obvious computational challenge. The orientational contribution is simple (if sometimes tedious) to obtain. However, the main contributions come from the effect of the presence of the defect on the lattice vibrations and the effect of the lattice thermal expansion on the internal energy of formation of the defect.

Second, there is the question of the propriety of comparing calculations of internal energies (done at constant volume) with experimentally derived enthalpies that refer to conditions of constant pressure. This is a question about the *meaning* of what is calculated.

These questions are, as we shall see, closely linked. The plan of this chapter is therefore as follows. In the next Section we shall consider the methods used to calculate the constant volume parameters. This enables us to obtain the Helmholtz free energy for defects in a solid. This of itself is not enough; what we require is the Gibbs energy in a solid where the thermal expansion is properly taken into account. In Section 2.3 we therefore consider two matters. First, can we calculate the lattice thermal expansion, and in particular, is the quasi-harmonic approximation (which is easy to implement) up to the job? Second, we need to consider the relationship between constant volume parameters, calculated by the methods of Section 2.2, and constant pressure parameters, which is what almost all experiments provide. We then turn to the theoretical apparatus that links these defects parameters to the measured properties that depend on the presence of defects. This entails a discussion of the mass-action approximation and the various methods that attempt to correct for the effects of high defect concentration. In Sections 2.4 to 2.6 we turn to consider transport properties and the theories that link the macroscopic observations to the microscopic defect processes. Here we review progress made in recent years in understanding the basis of the reaction rate theories first put forward by Slater and Eyring (see Glasstone *et al* (1941) for a list of the early references) and systematised by the work of Vineyard (1957). In particular, we shall consider the recent comparisons made between reaction rate theory and molecular dynamics simulations. In the light of all this we shall be able to see what progress has been made in the last few years and what the problems still are.

2.2 CALCULATION OF THE CONSTANT VOLUME PARAMETERS.

2.2.1 The crystal potential. The first necessity in calculating defect parameters is to have a model of the crystal forces. For the purposes of this article it will be assumed that the crystal is ionic. This is not so restrictive an assumption as might at first appear. The question of what *ionicity* means has been much discussed; those interested in the current state of the debate should consult the review article of Catlow and Stoneham (1983) and the recent discussion of the fundamental principles by Pyper (1987). For current purposes, a crystal is deemed to be *ionic* if (a) the crystal potential may be approximated as a sum of pair potentials between the components of the unit cell and (b) the Coulomb term dominates the long-range contributions. Such a definition is purely operational and while systems conventionally considered ionic (such as sodium chloride) certainly fall within its scope, a large number of systems not normally considered ionic do so also provided one restricts oneself to certain types of calculation. Thus the work of Catlow and Parker (1982), for example, shows the success of simple ionic models in calculating the structures of silicates. Provided one restricts oneself to calculating crystal or defect properties where the energies are dominated by the polarisation and distortion of the crystal, reasonable results may be obtained for systems that are, on the face of it, unpromising candidates for an ionic model. The point is discussed in detail by Harding and Stoneham (1982).

This still leaves the question of how we are to determine the crystal forces. If we are content to assume that the formal charges one might expect from simple chemical arguments give a reasonable representation of the long-range forces, we only require a model of the short-range forces (usually assumed to comprise the Pauli repulsion, dispersion and 'covalency' contributions). Most accurate calculations of defect properties also model the electronic polarisability of the ions. This is usually done using the shell model due originally to Dick and Overhauser (1964). This simple, yet very successful, model assumes that the ion may be considered as a massive core linked to a massless shell by a harmonic spring. This gives two fitting parameters: the shell charge Y and the spring constant K. This is very naive yet it does succeed in linking the short-range forces and the ionic polarisability in a physically sensible way especially if, as is nearly universal, the short-range forces are assumed to act between the shells only. Various attempts have been made to 'derive' the shell model from more fundamental considerations (for an example see Sinha 1973). These are interesting but not, in the end, convincing.

The question of potentials has recently been reviewed by Stoneham and Harding (1986); only the main points of that discussion are made here. In the past, potentials have been obtained by fitting simple analytic forms such as the Buckingham potential

$$V_{\alpha\beta}(r) = A_{\alpha\beta} \exp(-r/\rho_{\alpha\beta}) - C_{\alpha\beta}/r^6 \tag{2.2.1}$$

to perfect lattice properties. This is frequently satisfactory, but suffers from the problems that it ensures the adequacy of the potential only in the region of the interionic separations of the perfect lattice (see Figure 2.1) and also, clearly, relies on there being data available for fitting. The first problem may be partially alleviated by using the fitted potential to calculate properties of the lattice at other lattice parameters. Potentials have therefore been used to calculate the pressure derivatives of crystal properties (Cox and Sangster 1982), the Hugoniot equation of state (Harding and Stoneham 1984) and the lattice thermal expansion. This last is of more importance than simply as a test of potentials. We shall require values of the lattice thermal expansion to obtain free energies of defect processes and it is highly advisable to perform calculations entirely within the model we assume. We shall consider the lattice expansion further when we come to consider the question of the thermodynamics of defect processes.

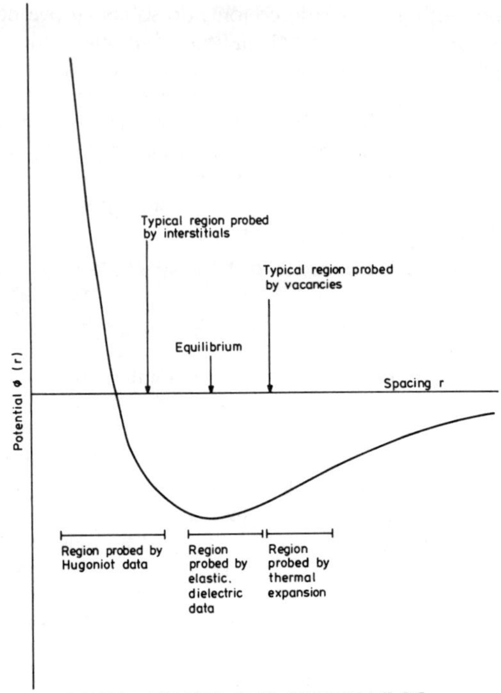

FIGURE 2.1 THE SHORT-RANGE CONTRIBUTION TO THE INTERIONIC POTENTIAL (SCHEMATIC)

However, there are cases where one must consider the question of calculating potentials. Two cases in point are where there are no data to fit to (the behaviour of different charge states in transition metal oxides for example) or where the defect configuration is so far from anything that can be investigated by fitting that there is serious doubt whether the fitted potential can be relevant (some highly distorted complexes or transition states). In addition to these cases where calculation might be considered necessary, there are some cases where there are obvious advantages over empirical parametrisations. If one wishes to consider a series of compounds it is useful to know precisely what approximations have gone into the potential so that one can be certain that trends in the calculated values are due to changes across the series and not the vagaries of fitting. The problem with fitting simple functional forms to experimental data is that it is never quite clear what approximations are being made.

Calculations of interionic potentials have, in the main, employed simple schemes like the electron gas approximation due originally to

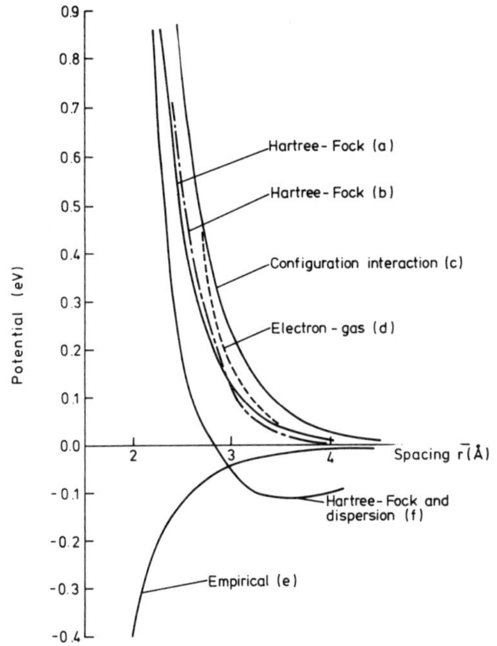

FIGURE 2.2 SHORT-RANGE INTERACTION POTENTIAL FOR O^{2-}- O^{2-} ((a,c) KENDRICK AND MACKRODT 1983; (b) HARDING AND HARKER 1985; (d) MACKRODT AND STEWART 1979; (e) CATLOW et al 1978; (f) PYPER 1987.

Wedepohl (1967) and popularised by Gordon and Kim (1972) (e.g. Mackrodt and Stewart 1979). These give results for the lattice parameter that typically are about 5% too large and the calculations of defects must be judged in the light of this. Much of this error may be ascribed to the neglect of dispersion. It is also important to ensure that the electron density is calculated in an appropriate crystal environment (Muhlhausen and Gordon 1981). If one is interested in trends rather than absolute values these methods are particularly useful. A few attempts have been made to calculate potentials by Hartree-Fock theory (Kendrick and Mackrodt 1983, Harding and Harker 1984, Pyper 1986, Pyper *et al* 1987). These methods are much less well developed and the potentials do not always agree as may be seen from Figure 2.2. The recent developments of Pyper do, however give a general framework for the calculations of such potentials, although the computer time required is rather large. More developments are needed in this area since the potentials are the foundation of any calculation of defect properties.

When we have obtained a satisfactory model of the crystal forces, we may proceed to calculate the defect parameters. Over the last fifteen years, standard programs have been written to do this. Here we briefly describe the physics that lies behind them. Reference to where the interested reader can find a more detailed account are given where appropriate.

2.2.2 Defect energies. The methods of calculating the internal energies of defect processes assume that the defect energy is minimised by relaxation of the surrounding ions. The amount of relaxation of these ions falls off rapidly the further from the defect they are. This suggests the possibility of dividing the crystal notionally into two regions; an inner region where the ion displacements are explicitly calculated and an outer region where some suitable approximation is used. This is the basic idea of the Harwell HADES program. We write the energy of the defective crystal as

$$E = E_1(\mathbf{x}) + E_2(\mathbf{x},\xi) + E_3(\xi) \tag{2.2.2}$$

where $E_1(\mathbf{x})$ is the energy of the inner region, $E_3(\xi)$ is the energy of the outer region and $E_2(\mathbf{x},\xi)$ is the interaction energy between the two regions. We assume that the displacements of the outer region, ξ, may be calculated by some continuum approximation and that they are an implicit function $\xi(\mathbf{x})$ of the inner region displacements $\mathbf{x}$. We define $E_3(\xi)$ to be a quadratic function of ξ;

$$E_3(\xi) = \tfrac{1}{2}\xi.\mathbf{A}.\xi \tag{2.2.3}$$

and, assuming the equilibrium condition,

$$\frac{\partial E(\mathbf{x},\xi)}{\partial \xi} = \frac{\partial E_2(\mathbf{x},\xi)}{\partial \xi} + \mathbf{A}.\xi = 0 \tag{2.2.4}$$

for $\xi = \xi_0$. We thus obtain an expression for $E_3(\xi)$ and hence for E;

$$E = E_1(\mathbf{x}) + E_2(\mathbf{x},\xi) - \tfrac{1}{2}\left.\frac{\partial E_2(\mathbf{x},\xi)}{\partial \xi}\right|_{\xi=\xi_0} \cdot \boldsymbol{\xi_0} \tag{2.2.5}$$

Here the outer displacements, ξ_0 are those corresponding to arbitrary values of $\mathbf{x}$. Formally at least, we may find the energy by solving the equations

$$\frac{dE}{d\mathbf{x}} = 0 \tag{2.2.6}$$

in which the derivative implies an implicit differentiation with respect to the coordinates ξ. Although this can and has been done, it is simpler to find the defect energy by requiring that the forces on each ion in the inner region be zero. This, the force balance condition, is equivalent to the direct minimisation of E provided that the outer region is in mechanical equilibrium. We shall in future denote the defect formation energy calculated by this method (or others like it) by the symbol u_V as it is an internal energy at constant volume.

This formal scheme has been turned into a computer program (Norgett 1974) by assuming the ionic model as defined above and using the Mott-Littleton approximation (Mott and Littleton 1938) for the outer region. This assumes that the only important feature of the defect for the outer region is its net charge. As originally implemented, it assumed that the crystal structure was diagonally cubic, so that the response of the outer region was isotropic. The effects of introducing anisotropy have been investigated. So far no system has been found where such effects are important. More recently, the pair potential approximation has been relaxed, and later versions of the CASCADE program (Leslie 1982) can handle certain types of bond-bending forces. The defect parameters calculated by these methods frequently agree very closely with the constant pressure enthalpies obtained from experimental data (for a review see Mackrodt 1982).

2.2.3 Defect entropies. The accurate calculation of entropies is more demanding. It requires knowledge not only of the potential $V_{\alpha\beta}(r)$ but also of its derivatives. A number of simple approximations have been used to attempt to calculate entropies. For example, it has been assumed that the perturbations of the defect on the vibrational spectrum of the lattice is small and a perturbation expansion may be used (Thiemer 1958) or that only a few lattice vibrations localised around the defect contribute significantly to the entropy and all the rest may be ignored (Agrawal and Garg 1973, Benière 1976). All such simplifying assumptions are incorrect. The three methods to be discussed here all acknowledge that there is no alternative to calculating the full lattice phonon spectrum of the defective solid. However, one may make two assumptions. First, one may note that the solid surrounded by relaxed lattice ions may still be considered as a harmonic solid. Second, as we did for the energy calculation, we shall assume that the defect is surrounded by distorted perfect lattice; the concentration of defects is assumed to be low. With these assumptions we may write the entropy of formation of a point defect as

$$\Delta s = S \; (defect \; crystal) - S \; (perfect \; crystal) \tag{2.2.7}$$

and so write, using the well-known formula for the entropy of a harmonic solid in the high temperature limit

$$S = k \sum_{i=1}^{3N} (1 - \ln \frac{\hbar\omega_i}{kT}) \tag{2.2.8}$$

$$\Delta s = -k \ln \frac{\prod_{i=1}^{3N'} \omega_i'}{\prod_{i=1}^{3N} \omega_i} + 3k\,(N' - N)\,(1 - \ln \frac{\hbar}{kT}) \tag{2.2.9}$$

where ω_i, ω_i' are the lattice frequencies for the perfect and defective lattices respectively and k is Boltzmann's constant. The problem, of course, is how to reduce the summation over 10^{23} modes to something more reasonable. The oldest method of doing this employs lattice Green functions. The technique had been much used in the calculation of local modes of defects (see Maradudin *et al* 1971, for a complete discussion and review) and it seemed natural to adapt this to this closely related problem. This method is based on the fact that the first term in (2.2.9) may readily be written as the ratio of the product of the squares of the lattice frequencies in the defect and perfect lattices; i.e. the ratio of the eigenvalues of the dynamical matrices. The method thus comprises three steps (Jacobs *et al* 1982). First, express the product of frequencies as the determinant of a Green matrix; second, write the Green matrix as the sum of the perfect lattice Green matrix and a set of force constant changes between the perfect and defect lattice; third, assume that these changes affect only a small number of ions around the defect so that a partitioning technique may be used. Formally, we may write the dynamical equations of motion of a harmonic crystal as

$$\mathbf{L.u} = \mathbf{M}^{\frac{1}{2}} \, (\, \mathbf{D} - \omega^2 \mathbf{I} \,) \, \mathbf{M}^{\frac{1}{2}} \mathbf{u} = 0 \tag{2.2.10}$$

where $\mathbf{D}$ is the dynamical matrix. It follows from this that

$$\prod_i \omega_i^2 = |\mathbf{D}| \tag{2.2.11}$$

and it may also readily be shown that

$$\prod_i \omega_i^2 = \lim_{\omega \to 0} \prod_i (\omega_i^2 - \omega^2) = \frac{1}{M} \lim_{\omega \to 0} |\mathbf{L}| \tag{2.2.12}$$

Since the defect crystal may still be considered as a harmonic system we can write

$$(\mathbf{L} + \boldsymbol{\delta}\mathbf{L}).\mathbf{u} = 0 \tag{2.2.13}$$

or equivalently,

$$(\mathbf{I} + \mathbf{G}.\boldsymbol{\delta}\mathbf{L}).\mathbf{u} = 0 \tag{2.2.14}$$

where $\mathbf{G}$, the inverse of $\mathbf{L}$, is the static Green matrix for the perfect crystal and $\boldsymbol{\delta}\mathbf{L}$ contains all the differences between the defect and imperfect crystal. We now define $\mathbf{g}$, $\boldsymbol{\delta}\mathbf{l}$ as the matrices containing only the elements of $\mathbf{G}$, $\boldsymbol{\delta}\mathbf{L}$ for the sites where $\boldsymbol{\delta}\mathbf{L}$ is significant. It may now readily be shown that the entropy given by (2.2.9) is

$$\Delta s = -\tfrac{1}{2} k \lim_{\omega \to 0} \ln \frac{|\mathbf{M}'|}{|\mathbf{M}|} \, (\mathbf{I} - \mathbf{g}.\boldsymbol{\delta}\mathbf{l}) \tag{2.2.15}$$

This equation constitutes the Green function method of calculating entropies. If we wish to consider vacancies or interstitials, we must add or delete rows and columns to $\mathbf{g}$ and $\mathbf{l}$. The details are given in the two references cited above. This method has two main problems. First, for technical reasons (see Sangster and Rowell 1982) it is difficult to use the shell model. The basic problem is that the Green matrices are calculated for the cores. Thus the force constant difference matrix must be calculated with respect to the cores also. Recalculating core-shell and shell-shell forces as effective core-core forces is exceptionally messy. Thus all such Green function calculations to date have been done with the rigid ion model, with its well-known inadequacies in calculating the optic modes. Second, a great deal of time and labour is necessary to check the accuracy of the Green matrix elements. Further, it is difficult to see how the method could be the foundation of a general purpose computer code (for a discussion of the practical problems of these calculations see Gillan and Jacobs 1983).

There is also a particular problem with charged defects (common to one of the other methods of calculating entropies that we shall discuss). In both neutral and charged defects, the displacements of the ions around the defect fall off with distance as R^{-2}. However, whereas for neutral defects there is only an elastic term, for charged defects there is also a dielectric term which will dominate. It follows that the changes in the *force constants* for the ions surrounding a neutral defect fall off as R^{-3} but only as R^{-2} for the charged defect. Since the number of ions near the boundary beyond which we ignore changes in the force constant matrix goes up as R^2, it follows that there is a surface effect that will not go away no matter how large the region where force constant changes are considered is made. This problem is revealed by the lack of convergence of Δs with increasing region size. The solution to this problem lies in recognising that it is not the defect charge that is the problem but the distortion field caused by it. We consider a perfect crystal with lattice displacements scaled down so that only the contribution to the entropy change linear in the displacements is retained. We denote this by S_{corr}, and so write

$$S_{corr} = S'(\lambda) - S(perfect\ lattice) \tag{2.2.16}$$

where λ is the scaling factor for the displacements. It may then be shown that the true entropy of the defect process, s_V, is given by

$$s_V = \Delta s - S_{corr} \tag{2.2.17}$$

where Δs is the result obtained from (2.2.15) with a finite region size. The value of s_V now converges to a value which is the entropy of the defect process in an infinite crystal and so corresponds to the definition of defect energies used earlier in the chapter. The subscript denotes the fact that this too is a constant volume calculation.

The reason why the correcting expression (2.2.16) takes the form that it does is the interaction of the defect with the surface. In a finite crystal, the surface relaxations produce an electrostatic potential in the bulk which alters individual defect energies but not the neutral combinations usually considered (Duffy *et al* 1984). Since these relaxations are temperature dependent, they also affect the entropy. However, they only matter when individual charged defects are considered (as for example in the analysis of thermopower experiments or the very recent determinations of the defect parameters of individual defects by analysis of their response to the surface potential). In a Green function calculation, this effect is responsible for the convergence problems noted earlier. This is because the expansion of the region size produces different (internal) ionic surfaces for different radii of the region. Further, since such surfaces are artifacts of the calculation, they are not normally physically sensible. The effect of subtracting S_{corr} is to remove such artifacts.

The two other methods of calculating defect entropies are 'brute force' methods. We first consider the large crystallite method. Here we divide the crystal into two regions; an inner region containing the defect and surrounding ions where the ions are allowed to vibrate and an outer region where the ions are held fixed in the positions calculated by the static lattice calculation. We may now calculate the entropy of a defect process using (2.2.9) where the products are taken only over the inner region. The ions in the outer region contribute only to the diagonal elements of the dynamical matrix. As with the Green function method, we must subtract the surface correction (2.2.16). Using this method it is possible to write a general code that will calculate entropies of defects. As with the static lattice case, it is possible to use symmetry to reduce the amount of calculation required. The method used is based on those developed for the vibrational analysis of large molecules and is described in a report of Ball and Harding (1983).

The second method again uses (2.2.9) but solves the problem of calculating the macroscopic number of modes that equation implies by using a large unit cell. The defect and surrounding ions are enclosed in a box and periodic boundary conditions imposed (Harding and Stoneham 1981). For consistency, the static lattice calculation should also be performed using periodic boundary conditions. Methods for doing this have been developed (Leslie and Gillan 1985). The original calculations using this method evaluated the phonon spectrum of the large unit cell only at zero wave vector, relying on the effect of the folding back of the dispersion curves onto the Γ point of the reduced Brillouin zone. More recent work (Allan, Mackrodt and Leslie 1987) has shown that better convergence with small region sizes may be obtained either by taking a mesh of points or by using special points within the Brillouin zone. However, to include the effect of local modes, it is necessary always to ensure that the Γ point is one of those chosen.

It is important to ensure that the calculation of the entropy has converged with respect to increasing the region size. In some early work this was not done, and the results are unreliable. It is therefore of interest to consider the relative speed of convergence. Gillan and Jacobs (1983) have discussed the convergence of the Green function and large crystallite methods and have shown that they converge to the same answer, the Green function method converging the faster. Figure 2.3 shows the convergence of the 'brute force' methods for the anion

Frenkel defect in calcium fluoride as a function of the number of ions N in the supercell. The first satisfying point to note is that again they converge to the same answer. It is interesting to note that the convergence of the large crystallite method is the poorest. The figure also shows the effect of various ways of constructing the large unit cell on the convergence of the problem.

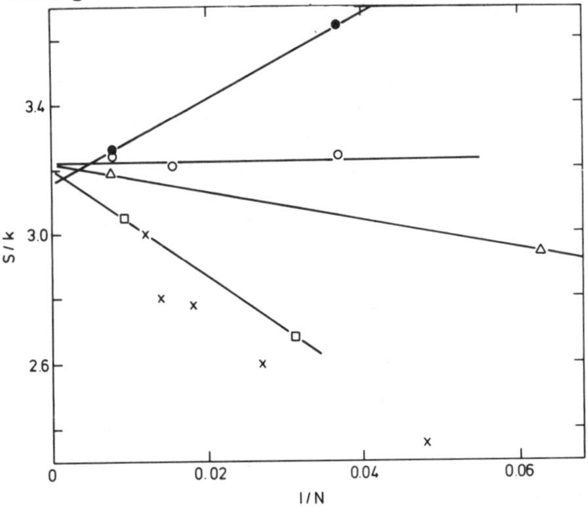

FIGURE 2.3 CALCULATED FRENKEL ENTROPY FOR CALCIUM FLUORIDE CROSSES SHOW THE LARGE CRYSTALLITE CALCULATIONS, OTHER SYMBOLS SHOW SUPERCELL CALCULATIONS (FOR DETAILS OF THE MEANING SEE THE TEXT)

The open symbols refer to a simple cubic cell (circles), a body- centred cell (triangles) and a face-centred cubic cell (squares). The filled circles show the effect of doing the calculation with the two defects in the same cell (thus producing a neutral cell) rather than doing two separate calculations with the isolated defects (and so using charged cells). It is clear that a careful choice of cell can signficantly enhance convergence. However, the main advantage of the 'brute force' methods does not lie in how rapidly they converge. They have two great advantages over the Green function methods. First, it is straightforward to use the shell model and so get a good representation of the phonon spectrum. Second, as noted, it is much simpler to write efficient code for these methods. Although the large crystallite method has the slowest convergence, it does not follow that it should be discarded. Its advantage is that, although it requires larger region sizes to ensure convergence than the periodic boundary condition method, it is much cheaper on computer time. This is for two reasons. First, the defect calculation enclosed in a sphere converges significantly faster than the same calculation enclosed in a box. Second, the main step of the entropy calculation requires only the diagonalisation of large matrices, a task that modern vector processor are extremely good at. This is the basic reason why the 'brute force' methods are much more efficient than the Green function method, despite the greater mathematical elegance of the latter technique. It is worth noting that while the large crystallite method is fast, it requires a fair amount of space, even when the dynamical matrix is symmetrised. For that reason, entropy calculations done in a periodic repeating cell, but using a mesh of points to reduce the size of the repeating unit needed, may be preferable although they take slightly more time.

Finally, we note that in all our methods of calculating entropies, we have necessarily solved the problem of calculating local modes and resonances of defects. These were obtained in the past by Green function calculations, frequently fitting elements of the force constant matrix to the experimental data (see Maradudin *et al* 1971). The discussion in this section suggests that the large crystallite method should be used for such calculations. This has been tried for the case of alkali halides and promising results obtained for gap modes, local modes and isotope shifts (Harding and Sangster 1986, 1987).

2.3 THERMODYNAMICS OF POINT DEFECTS.

2.3.1 Basic principles. We have already noted that the aim of defect simulations must be to calculate the Gibbs free energy of defect processes, g_P. A simulation of the type discussed in the previous section calculates the Helmholtz free energy at constant volume, f_V. (Strictly speaking, the calculation is done at constant *lattice parameter* but the difference is negligible for our purposes). The relation between the constant volume parameters that we calculate and the constant pressure parameters that we normally require has been much discussed in the literature. A convenient summary and appraisal of the position may be found in the paper of Catlow *et al* (1981). The fundamental result is that, to first order in the defect volume, v_P, (that is to say, for low defect concentrations) the Gibbs free energy at constant pressure is equal to the Helmholtz free energy at constant volume. A derivation of this result is given by Gillan (1981). The relationship has recently been tested by direct calculation (Allan, Mackrodt and Leslie 1987) and found to hold up to quite high defect concentrations. Catlow *et al* also quote the relationships between the constant pressure enthalpy, h_P and the internal energy, u_V and between the constant pressure entropy, s_P and the constant volume entropy, s_V. These are

$$h_P = u_V + \frac{v_P \beta T}{\kappa_T V_m} \qquad (2.3.1)$$

$$s_P = s_V + \frac{v_P \beta}{\kappa_T V_m} \qquad (2.3.2)$$

where β is the thermal expansivity, κ_T the isothermal bulk modulus and V_m the molar volume. These formulae are useful if one wishes to calculate the correction to the constant volume parameters using experimental data, but since we want to calculate all the terms, it is a little cumbersome to proceed by way of the defect volume. We therefore use an alternative formulation of (2.3.1) and (2.3.2) (see Harding 1985) which expresses the correction to constant pressure directly in terms of the defect quantities we calculate.

$$h_P = u_V - T\left(\frac{\partial V}{\partial T}\right)_P\left(\frac{\partial f_V}{\partial V}\right)_T \tag{2.3.3}$$

$$s_P = s_V - \left(\frac{\partial V}{\partial T}\right)_P\left(\frac{\partial f_V}{\partial V}\right)_T \tag{2.3.4}$$

In many cases, we can plot f_V against temperature and identify the slope with s_P and the intercept with the energy axis with h_P. The results for the Schottky defect in potassium chloride are shown in Figure 2.4. The free energy is not a perfectly linear function of temperature, but the deviation is not detectable in conductivity experiments.
The first two calculations are performed using the calculated lattice parameter (i) and the experimental one (ii). The results are very similar. However, the use of the experimental parameter is not consistent with the potential, and the lattice is stressed. For consistency this effect should be added to the calculated defect energy. When this is done we obtain curve (iii). This illustrates the point that for accurate self-consistent calculations the *calculated* lattice parameter should always be used.

With equations (2.3.3) and (2.3.4) we can also discuss the common assumption that the zero temperature constant volume calculation of the defect energy may be identified with the experimental enthalpy i.e. that $h_P(T) = u_V(0)$. This equation apparently ignores two important effects; first, the difference between constant pressure and constant volume and second the effect of temperature. The correction term to go from constant volume to constant pressure is given in equation (2.3.3) above. To first order in a Taylor expansion we may write the correction due to finite temperature as

$$u_V(T) = u_V(0) + \left(\frac{\partial u_V}{\partial V}\right)\delta V \tag{2.3.5}$$

where δV is the volume change. Assuming that the coefficient of thermal expansion is reasonably constant from zero kelvin up to the temperature of interest, we may write δV as $T(\partial V/\partial T)_P$ and so obtain

$$u_V(T) = u_V(0) + T\left(\frac{\partial u_V}{\partial V}\right)\left(\frac{\partial V}{\partial T}\right)_P \tag{2.3.6}$$

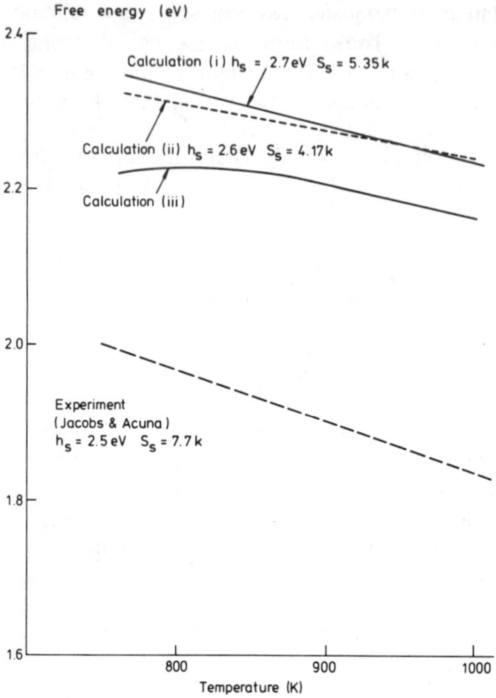

FIGURE 2.4 FREE ENERGY OF FORMATION OF SCHOTTKY DEFECT IN POTASSIUM CHLORIDE

If we compare (2.3.6) with (2.3.3), it is clear that (2.3.6) cancels out the largest part of (2.3.3), that involving the variation in the internal energy. We are left with

$$h_P(T) = u_V(0) + T^2\left(\frac{\partial s_V}{\partial V}\right)_T\left(\frac{\partial V}{\partial T}\right)_P \tag{2.3.7}$$

i.e. the assumption that $h_P(T) = u_V(0)$ is correct provided that the variation of s_V with volume is small and that the thermal expansion coefficient is constant down to zero kelvin. Whether or not the first condition is met is a matter for detailed calculation. It turns out that the effect of variation in s_V usually does not give rise to corrections of more than about 0.1eV; it is only of real importance when the defect enthalpy is low. Thus it is more likely to matter when enthalpies of migration are being discussed.

The condition of constant thermal expansion is more important and it is failure to meet this condition that is responsible for gross deviations of

$u_V(0)$ from the experimental enthalpy. This point also touches an issue directly relevant to experiment; the assumption that the parameters h_P and s_P are independent of temperature. This assumption is made almost universally as a way of cutting down the number of variables in the fitting process. There are no grounds for supposing this assumption to be universally true and from the above argument one would expect it to break down for solids with anomalous thermal expansion. This has been observed for the case of the silver halides both in experiment (Abroagye and Friauf, 1975; for a review of the data see Laskar 1984) and calculation (Catlow *et al* 1987).

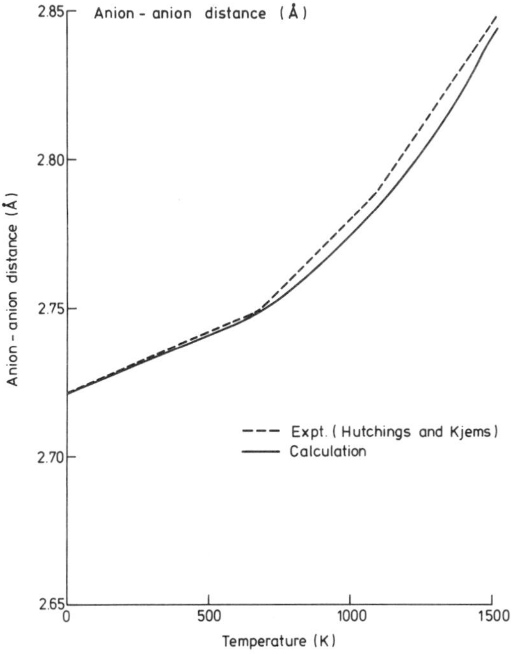

FIGURE 2.5 LATTICE THERMAL EXPANSION OF CALCIUM FLUORIDE

If we follow through the same argument for the entropy s_P we obtain the following

$$s_P(T) = s_V(0) - \left(\frac{\partial V}{\partial T}\right)_P \left\{ \left(\frac{\partial u_V}{\partial V}\right)_T - 2T\left(\frac{\partial s_V}{\partial V}\right)_T \right\} \qquad (2.3.8)$$

A significant, frequently the dominant, contribution to s_P comes not from the lattice vibrations but from the volume dependence of the internal energy. This point is of particular importance when we come to consider diffusion. For now, it is sufficient to realise that $s_V(0)$ will never, except by chance, be a good approximation to $s_P(T)$. The quantities may even be of the opposite sign. However, if the effect of change in s_V is small and the thermal expansion is constant, s_P may still be independent of temperature provided that u_V varies linearly with temperature. This is usually approximately true, at least to the extent to which fitting can fix values of s_P (normally to within about $1k$).

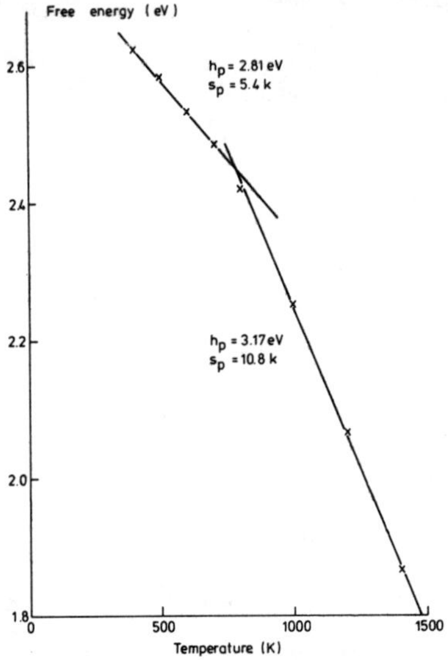

FIGURE 2.6 FREE ENERGY OF FORMATION OF THE ANION FRENKEL DEFECT IN CALCIUM FLUORIDE

The points made in this section are well illustrated by a particular example. We mentioned in Section 2.2 the necessity of calculating the lattice thermal expansion. This may be done by using the crystal potential to calculate the internal energy and entropy of the crystal as a function of volume. This gives the Helmholtz free energy, $\mathscr{F}$, and we then use the identity

$$P = -\left(\frac{\partial \mathscr{F}}{\partial V}\right)_T \simeq 0 \qquad (2.3.9)$$

to obtain the lattice expansion since atmospheric pressure is negligible here. (For another method of calculating the lattice expansion from the crystal potential see Allan, Mackrodt and Leslie, 1987.) An example of such a calculation is shown in Figure 2.5. It suggests that the potential, fitted at room temperature, performs well at higher temperatures. Also, the voiume-temperature relationship is far from linear. This is reflected in the calculation of the free energy of formation of the anion Frenkel defect shown in Figure 2.6. This curve fits quite well to two straight lines, one above and one below 800K. The values for h_P and s_P are shown in Table 2.1 together with the anion Frenkel energies extracted from a variety of experiments.

Temperature Range	h_P (eV)	$s_P(k)$	Reference
500-1000	2.71	5.5	Jacobs and Ong, 1976
300-1200	2.74	13.4	Bollmann and Henniger, 1972
910-1100	2.82	13.5	Ure, 1957
870-1100	3.0±0.16		Oberschmidt and Lazarus, 1980
400-800	2.81	5.4	calculation
800-1500	3.17	10.8	calculation

Table 2.1. Anion Frenkel defects in calcium fluoride

If we perform the calculation at zero temperature, we obtain a value of $u_V(0)$ of 2.74eV. This is close to the calculation for h_P below 800K where the volume temperature relation is linear to low temperature. The value of h_P above 800K is quite different. This shows the importance of the effect of non-linear thermal expansion. If we look at the experimental values, we see there also the evidence of two sets of values below and above about 800K.

The way these calculations are performed is summarised in Figure 2.7. From the calculations performed up to now, it appears that the assumptions of the quasi-harmonic approximation and $g_P = f_V$ are remarkably accurate. The likely main source of error is in the interionic potentials used. It would, however, be of interest to see how accurate the assumptions were. Ball discusses in another chapter how theories beyond the quasi-harmonic approximation can be constructed. Such theories have been used to discuss quantum crystals but only within the rigid ion approximation, not the shell model. These theories are most

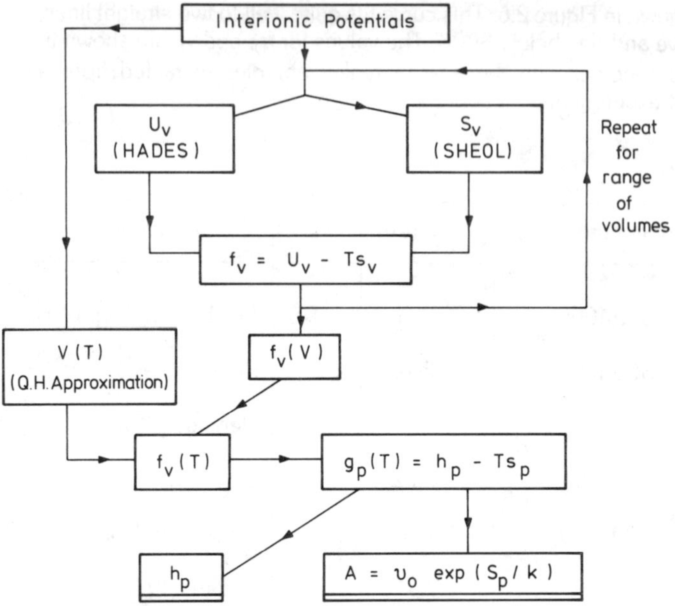

FIGURE 2.7 FLOW CHART FOR ENTROPY CALCULATIONS

likely to be of use in ferro-electrics and other solids where highly anharmonic behaviour may be expected. A particular case where such theories may be of use is in the study of surfaces, where the existence of the free surface may cause problems within the quasi-harmonic approximation. This approximation ensures the correct thermodynamics, but does not ensure mechanical stability. In a calculation performed in the bulk, this may not be of great importance, since the boundary conditions will suppress undesirable effects. This may not be the case at a free surface.

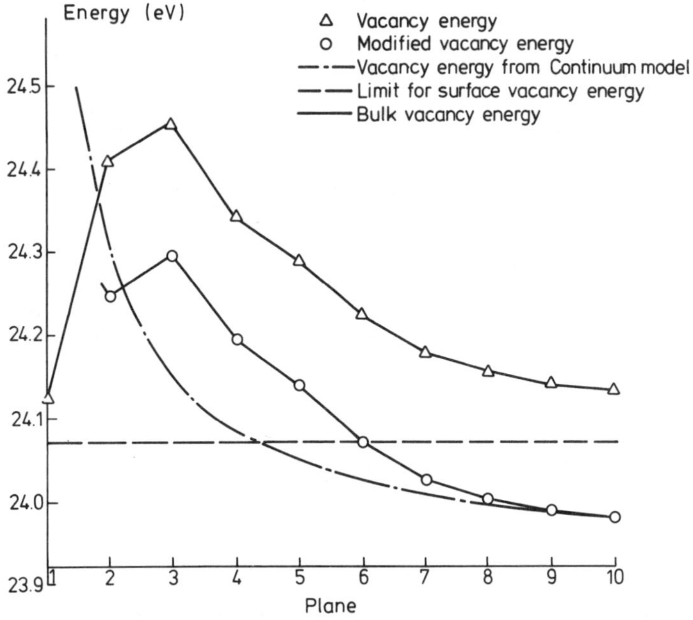

FIGURE 2.8 VARIATION OF THE ENERGY OF FORMATION OF A CATION VACANCY IN MAGNESIUM OXIDE AS A FUNCTION OF DISTANCE FROM THE SURFACE (AFTER DUFFY et al 1984)

Finally in this section we note one more hidden approximation: real crystals stop somewhere. Naively, one might hope that the effect of the surface could be neglected once the defect was a reasonable distance (say, a few tens of planes) away from it. As Duffy *et al,* (1984) have pointed out, matters are not so simple. Figure 2.8 shows the energy of formation of a vacancy as a function of distance from the surface. There are three effects that need to be considered. First, there is the variation in the Madelung term close to the surface. Second, there is an image term arising from the change in dielectric constant at the surface. This is given by

$$E_{image} = \frac{Q^2(\varepsilon - 1)}{4\pi\varepsilon_0 \varepsilon(\varepsilon + 1)R} \qquad (2.3.10)$$

where Q is the defect charge, R the distance from the image plane and ε the dielectric constant of the solid. This term will obviously fall off only slowly with distance. Finally, for many surfaces, there is a dipolar term, that is in principle of infinite range, arising from the surface rumpling. In practice, the last two terms will be screened by a space-charge layer. Further, since at reasonable distances all that matters is the defect charge, Q, when we consider defect reactions that are, overall, neutral these terms cancel. However, in thermopower measurements it *is* possible to obtain the energies of formation of isolated charged defects and recent experiments (e.g. Hudson *et al* 1987) have managed to extract values for the enthalpies and entropies of individual charged defects in silver halides and sodium chloride. The proper calculation of these parameters would be of some interest.

2.3.2 Clusters and non-stoichiometry. Most calculations of defect parameters start from the assumptions that the defect population is dilute and that the defect-defect interactions are negligible. In such a case, the Gibbs free energy for the defective solid may be written (Lidiard 1974)

$$\mathcal{G} = \mathcal{G}_0(P,T) + \sum_{i=1}^{N} n_i g_i - kT \ln W(\ldots. n_i \ldots.) \tag{2.3.11}$$

where $\mathcal{G}_0(P,T)$ is the Gibbs free energy of the perfect solid containing the same number of atoms (though not necessarily the same number of sites) as the perfect solid; g_i the free energy of formation of an individual defect and W the combinatorial function giving the number of ways of distributing n_i defects over the lattice. The values of n_i at equilibrium are found by minimising $\mathcal{G}$ subject to any applied constraints, i.e.

$$d\mathcal{G} = \left(\frac{\partial \mathcal{G}}{\partial n_i}\right)_{T,P,n_j \neq n_i} dn_i = 0 \tag{2.3.12}$$

The term $(\partial\mathcal{G}/\partial n_i)_{T,P,n_j \neq n_i}$ is called the defect chemical potential by analogy with the standard definition of chemical potentials.

These methods may be used to generate the standard equilibrium formulae (Howard and Lidiard 1964). A frequently-used approximation to this approach is the mass-action approximation. This involves writing the formation processes of defects as chemical reactions and then writing the expression for the equilibrium constants for such reactions in the normal way. Thus for the general reaction

$$n_1 \mathrm{A}_1 + \ldots + n_i \mathrm{A}_i + \ldots \leftrightarrows m_1 \mathrm{B}_1 + \ldots + m_j \mathrm{B}_j + \ldots \tag{2.3.13}$$

We can write

$$\sum_i n_i \mu(\mathrm{A}_i) = \sum_j m_j \mu(\mathrm{B}_j) \tag{2.3.14}$$

Since the chemical potentials are conventionally written

$$\mu(\mathrm{A}_i) = \mu^0(\mathrm{A}_i) + kT \ln a_i \tag{2.3.15}$$

where a_i is the activity of A_i, we can obtain the mass action law

$$K_{eq} = exp(-\Delta\mathcal{G}/kT) = \frac{\prod_{i=1}^{k} b_i^{m_i}}{\prod_{j=1}^{l} a_j^{n_j}} \qquad (2.3.16)$$

where $\Delta\mathcal{G}$ is the free energy absorbed in going from left to right and a_i, b_i the activities (usually approximated as the concentrations) of A,B. It should be noted that (2.3.14) may contain temperature-dependent terms in addition to those of the form $kT \ln a_i$. A common example is when we consider a non-stoichiometric solid in equilibrium with a gas. There we must take into account the chemical potential of the gas which, for a diatomic, is given by

$$\mu(A_2) = -kT \ln\left\{\left(\frac{mkT}{2\pi\hbar^2}\right)^{\frac{3}{2}} \frac{V}{N} \frac{2\pi\mathcal{I}kT}{\hbar^2} \mathcal{Q}_{vib}\right\} \qquad (2.3.17)$$

where $\mathcal{I}$ is the moment of inertia of the molecule and $\mathcal{Q}_{vib}$ is the vibrational partition function. There may also be contributions from the electronic partition function. We may assume the perfect gas law $PV = NkT$ and thus a $T^{\frac{7}{2}}$ term appears in the pre-exponential term of (2.3.16).

In ionic solids, the most obvious term that these procedures neglect is the Coulomb interaction between the defects. At low defect concentrations this can be included by use of the Debye-Hückel approximation used in the theory of dilute electrolyte solutions (Teltow 1949, Lidiard 1957). This represents the effect of the Coulomb interactions as screening the defect charge. The electric potential around a defect of charge qe becomes

$$\phi(r) = \frac{qe\ exp(-\mathcal{K}r)}{4\pi\varepsilon_0\varepsilon} \qquad (2.3.18)$$

where $\mathcal{K}$ is the screening length given by

$$\mathcal{K}^2 = \sum_{i=1}^{i=N} n_i q_i^2 e^2/\varepsilon_0\varepsilon kT \qquad (2.3.19)$$

where the summation is over the species i. We may then write the activity coefficient for a species, γ_i, as

$$\ln \gamma_i = \frac{-q_i^2 e^2 \mathcal{K}}{8\pi\varepsilon_0\varepsilon kT(1 + \mathcal{K}R)}. \qquad (2.3.20)$$

The last term in the denominator of (2.3.20) defines a 'distance of closest approach', R, inside which the two defects cannot be considered as separate entities. Allnatt and co-workers have discussed the validity of this theory, both from the point of view of a more general theory (the Mayer cluster expansion; Allnatt and Cohen 1964) and by direct calculation of the Coulomb interactions using the hypernetted chain approximation (Allnatt and Loftus 1979; Allnatt and Allnatt 1982). They conclude that it should be acceptable up to defect concentrations of up to about 0.1%.

The simplest way of recognising the effects of defect-defect interactions beyond the Debye-Hückel approximation is to admit the possibility of clustering. We then write down chemical equations representing the formation of clusters from the primitive vacancies, interstitials or substitutionals as a chemical reaction and use the mass-action equation (2.3.16) on the result. This method has been used with great success and a large variety of clusters considered. Sometimes it is possible to detect these clusters directly by such techniques as small-angle neutron scattering. This is discussed in the chapter by Hayes and Hutchings. More frequently the existence of clusters must be inferred from a number of experiments. In this case, calculations may frequently be of use in the interpretation (see, for example, the study of clustering in anion-deficient fluorites by Bendall *et al* 1984).

Clustering is of particular importance in non-stoichiometric compounds, especially in oxides. Here high defect concentration, in extreme cases ten percent or more, are readily obtained. Many reviews of the experimental data are available. In particular, we note the book edited by Sørensen (1981) and the proceedings of the NATO research workshops at Alenya (edited by Petot-Ervas *et al* 1984) and Keele (edited by Catlow 1987). Some oxides achieve stability by eliminating whole planes of vacancies to form shear planes. The classic example of this is TiO_{2-x} . The structure of such planes may be calculated by methods similar in principle to the static lattice methods discussed in section 2. Details of the methods used may be found in Tasker and Bullough (1981). Here, however, we shall consider only those oxides that reduce the effective defect population by clustering. A particularly instructive set of examples is found in the oxides MnO, FeO, CoO and NiO. All these oxides have the rock-salt structure. The degree of non-stoichiometry varies from FeO, which cannot be obtained closer to stoichiometry than $Fe_{0.97}O$ to NiO where the departure from stoichiometry is about 10^{-3} in air at high temperatures. The oxides are in fact better considered as two groups; the three oxides MnO, CoO, NiO and FeO on its own. In the first group, the clusters inferred from experiment are all comparatively simple, being either the cation vacancy in its various charge states or at most the 4:1 Koch-Cohen cluster (see Figure 2.9; squares are vacant sites, black circles are interstitials)

In FeO, on the other hand, a large number of clusters have been postulated. These all use the 4:1 cluster as a basic unit and are obtained by sharing faces, edges or corners (for some examples from the zoo see Figure 2.9). The calculations on these clusters are reviewed by Catlow (1981). That this defect structure is not due simply to the greater non-stoichiometry is clearly shown by the work of Sykora and Mason (1987) done at very high oxygen partial pressures. Under these conditions, it is possible to raise the defect concentrations in CoO and MnO to levels typical of FeO. However, even under such conditions they do not show the clusters typical of FeO. The reason for the different stoichiometries of the oxides under ambient conditions lies in the different ionisation potentials of the divalent cations. The ionisation of Fe^{2+} is particularly low because it has one electron outside a full half-shell. The Fe^{3+} ion is stabilised by Hund's rule. Furthermore, this ion is spherically symmetric and so there will not be crystal field terms opposing the formation of the cluster with the central ion in a tetrahedral environment. Calculations on the crystal field terms for these clusters have been done (Grimes *et al* 1986) which make useful qualitative points but cannot be considered quantitative because they neglect important electrostatic and relaxation terms.

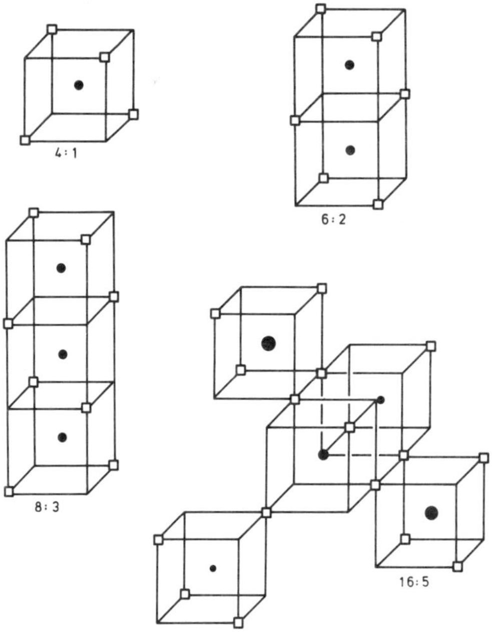

FIGURE 2.9 VACANCY CLUSTERS IN THE TRANSITION METAL OXIDES

If we can calculate the energies and entropies of all the defect clusters that are formed as the oxide departs from stoichiometry, it is obviously possible to calculate the non-stoichiometry as a function of such external constraints as oxygen partial pressure or temperature. The results can then be compared directly with the experimental data. This has been done for MnO (Stoneham *et al* 1986) and for CoO (Harding and Tarento 1987). The results for MnO are shown in Figure 2.10 and for CoO in Figure 2.11. In both cases the agreement with experiment is reasonable. In MnO, the calculations show the presence of charged 4:1 clusters in significant concentration, but in CoO these clusters are of no importance unless one is close to the CoO/Co_3O_4 phase boundary. The results for MnO should, however, be treated with some caution as the calculation was performed only within the harmonic approximation. More work is currently in progress, which shows the necessity to take deviations from the dilute approximation into account as well as clustering.

The calculated populations for the different charge states of the cation vacancies in CoO are shown in Figure 2.12. As suggested also by fitting to the experimental thermodynamic and diffusion data (Dieckmann 1977) the dominant defect over most of the experimental range is the single charged cation vacancy (V'_{Co}). The calculations treat the hole as a small polaron. The question of whether or not the hole *is* localised and if so, under what conditions, has been much discussed and different authors come to different conclusions on the point. We refer the reader to the careful review of Hönig (1984) for a discussion of the issues.

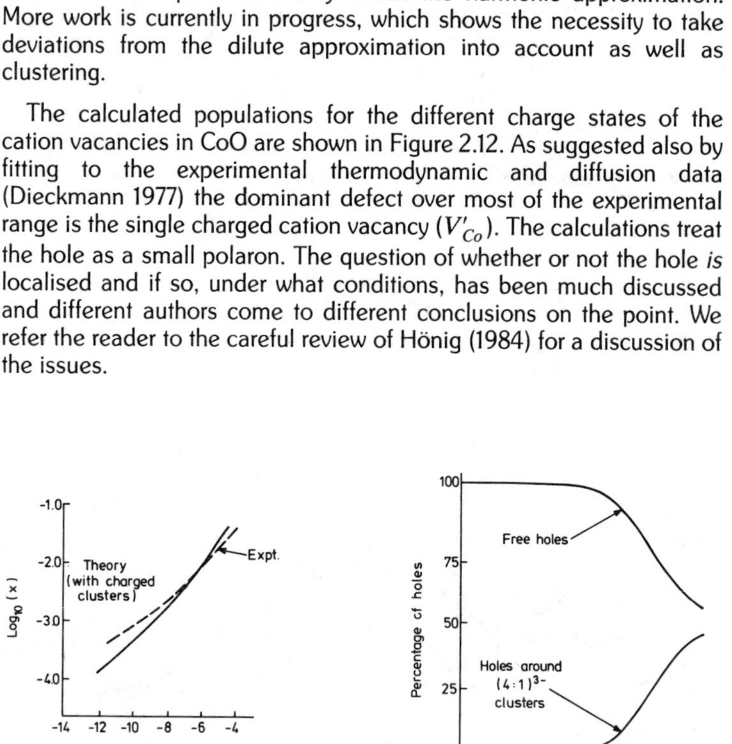

FIGURE 2.10 CALCULATION OF THE NONSTOICHIOMETRY OF MANGANESE OXIDE ($Mn_{1-x}O$)

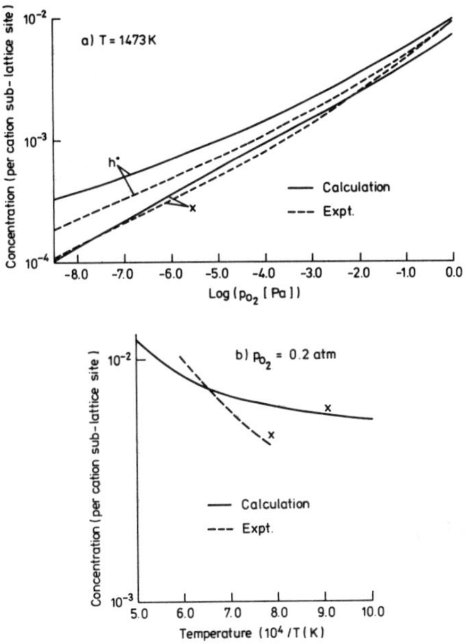

FIGURE 2.11 CALCULATION OF THE NONSTOICHIOMETRY OF COBALT OXIDE (Co_{1-x})

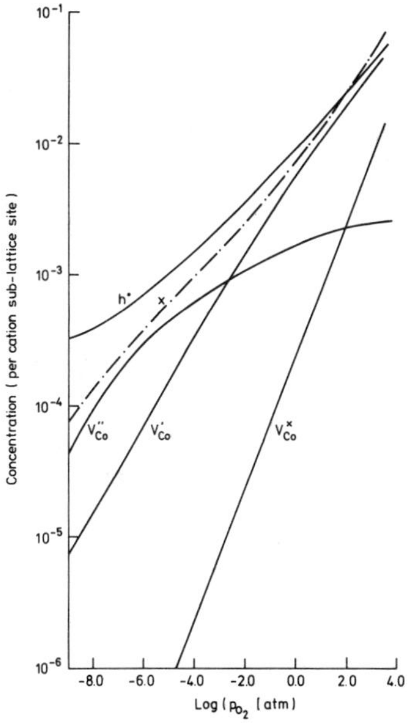

FIGURE 2.12 CONCENTRATIONS OF DEFECTS IN COLBALT OXIDE AS A FUNCTION OF OXYGEN PARTIAL PRESSURE AT 1473K.

All the calculations discussed so far have been performed within the mass-action approximation. At high defect concentrations this will no longer be adequate. In extreme cases, it is no longer sensible to try and identify individual defect clusters. In such circumstances one may try to evaluate the partition function directly by such techniques as Monte-Carlo (see the work of Boureau 1981, 1985). Even when individual defects can plausibly be identified, direct methods may still be the most fruitful approach. Gillan discusses the evaluation of defect-defect correlation functions by molecular dynamics elsewhere in this volume.

If one still wishes to use the idea of defect clusters there remain a number of approaches. The simplest is to recognise the fact that a defect cluster may occupy more than one lattice site. Even when it does not, it may still exclude other defects from the sites around it because of strong elastic or Coulomb interactions. We can incorporate this into the statistical mechanics of equation (2.3.11) by altering the combinatorial function $W(\ldots n_j \ldots)$. If we consider that a defect n_j blocks $C - 1$ other sites, then the function becomes

$$W(\ldots n_j \ldots) = \frac{C^{n_j}(N/C)!}{n_j!\,(N/C - n_j)!} \qquad (2.3.21)$$

This reduces to the standard combinatorial formula for $C = 1$. The effect of using an equation like (2.3.21) is to decrease the configurational entropy and so steepen the slope of the chemical potential as a function of non-stoichiometry. Qualitatively, this is correct, but the method has many problems. In effect, it represents a very simple interaction potential; the hard sphere model. In practice, if one considers (2.3.21) as modelling not simply the effect of not being able to have two clusters on top of one another, but also (crudely) the effect of defect-defect repulsion, the quantity C becomes merely a fitting parameter without any clear physical meaning. More sophisticated versions of (2.3.21) do manage to model the fact that at high defect concentrations the defect energy is a function of the local environment. In particular we would mention the 'spacing statistics' of Atlas (1970), reviewed and extended by Manes (1981). This method still suffers from the short-coming that it considers the cluster to be spherical. It is, indeed, very difficult to go beyond this.

Mitra and Allnatt (1979) have calculated the configurational entropy term for UO_{2-x} taking into account the asphericity of the defect clusters. However, the method, involving as it does the summation over all configurations using the Mayer cluster integrals (Mayer 1950) requires the solution of large numbers of unpleasant integrals. Allnatt and Lidiard (1987) have recently suggested that the first order term of the Mayer expansion (which is simple to calculate) may be sufficient for many purposes. This would be worth investigating. Also, the standard techniques used in discussing alloys; the Bragg-Williams and quasi-chemical methods may repay more detailed study than they have received to date (for a review of their use in this field see Murch 1981) but one may in the end be forced to the conclusion that the only answer to the general problem is to find efficient algorithms for a direct simulation of the partition function as pioneered by Boureau.

2.4 THEORIES OF DIFFUSION IN IONIC SOLIDS.

Most of the experimental data we have discussed in previous Sections comes from studies of diffusion or conductivity. With the methods we have discussed it is possible to calculate the diffusion coefficient, D, or the conductivity, σ, directly, and so bypass the analysis of the data into defect parameters. This does not, however, avoid the problem inherent in all static simulations; that you must think of and calculate all relevant defects and migrations pathways. In simple cases this is not a problem. However, in complex situations there is no definite answer to the question of whether all relevant defects have been considered and, if so, whether all possible ways of moving them have been considered also. It might be thought that molecular dynamics would offer a way out of this difficulty since it does generate all possible configurations given time. The snag lies in the phrase 'given time'. Many defect processes important in ionic crystals have high activation energies, so high that no molecular dynamics simulation would generate them spontaneously sufficiently often to give reliable statistics within the time that such simulations can be run for even on modern computers. It is possible to bias the simulation to investigate regions of configuration space of particular interest and we shall discuss ways of doing that. However, this returns us to the original problem. Once we bias the simulation in particular directions, we can never be certain we have biassed it in the right way. The cases we shall discuss are sufficiently simple that this problem is not serious, but it should always be born in mind when the defect structure of complex solids is discussed.

2.4.1 Phenomenology of diffusion and conductivity. The basic phenomenological equations for diffusion have been greatly discussed in the literature. The reader is referred to such volumes as Flynn (1972), Kroger (1974) and Schmalzried (1981). The relationship between the traditional formulation in terms of Fick's laws and the irreversible thermodynamics of Onsager is discussed in the reviews of Howard and Lidiard (1964) and Allnatt and Lidiard (1987). We present here a brief summary of the main points required in this review.

In an inhomogeneous single-phase system there will be non-convective flows that tend to eliminate the chemical potential gradients existing within the system. One of the first formulations of this was Fick's first law of diffusion that relates the flow, J_i of particles of type i due to a concentration gradient dc_i/dx

$$J_i = -\tilde{D}_i \frac{dc_i}{dx} \tag{2.4.1}$$

where $\tilde{D}_i$ is the individual partial chemical diffusion coefficient of species i. As already noted, the fundamental gradient involved is not the concentration gradient but the chemical potential gradient. We should therefore replace (2.4.1) by

$$J_i = -L_{ii} \frac{d\mu_i}{dx} \tag{2.4.2}$$

where L_{ii}, the transport coefficient, is given by

$$L_{ii} = \tilde{D}_i \left(\frac{dc_i}{d\mu_i} \right) \tag{2.4.3}$$

This coefficient is, in fact, the diagonal element of the matrix of coefficients in the Onsager formalism. If we recall the definition of chemical potential from equation (2.3.15), we have

$$L_{ii} = \frac{\tilde{D}_i c_i}{kT} \frac{d \ln c_i}{d \ln a_i} \tag{2.4.4}$$

It is usual to define a *particle* diffusion coefficient, D_i, such that

$$D_i = \tilde{D}_i \frac{d \ln c_i}{d \ln a_i} = \tilde{D}_i \left(1 + \left(\frac{d \ln \gamma_i}{d \ln c_i} \right) \right)^{-1} \tag{2.4.5}$$

where we write $a_i = \gamma_i c_i$, which defines the activity coefficient γ. This ignores any effect of the concentration on the molar volume (which is usually negligible). We may therefore write

$$J_i = -\frac{D_i c_i}{kT} \frac{d\mu_i}{dx} \tag{2.4.6}$$

which is another common way of writing (2.4.2). In the dilute limit, γ_i tends to unity, D_i and $\tilde{D}_i$ become equal and (2.4.6) tends to (2.4.1).

One very common method used to determine D_i is by the use of tracers. The details of how the experiments are carried out are discussed in an article by Rothman (1984). Here we note that the method relies on the assumption that the drift velocity of the tracer isotope differs negligibly from that of the host species, that is

$$J_{i^*} \left(c_{i^*} \frac{d\mu_{i^*}}{dx} \right)^{-1} = J_i \left(c_i \frac{d\mu_i}{dx} \right)^{-1} \tag{2.4.7}$$

and so tracer and particle diffusion constants are equal. This is not as obvious as it might seem. The two coefficients are only equal if the tracer ion moves by uncorrelated elementary steps. We shall discuss this in more detail in Section 2.5.2.

All the discussion up to now has made one further significant assumption; that the chemical potential gradients of *other* particles, $d\mu_k/dx$, say, do not affect the flux J_i. The more general expressions where coupling of the fluxes is taken into account are discussed by Howard and Lidiard (1964) and Allnatt and Lidiard (1987). We consider the question of how to calculate some of these coefficients for a simple kinetic model in Section 2.5.2

When we consider charged particles, we must consider the coupling of the ion fluxes by the requirement of local charge neutrality. If the defects causing ion motion have different velocities the resulting charge separation gives rise to a diffusion potential (often called the Nernst potential) which retards the faster-moving species and accelerates the slower-moving one. If we consider, for example, the case of vacancies of charge Z_v and holes diffusing in a non-stoichiometric oxide, we know from the electroneutrality condition that

$$J_v Z_v = J_h \text{ and } c_v Z_v = c_h \tag{2.4.8}$$

We can then show from this and the definitions of the fluxes that the chemical diffusion coefficient of the vacancies is given by

$$\tilde{D}_v = D_v(1 + |Z_v|) \tag{2.4.9}$$

In general when we consider the motion of charged particles we should use the electrochemical potential η_i defined as

$$\eta_i = \mu_i + Z_i|e|\phi \tag{2.4.10}$$

where Z_i is the particle charge and ϕ the electrical potential. In ionic solids therefore the transport equation should be written

$$J_i = \left(\frac{D_i c_i}{kT}\right)\frac{d\eta_i}{dx}. \tag{2.4.11}$$

If we define the mobility u_i as the drift velocity per unit field strength, then it is related to the particle velocity v_i by $v_i = u_i Z_i|e|$ and the particle diffusion coefficient, $D_i = kTv_i$, is related to the mobility by

$$D_i = \left(\frac{kT}{Z_i|e|}\right) u_i \tag{2.4.12}$$

The partial electrical conductivity σ_i is related to the mobility by

$$\sigma_i = Z_i|e|c_i u_i \tag{2.4.13}$$

and so the diffusion coefficient and the electrical conductivity are related by

$$c_i D_i = \sigma_i \left(\frac{kT}{Z_i^2 e^2}\right) \tag{2.4.14}$$

This, the Nernst-Einstein relation, is subject to two caveats. First, in applying it to experimental data, one must be sure that neutral defects are not important since they will contribute to the diffusion but not the conductivity. Second, if we use the tracer diffusion coefficient D_{i*} we must note the point made above about the correlation of the elementary jumps. The conductivity does not involve correlated ion jumps; this assumption is made implicitly when we define the mobility as the drift velocity per unit field strength. The tracer diffusion coefficient frequently does contain terms resulting from correlation.

2.4.2 Microscopic theories of diffusion. The above discussion describes diffusion and conductivity in terms of macroscopic fluxes and chemical forces but gives no clue as to how these phenomena might be calculated from more fundamental considerations. One way of providing such a link comes from random walk theory and the Einstein-Smoluchowski equation. For an isotropic solid, it may be shown that the diffusion constant D_i is given by

$$D_i = R_i^2 / 2z\tau \qquad (2.4.15)$$

where R_i^2 is the mean square displacement of the moving particle i in time τ and z is the lattice dimension. The equation may easily be generalised to the anisotropic case (Flynn 1972). If the ion moves randomly over a lattice such that each jump distance is r, then we have $R_i^2 = nr^2$ where the ion makes n jumps in time τ. If we therefore define the jump rate $\Gamma = n / \tau$, we arrive at our fundamental result

$$D_i = \Gamma r^2 / 2z \qquad (2.4.16)$$

where we have related the macroscopic quantity D_i to a microscopic quantity; the ion jump rate. The assumption of randomness does not always hold. For example, in the case of transport by vacancies an ion which has just exchanged sites with a vacancy is much more likely to reverse that jump than to move off in other directions because of the arrival of another vacancy. If this ion is distinguishable (for example by being a radioactive tracer), its motion is no longer random despite the fact that the motion of the vacancy remains random. The result of this *correlation* between successive jumps is to modify equation (2.4.16) by a multiplicative factor; the correlation factor f. The calculation of this quantity is considered in great detail in a review by Le Claire (1970) to which the interested reader is referred. In simple cases f is fixed by the geometry of the problem, but in general it depends on the details of the ion jump rates.

It has long been known that a useful expression for the jump rate is of the form

$$\Gamma = A\, exp(-E_A/kT) \qquad (2.4.17)$$

where A has the dimensions of a frequency and E_A is called the activation energy although in most cases where (2.4.17) is used it is in fact an enthalpy.

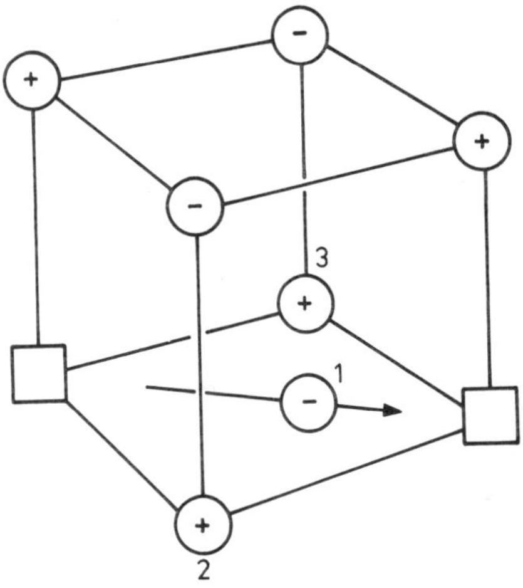

FIGURE 2.13 HOPPING GEOMETRY OF THE ANION VACANCY IN THE ROCK-SALT LATTICE

A simple argument for the rate expression goes as follows. The ion jumping from lattice site to lattice site (or interstitial site to interstitial site) must squeeze between the surrouding ions to do so (Figure 2.13). There is therefore a barrier to motion. Ions can surmount this barrier because of the existence of thermal fluctuations which supply the local energy and momentum required. The frequency of fluctuations large enough to surmount a barrier of height E_A is governed by the Boltzmann factor $\exp(-E_A/kT)$ and so we obtain the form of equation (2.4.17). In cases of high symmetry, a unique saddle point may be identified and so the programs discussed in Section 2.3 may be used to calculate the energy difference between a defect in the ground state and the saddle configuration where the ion is migrating. (In the case of vacancy diffusion, of course, this corresponds to the defect moving in the opposite direction.) Such calculations have been performed and the results compared with the enthalpies obtained from diffusion experiments . As may be seen from Table 2.2, the agreement is frequently very satisfactory.

Yet the simple account given above begs many questions as will be seen when we discuss the matter in more detail below. These can be given satisfactory answers, and for a wide range of defect processes the picture outlined above is correct, in outline if not in detail. There are, however, a number of diffusion phenomena, mainly those involving quantum effects, where the argument of this paragraph is quite wrong and although equation (2.4.17) is still correct, the meaning of the terms is different. We shall not be much concerned here with quantum diffusion, but will indicate the main points at issue later on.

defect	experiment (eV)	calculation (eV)
NaCl; cation vacancy	0.5-0.8	0.66
CaF_2; anion vacancy	0.5-0.6	0.35
CaF_2; anion interstitial	0.9-1.0	0.91
MgO; cation vacancy	2.0-2.3	1.8-2.2
NiO; cation vacancy	1.5	1.86

Table 2.2 Comparision of experimental and calculated migration energies (see Catlow (1986) for references).

The style of argument in the preceding paragraph bears a strong resemblance to the reaction rate theory (see Glasstone *et al* 1941) that was first applied to diffusion problems by Wert and Zener (1949). This was most elegantly formulated by Vineyard (1957) whose analysis brought out the many-body nature of the diffusion process. The basic assumptions made may be seen by considering a simple one-dimensional model (Waldram 1985). Consider an atom A next to a vacancy (using the nomenclature of Figure 2.14). We may construct a saddle plane that passes through the saddle point P and perpendicular to the contours of constant potential. We consider an ensemble of such systems each with the atom A to the left of the saddle plane. Once the atom is on the right-hand side of the saddle plane the system is removed from the ensemble. The jump rate, Γ, for this system is the net flux of atoms moving left to right. Since the atoms jump because of random thermal forces, they are also subject to damping from these same forces as they move about on the potential surface. We assume that this damping is such that it does not cause appreciable deviations of atoms moving left to right in the vicinity of the saddle plane but does ensure that if atoms fail to cross it and fall back, they equilibrate with the rest of the ensemble in the ground state before they return to the saddle plane region. What this amounts to saying is that the barrier is high enough to ensure that the amount of time the atom spends in the saddle plane region is small compared to the amount of time it spends in the ground state.

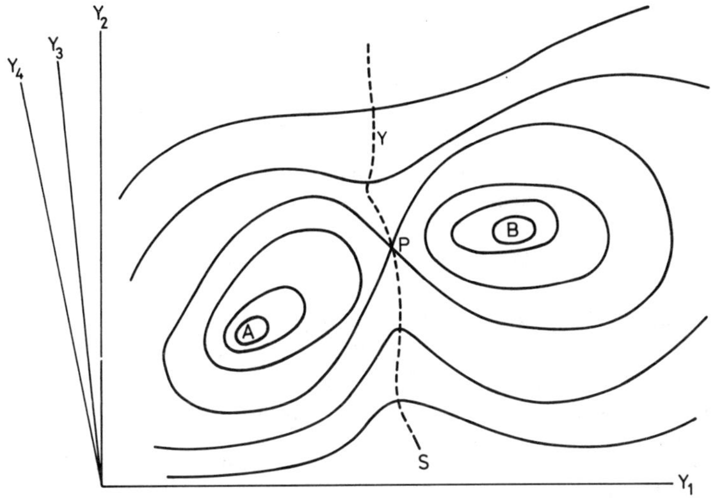

FIGURE 2.14 SCHEMATIC POTENTIAL SURFACE FOR A SIMPLE VACANCY MECHANISM (AFTER VINEYARD 1957)

These assumptions allow us to calculate the ion flux through the saddle plane. The flux density across the plane moving left to right is $\frac{1}{2}\bar{v}n$ where $\bar{v}$ is the mean value of the velocity component v_x for the one-component Maxwell speed distribution $\sqrt{2kT/\pi\hbar m}$ and n is the number of atoms per unit area in the saddle plane. This number is obtained from the fluctuation distribution of the atoms on the left-hand side of the saddle plane. By assumption, all atoms in the ground state are in thermal equilibrium and so the fluctuation distribution is identical to the equilibrium ensemble distribution. We can therefore write the probability that atom A, in the course of its fluctuations, finds itself in the saddle plane as $n_g \exp(-E/kT)$, where n_g is the population of particles in the ground state and E is the barrier height. Since the particle need nto cross the barrier at precisely the lowest point, it is convenient to write $E = E_A + \delta E(y)$ where $\delta E(y)$ is the extra height of the barrier a distance dy from the saddle point in a direction within the saddle plane. We may therefore write the flux of atoms as

$$\tfrac{1}{2}\bar{v}n_g \exp(-E_A/kT) \int_{-\infty}^{\infty} \exp(-\delta E(y)/kT)\, dy \qquad (2.4.18)$$

It should be carefully noted that this does *NOT* imply that the saddle plane configuration is a thermodynamic state in thermal equilibrium with the ground state. The temperature in (2.4.18) is the temperature of the fluctuation ensemble about the ground state. This is well defined because of the assumption that we make about the system always having time to come to equilibrium between each attempt of the particle to jump.

The claim about thermal equilibrium with the saddle point is still (unfortunately) found in textbooks; it is even claimed that the assumption of thermal equilibrium of the saddle plane configuration with the ground state is fundamental to (2.4.18) and hence to reaction rate theory. This assertion is not merely untrue, it is the exact reverse of the position. If the saddle plane configuration were in thermal equilibrium with the ground state, it would be impossible to ignore jumps that went from left to right and then immediately reversed themselves because we should have detailed balance.

We may evaluate the integral in (2.4.18) by assuming that $\delta E(y)$ varies quadratically with distance within the saddle plane. We then obtain

$$\tfrac{1}{2}\bar{v}\exp(-E_A/kT) \,/\, 2\omega'_y \qquad (2.4.19).$$

The total number of atoms in the ensemble is given by the volume integral

$$n_g = \int\int exp(-\delta E(x,y)/kT)\, dxdy \qquad (2.4.20)$$

where the integral is taken over the plane near A. We may write this as

$$n_g/\omega_x \omega_y \qquad (2.4.21)$$

The rate is then

$$\left(\frac{\omega_x \omega_y}{\omega'_y}\right) \exp(-E_A/kT) \qquad (2.4.22)$$

If we compare with the Arrhenius expression quoted above we find that the prefactor A is to be identified with the product of frequencies shown. It should again be noted that these appear as frequencies because of assumptions that we have made about the shape of the potential surface near the ground state and the saddle point. In the case of the ground state, modes are observable; the local modes or resonances in the phonon spectrum of the defective crystal. The saddle plane 'modes' are not observable since the saddle plane configuration does not exist for long enough to permit observation. For the purposes of reaction rate theory and equation (2.4.22), this is irrelevant. The remarkable thing about reaction rate theory is the manner in which it evades discussion of the details of the jumping trajectory. It can only do this because of the special assumptions it makes about the jumps.

Vineyard (1957) applied arguments fundamentally similar to those above to the many-body problem of defect migration in a crystal. He obtained the expression for the rate

$$\Gamma = \left[\frac{\prod_{n=1}^{3N} \omega_n}{\prod_{n=1}^{3N-1} \omega'_n} \right] exp(-E_A/kT) \tag{2.4.23}$$

where the term in brackets is the ratio of the frequencies in the saddle hyperplane (primed) to those in the ground state and N is the number of ions in the system. Two points are worth noting in addition to those already made above.

First, this is a calculation performed using a constant volume ensemble and so E_A is an internal energy. If we now perform an analysis along the lines of that discussed in Section 2.3 to obtain constant pressure parameters, we will obtain a new type of term in the pre-exponential factor. As can be seen from equation (2.3.8), the constant pressure entropy contains a term arising from the temperature dependence of the internal energy. Thus one contribution to the pre-exponential factor has nothing to do with the lattice vibrations, but comes rather from the temperature dependence of the barrier height. This term may be very large, resulting in pre-exponential factors that a much larger than any lattice frequency. This need not be cause for any surprise.

The second point concerns the definition of the saddle hyper-surface. At the beginning of his calculation, Vineyard gave a definition equivalent to that of the simple case discussed above. The $(3N-1)$ dimensional hyper-surface is defined as that surface that passes through the saddle-point and is perpendicular to the contours of constant potential everywhere else. However, by the approximation of expanding the potential to quadratic order at the saddle-point and so obtaining the primed frequencies, Vineyard implicitly redefined the saddle-surface . This new definition is provided by the eigenvectors of the dynamical matrix used to obtain the primed frequencies and the final Vineyard hyper-surface is that surface spanned by those eigenvectors. The choice of a dividing surface is not important in an exact theory, but reaction rate theory is not exact and the surface should be chosen to minimise the number of return jumps. The importance of the geometry of the saddle-surface has been emphasised in a number of publications by Flynn (1975, 1986) and Flynn, Jacucci and co-workers (1975, 1984, 1985). These authors defined a curved watershed saddle-surface obtained by passing from the saddle-point up the maximum gradient. This may be specified in terms of displacements, ξ, of the saddle-surface normal to the Vineyard hyper-plane; i.e.

$$\xi = \xi(x) \tag{2.4.24}$$

The saddle-surface and the hyperplane coincide at the saddle-point but nowhere else. Their method then consists of evaluating the radii of curvature of the saddle-surface and comparing them with the radii of curvature of possible trajectories. The objective here is to retain the basic framework of the reaction rate theory, but to correct for an inadequacy in its basic assumptions. To this we now turn.

It should be clear from this discussion that the critical assumption of reaction rate theory is that there are no return jumps. This is the 'fundamental assumption' set out by Wigner in 1938 and debated by many authors since then. In particular, we mention the work of Chandler (1978) who showed the connection between the 'fundamental assumption' and the time-scales relevant to the diffusion process and showed how to put the rather qualitative assumptions we made about damping on a firm footing. In particular, as argued above, we must assume that between each successful jump the lattice has time to come to equilibrium in the new configuration. It is because of the assumptions about timescales that reaction rate theory in the solid state is a theory about defect hopping, where the defect spends most of its time on one site and then rapidly moves from one site to the next. If the residence time becomes comparable to the flight time, the reaction rate theory will break down. It is clearly of interest to see how good this 'fundamental assumption' is. Since it is an assumption about dynamics, we must turn to a technique that can consider the details of the ion trajectories during the jump process; that is to say we must turn to molecular dynamics.

Two detailed simulations have been carried out to investigate this point; those of DeLorenzi *et al* (1986) for rare-gas solids and Gillan *et al* (1987) for ionic solids. They come to very similar conclusions, despite the wide difference in interatomic force law used, which gives rise to the hope that the conclusions may be of general utility.

The effect of return jumps is to reduce the number of jumps from that calculated by the Vineyard formula. We define a transmission coefficient, $\langle S \rangle$, as the ratio of successful jumps, Γ, to the Vineyard result, Γ_0.

$$\Gamma = \Gamma_0 \langle S \rangle \tag{2.4.25}$$

This does not answer the question of what counts as 'successful'. How soon does a return jump have to occur for it to count as a jump negating the first one and not just another independent hopping event that happens to move in the opposite direction to the first? This is again a question about timescales; whether it is possible to separate off the short-time dynamical motion associated with one jump from the diffusive motion of the random walk. In practice this can indeed be done as has been shown by some direct simulations of da Fano and Jacucci (1977). A formal criterion for performing this separation and so giving a (reasonably) precise meaning to $\langle S \rangle$ is discussed by Chandler (1978) and Gillan *et al* (1987).

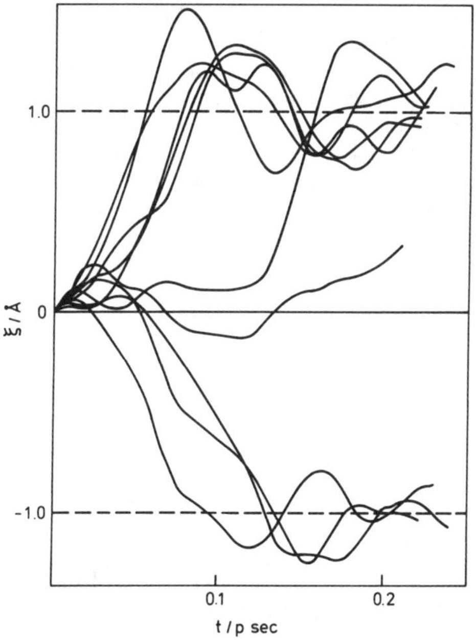

FIGURE 2.15 TIME EVOLUTION OF THE REACTION COORDINATE ξ FOR A SELECTION OF THE HOPPING TRAJECTORIES AT 1368K; DASHED LINES SHOW THE VALUES OF ξ FOR THE EQUILIBRIUM POSITIONS OF THE VACANCY

We thus have two non-trivial factors in our diffusion simulation; the probability of arriving in the saddle-plane (which is equivalent to Γ_0) and the transmission coefficient $\langle S \rangle$. The evaluation of this probability for ionic solids suffers from a particular difficulty because of the height of the barrier. A direct simulation would spend almost all its time executing the (uninteresting) vibrational motion about the ground state minimum and its would be impossible to get adequate statistics on the hopping system. It is, however, possible to avoid this problem by adapting an idea due to Bennett (1975). We cannot calculate the true probability distribution $P(\xi)$ for the particle where ξ is the distance from the saddle-point. We therefore calculate a distribution $P'(\xi)$ for the system in an external potential $V_{ext}(\xi)$ chosen so that $P'(0)$ has a reasonable value. It is then possible to obtain $P(\xi)$ and hence $P(0)$ from the relation (Valleau and Whittington 1977)

$$P'(\xi) = A\exp(-V_{ext}(\xi)/kT)\ P(\xi) \qquad (2.4.26)$$

To estimate $\langle S \rangle$ we need to generate a set of statistically representative trajectories for the system as it crosses the barrier. We do this by performing a molecular dynamics simulation constrained to move exactly within the saddle hyperplane. We can thus generate a constrained trajectory within this plane which is typical of thermal equilibrium. We use a sequence of configurations from this trajectory as initial conditions for unconstrained evolution, except that each initial configuration ξ' is given an initial velocity component v' normal to the saddle plane whose magnitude is random but whose direction is always the same. A set of hopping trajectories thus generated in shown in Figure 2.15 for vacancy motion in a rock salt oxide. The calculations show that a majority of the trajectories do not return, but a significant number do.

T/K	$P(0)/\text{Å}^{-1}$	n_+	n_-	$\langle S \rangle$	Γ/sec^{-1}	$\tilde{\nu}/10^{12}\ \text{sec}^{-1}$
995	9.6×10^{-9}	52	24	0.37	1.8×10^{4}	5.2
1368	1.6×10^{-6}	52	27	0.32	3.0×10^{6}	4.2
1520	7.8×10^{-6}	60	18	0.54	2.6×10^{7}	8.8
1760	2.3×10^{-5}	48	24	0.33	5.0×10^{7}	3.0

Table 2.3. Simulation results for the saddle-plane probability $P(0)$, the transmission coefficient $\langle S \rangle$ (numbers of successful and unsuccessful crossings n_+ and n_-), the effective crossing rate Γ and the frequency pre-factor $\tilde{\nu}$ at four temperatures.

It is possible to obtain values of $\langle S \rangle$ from the statistics of these trajectories as shown in Table 2.3. The variation for different temperatures is because of the difficulty of getting adequate statistics rather than any underlying physical reason. The Vineyard calculation gives a value for $\tilde{\nu}$ of $1.4\times10^{13}\text{sec}^{-1}$. Comparing with the numbers in the Table we see that they agree to within about a factor of two. It is important to compare the absolute rates rather than just look at the transmission coordinate since the molecular dynamics calculation used a different choice of saddle-plane to the Vineyard one. Only the product $\langle S \rangle \Gamma_0$ has any physical meaning. This agreement suggests that for practical calculations on systems where the assumptions of reaction rate theory apply, the errors due to return jumps are not serious. The inaccuracies in the potential function are likely to cause far larger errors than those found here. Nevertheless, these calculations do demonstrate important points of principle. First, there are return jumps and to the extent that there are, reaction rate theory must fail. Second, for many cases of interest it is possible to make a sensible distinction between short-term dynamical effects and long-term diffusive hops. Where this can be done, it is possible to calculate corrections to reaction rate theory. Indeed, it is possible to construct a theory that incorporates these dynamical fluctuations while remaining within the spirit of the reaction-rate approach. This has been done by Toller *et al* (1985) using the ideas of the geometry of the saddle-surface discussed briefly earlier.

There are, however, important cases where the reaction rate theory cannot apply. The theory relies on the possibility of specifying a saddle-point and a saddle-plane. In the important case of quantum diffusion such a specification is not possible. Further, the reaction rate theory assumes, as a classical theory is entitled to assume, that the distributions of the position and momentum coordinates of the ions are independent. Such an assumption violates the uncertainty principle in quantum mechanics. A proper quantum treatment of diffusion must take as its starting point the eigenstates of the system in which diffusion occurs. Here we encounter a problem, since the exact eigenstates of a crystal are not localised whereas point defects usually are. The exact states are linear combinations of the set of equivalent localised states that reflect the translational symmetry of the lattice. At low enough temperatures these exact states will propagate freely, but at higher temperatures the interaction of the moving particle with the lattice phonons transfers pseudomomentum between the particle and the lattice. It is no longer possible to construct a translationally invariant wavefunction for the particle alone; one would have to consider the particle and phonon system together. Hopping emerges when the interaction between the particle and the lattice is so strong that the free propagation of the particle eigenstate is less than an interatomic spacing. In this regime it is possible to calculate the transition probability between the (approximate) localised eigenstates under the crystal hamiltonian and so calculate a hopping rate. The details whereby this sketch is turned into a theory capable of giving numbers are discussed in Flynn and Stoneham (1970) and Flynn (1972).

One specific case is worth considering here ; that of small polaron motion. In this case , as Austin and Mott (1969) point out, it is possible to define a saddle configuration; this is the configuration where the lattice distortion is such that the two sites between which the migrating electron is to tunnel have the same energy. Thus it would seem to be possible to calculate the activation energy of this quantum process by a static lattice program. Such calculations have been done (see, for example, Catlow *et al* 1979). However, the results should be treated with caution as extra 'electronic' terms (crystal field, tunnelling integrals) must be added in by hand.

Recently, workers have investigated the possibility of using molecular dynamics to simulate quantum diffusion. Such approaches have been discussed for some years. The path integral method (Barker 1979, Chandler and Wolynes 1981, Parinello and Rahman 1984) exploits the isomorphism between the Schrodinger equation and the diffusion equation. We may simulate a quantum system of particles as a purely classical system of cyclic chains of beads coupled by harmonic springs. Each chain represents a particle; the chain may be considered to represent (in a pictorial sense) the delocalisation of the quantum particle required by the uncertainty principle. The path integral method cannot be used to simulate diffusion directly, but Gillan (1987) has recently pointed out that the probability of finding the centre of mass of the chain at the saddle point of the classical simulation is closely related to the transition rate of the theory of Flynn and Stoneham. In particular, the activation energies of the two should be the same. More investigations in this area would be of value, in particular in establishing the details of the transition from the quantum to the classical diffusion regime.

Finally in this section we turn to a wholly different approach to diffusion; the 'dynamical diffusion' theory of Rice (1958) and Slater (1959). This theory focusses attention on the fluctuations of a special *reaction* coordinate, and defines a jump to occur when this coordinate exceeds a specific value. This differs fundamentally from the reaction rate theory discussed above which considers fluctuations to reach a specific state (the saddle-point configuration) and defines a jump to occur when that state is reached. One might expect from this brief description that there nevertheless a relationship between these two approaches and Glyde (1967) has shown that the they are equivalent in the particular case when one can obtain the saddle-point configuration by expanding about the ground state configuration in a Taylor series to second order. The ideas that guide the theories are, however, quite different.

The reaction coordinate in the dynamical rate theory is written as a superposition of harmonic phonons

$$\mathcal{R}(t) = \sum_n A_n u_n \boldsymbol{\theta}_n \exp(i\omega_t) \tag{2.4.27}$$

where A_n are the (complex) coefficients and u_n, $\boldsymbol{\theta}_n$, ω_n describe the harmonic phonons in the usual way. The theory then considers how frequently $\mathcal{R}(t)$ will exceed a critical value $\mathcal{R}_{crit}$. This may be obtained from a result of Kac (1943) who showed that this frequency ν_{crit}

$$\nu_{crit} = \int_{-\infty}^{\infty}\int_{-\infty}^{\infty} \frac{d\xi\, d\eta}{2\pi^2\eta^2} \cos(\eta\mathcal{R}_{crit})$$

$$\left\{\prod_n J_0(|A_n u_n|\eta) - \prod_n J_0(|A_n u_n|(\xi^2 + \omega_n^2\eta^2)^{\frac{1}{2}}\right\} \tag{2.4.28}$$

The Bessel functions $J_0(z)$ are approximated as

$$J_0(z) = (1 - z^4/64)\exp(z^2/4) \tag{2.4.29}$$

and after some algebra the expression for v_{crit} may be found to be

$$v_{crit} = \left\{\frac{\Sigma_n(\omega_n A_n u_n)^2}{\Sigma_n|A_n u_n|^2}\right\}^{\frac{1}{2}} \exp\left(-\mathcal{R}_{crit}^2/\Sigma_n|A_n u_n|^2\right) \tag{2.4.30}$$

We now consider (Feit 1971) the system as a canonical ensemble. The ensemble average jump rate Γ is given by

$$\Gamma = \int_0^\infty v_0(\mathcal{R}_{crit}, \{u_n\}) \prod_n P_n(E_n)dE_n/kT \tag{2.4.31}$$

where

$$P(E_n) = \exp(-E_n/kT) = \exp(-m\omega_n^2 u_n^2/kT) \tag{2.4.32}$$

and gives the distribution of the amplitudes $\{u_n\}$. The resulting jump rate is

$$\Gamma = \left\{\frac{\Sigma_n|A_n|^2}{4\pi^2\sum_n\left(\frac{|A_n|}{\omega_n}\right)}\right\}^{\frac{1}{2}} exp(-E_m/kT) \tag{2.4.33}$$

$$E_m = \frac{\frac{1}{2}m\mathcal{R}_{crit}^2}{\sum_n\left(\frac{|A_n|}{\omega_n}\right)^2} \tag{2.4.34}$$

This, as before, is a constant volume calculation. If we consider the pre-exponential factor we may see (Flynn 1968, Feit 1971) that, whereas the prefactor in the reaction rate theory is the ratio between frequencies at the saddle-point and the ground state, the prefactor in the dynamical theory is a weighted average of the frequencies of the lattice vibrations contributing to the reaction coordinate. The dynamical theory has been used by a number of authors as a convenient analytical tool to investigate diffusion phenomena. In particular, since it does not focus attention on a saddle-plane configuration, it has been thought more suitable for considering cases where quantum effects may be involved. However, it suffers from two great disadvantages when compared with reaction rate theory. First, by its use of Kac's result, it is necessarily confined to fluctuations in a harmonic system. It cannot therefore provide a complete classical description of diffusion since a harmonic system cannot exchange energy between its modes and such an exchange is essential if the migrating particle is first to gain energy to migrate and then dissipate it. Second, despite the work spent on the approach, it has never proved possible to use it to calculate defect hopping rates. The main reason for this is that it is not possible to provide an unambiguous method for determining the critical value of the reaction coordinate, $\mathcal{R}_{crit}$. As we shall see in the next section, the reaction rate theory not only gives a clear prescription for calculating hopping rates, it gives one that can be carried out in a large number of systems of practical interest.

2.5 CALCULATIONS WITH THE REACTION RATE THEORY.

2.5.1 The hopping rate. In the previous section we examined the reaction rate theory and in particular Vineyard's reformulation of it for problems in defect physics. From equation (2.4.23) it is clear that the methods of Section 2.3 are well suited to the calculation of hopping rates. In this section, we give some examples of the accuracy that can be achieved.

Many calculations have been performed of migration enthalpies using static lattice programs and assuming that $h_P(T) = u_V(0)$. Such calculations in effect assume the validity of reaction rate theory but, since they do not attempt to calculate the prefactor, they do not offer a decisive test of how well the theory works in practice. Even in cases where we do attempt a complete calculation, it will be clear that in all cases so far considered the limitations on accuracy are not the fundamental assumptions of reaction rate theory but the more mundane difficulties of obtaining accurate enough potential functions and adequate convergence of the calculation as a function of the number of ions considered. That this is likely to be a problem can be seen by a simple calculation. Let us suppose we wish to get the defect jump rate accurate to within about an order of magnitude at 1600K. To do this, we must obtain the defect enthalpies affecting the activation enthalpy (i.e. any formation enthalpies plus the migration enthalpy) accurate to within 0.2eV assuming that the prefactor is correct. In practice, the prefactor is the more difficult to get right. Here the main error is not so much getting the vibration spectrum correct as getting the temperature dependence of the internal energy accurate enough. Because of this it is difficult to calculate constant pressure enthalpies to better than $0.5k$ and so the prefactor cannot be obtained to within better than a factor of three. Thus we require enthalpies to within 0.1eV to obtain the jump rate to within an order of magnitude. This can be done in favourable cases, but it is not possible to obtain reliable results to greater accuracy than that. Obviously, this is not as accurate as diffusion rates can be measured, but it is certainly accurate enough for the calculations to be of use as a predictive and interpretative tool. Also, it must be said that while diffusion coefficients can be measured to greater accuracy than the calculations, it does not follow that they agree amongst themselves to that kind of accuracy. The table of Wuench (1983) of the diffusion coefficients of the alkaline earth oxides is illuminating in that regard. The most glaring case is two determinations of the diffusion coefficient of BaO where, when expressed in the form $D = D_0 \exp(-h/kT)$, the values of D_0 differ by a factor 10^{40} and the values of h by 11.5eV! It must be said in fairness that these differences are such as to cancel against each other in the temperature range of the experiments, but here at least calculations ought to be able to sort something out.

Calculations have indeed been done on the cation hopping rate in the alkaline earth oxides, most recently by Harding *et al* (1987). The values of the pre- exponential factor ν and the activation enthalpy h obtained are shown in Table 2.4

oxide	ν(THz)	h (eV)
MgO	31.29	2.14
CaO	20.55	1.87
SrO	11.60	1.68
BaO	6.51	1.48

Table 2.4 Calculation of cation hopping rates for alkaline earth oxides.

Comparison with experiment is difficult, partially because the experiments are frequently in disagreement amongst themselves, partly because of problems in interpretation. For example, it is now generally agreed that no-one has ever measured intrinsic diffusion in MgO. The theoretical calculation of 7.5eV for the Schottky enthalpy and 4.0k for the entropy show that the intrinsic point defect concentration is far too low for intrinsic diffusion to be observable. Thus the measured coefficients are extrinsic, the varying values of h being due to association of the extrinsic vacancies with whatever impurity produced them. The experiments of Sempolinski and Kingery (1980) are particularly useful in that they give a value of the diffusion coefficient containing only the migration term. Although the calculated and experimental values of D_0 and h are somewhat different, the values of D differ by less than a factor of two over the entire measured range.

One general point worth noticing about these results is that if we plot the logarithm of the pre-exponential factor against the enthalpy of activation we observe an approximate correlation. Figure 2.16 shows such a correlation for a large variety of ionic conductors. Correlations of this kind have frequently been claimed for diffusion data that refer to a series of compounds of the same structure. The correlation is usually referred to as the Meyer-Neldel rule (Meyer and Neldel 1937; for a recent review of data see Almond and West 1986). If the apparent pre-exponential factors and enthalpies do represent the activation of a simple defect process (see Nowick and Lee (1987) for a discussion of the problems involved in correctly interpreting the data) we can advance a possible (and certainly the simplest) explanation of such a correlation. We recall that a major term in the pre-exponential factor is the temperature dependence of the barrier height. Thus the correlation observed is largely between the value of the enthalpy (approximately $u_V(0)$) and the derivative of the internal energy. If $u_V(0)$ is large, this means either that the defect formation or migration seriously distorts the lattice. As the lattice expands with temperature, the relief will be great because of the exponential form of the repulsive forces; therefore du/dV will be large also. There is also the obvious point that only large initial values of u_V can possibly have large reductions.

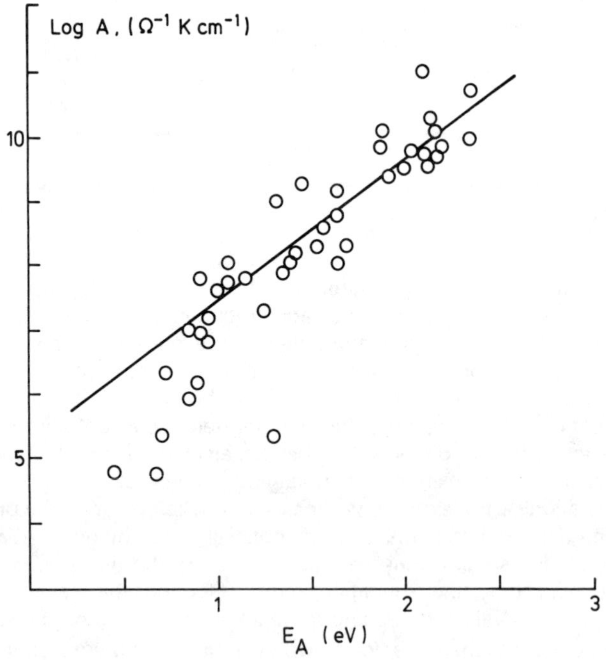

FIGURE 2.16 MEYER-NELDEL CORRELATION (AFTER UVAROV AND HAIRETDINOV)

It is clearly possible to treat the more complex problem when the diffusion coefficient is a function of non-stoichiometry. Using the mass-action equations we may obtain the concentrations of the various defects over the range of values of non-stoichiometry required. Examples of this were shown in Section 2.3. (see Figures 2.10 and 2.11). We must now calculate the hopping rates for all the relevant defects. In the case of the oxide CoO which we take as an example, this is simplified by the fact that the only relevant defects (except at high oxygen partial pressure) are the three possible charge states of the cation vacancy: V^x_{Co}, V'_{Co}, V''_{Co}. Further, all the charge states migrate at the same rate. This is suggested by both experiment and calculation. The calculation, assuming the small polaron model, shows that it is always more favourable to move the Co^{2+} ion than the Co^{3+} ion. Thus, in all the charge states of the vacancy, it is the Co^{2+} ion migration that is the rate determining step. If bound holes are present, these can readily adjust to the motion of the vacant site with negligible effect on the hopping rate. Figure 2.17 shows the dependence of the tracer diffusion coefficient for cobalt as a function of stoichiometry at 1473K. The agreement (to within a factor of five) is within the limits of expected accuracy as discussed above.

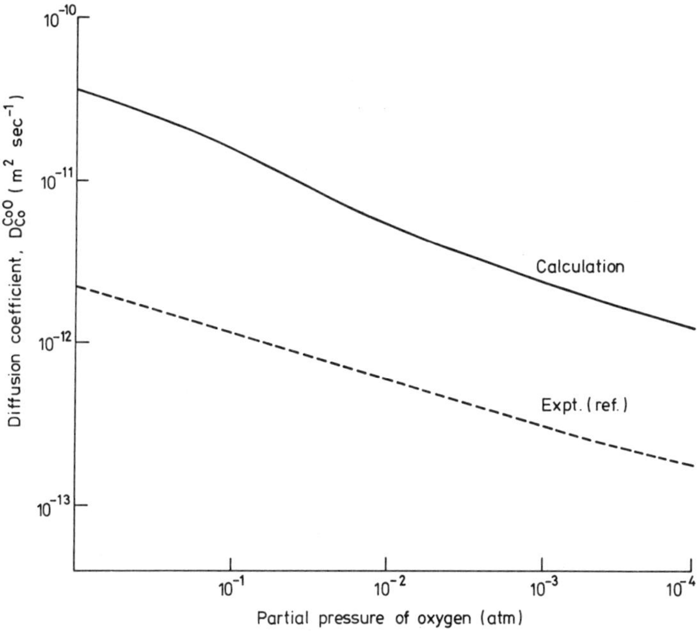

FIGURE 2.17 CALCULATION OF THE DIFFUSION COEFFICIENT OF COLBALT OXIDE AS A FUNCTION OF PARTIAL PRESSURE OF OXYGEN (1473 K)

2.5.2 The correlation factor. As noted in the previous section, the expression for the diffusion coefficient contains a term, the *correlation factor,* that expresses the departure of defect motion from a random walk. In general, this factor depends on the relative jump rates of the possible defect pathways. An expression was derived by Lidiard (1955) for the case of a vacancy hopping near an impurity. We consider an f.c.c. lattice (or sub-lattice). By convention, the impurity-vacancy exchange rate is labelled ω_2; the other vacancy jumps close to the vacancy are labelled as shown in Figure 2.18 and jumps beyond the second neighbour shell are deemed random and equal to the host lattice vacancy jump rate ω_0. It may then be shown that the correlation factor f defined above is given by

$$f = \frac{\omega_1 + \frac{7}{2}\omega_3}{\omega_1 + \omega_2 + \frac{7}{2}\omega_3} \tag{2.5.1}$$

This expression ignores the possibility that after a dissociation of the (nearest-neighbour) pair, the vacancy returns to the impurity from all directions with equal probability.

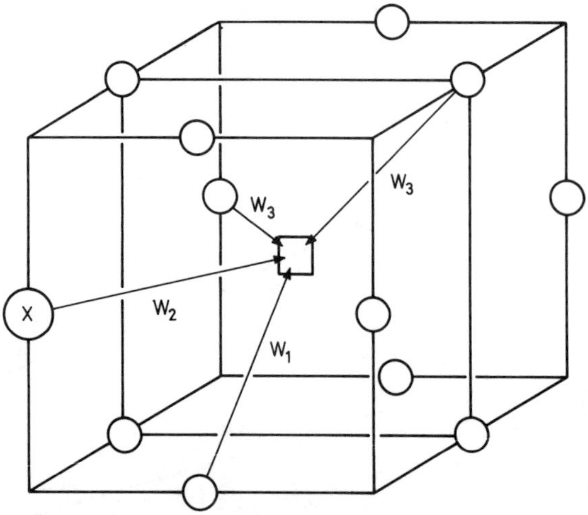

FIGURE 2.18 THE LIDIARD FIVE - FREQUENCY MODEL

To take account of this, it is necessary to consider all the possible paths whereby the vacancy can return to the impurity after dissociation. This has been done by Manning (1964) who showed that equation (2.5.1) should be amended by replacing the factor $\frac{7}{2}$ by the expression $\frac{7}{2}F(\omega_4/\omega_0)$. Expressions for $F(\omega_4/\omega_0)$ have been derived by Manning for a number of lattices. Since, as we have shown, it is possible to calculate defect jump rates, there are no problems of principle in evaluating expressions like (2.5.1). The practical problems are the number of calculations that are required and the size of each individual calculation if the jump takes place in a low symmetry direction. This is a particular problem if an attempt is made to evaluate Manning's formulae. Thus the only calculations that have been done are for very simple cases. The correlation factors for three such cases (Ni^{2+} and Fe^{2+} in CoO and Co^{2+} in NiO) are shown in Figure 2.19 and compared with the experiments of Chen and Peterson (1972) in Figure 2.20. The agreement is excellent given the errors that might reasonably be expected from the potentials used.

Recently, there have been attempts to calculate correlation factors beyond the dilute limit using Monte Carlo. These have been reviewed by Murch (1984). We will consider the problems of calculating defect hopping at high defect concentrations briefly in section 2.5.4.

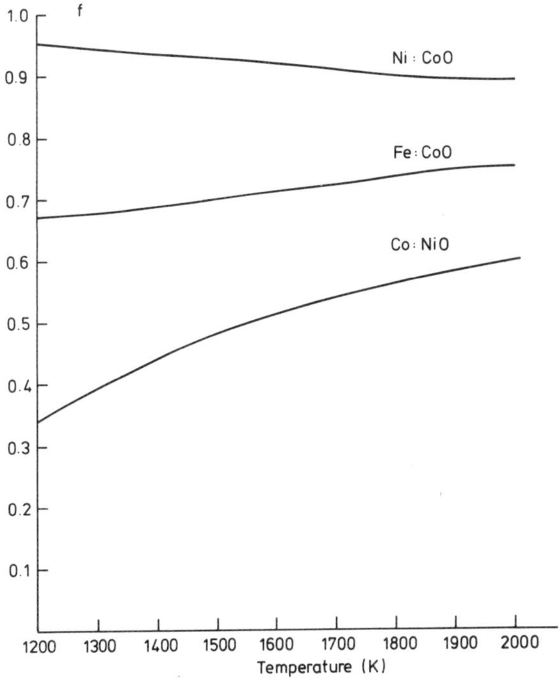

FIGURE 2.19 CORRELATION FACTORS IN SIMPLE OXIDES

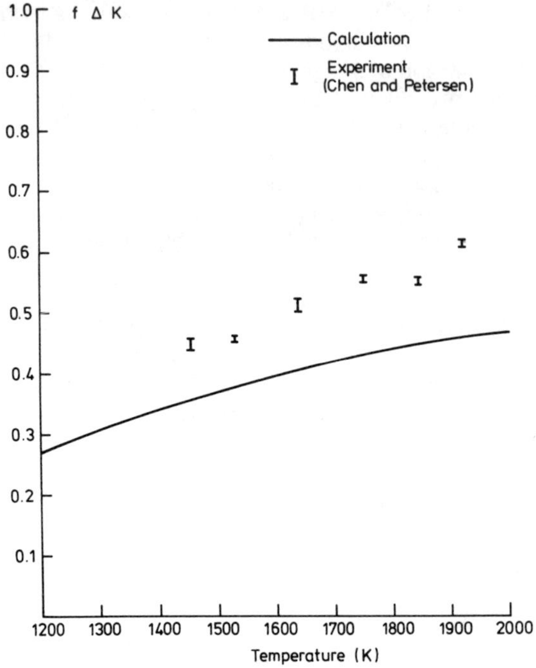

FIGURE 2.20 CORRELATION FACTOR FOR COBALT DIFFUSION IN NICKEL OXIDE

2.5.3 The isotope effect- This is important for two reasons. First, a value for it is needed in experimental determinations of f, the correlation factor. The simple expression used to obtain correlation factors such as those shown in Figure 2.20 is (Schoen 1959, Tharmalingam and Lidiard 1959)

$$\frac{D_\alpha}{D_\beta} = f_\alpha\left(\frac{\Gamma_\alpha}{\Gamma_\beta} - 1\right) \tag{2.5.2}$$

where α, β label the isotopes. This is correct provided that the migration of the diffusing species involves a unique jump frequency ω and the correlation factor may be written in the form

$$f_\alpha = u/(u + \Gamma_\alpha) \tag{2.5.3}$$

where u contains all the jumps not involving the tracer. The expressions quoted above indeed have this form. This leaves the problem of determining the ratio $\omega_\alpha/\omega_\beta$. In the absence of explicit calculation, it was usually assumed that

$$\frac{\Gamma_\alpha}{\Gamma_\beta} = \left(\frac{m_\beta}{m_\alpha}\right)^{\frac{1}{2}} \tag{2.5.4}$$

where m_α, m_β are the isotope masses. This expression assumes that one atom migrates while decoupled from the rest of the lattice. From the argument of Section 2.4 this is clearly not correct. This point was realised by Mullen (1961) and particularly by Le Claire (1966). Using the Vineyard expression (2.4.23) for the jump rates Γ_α, Γ_β and assuming that the change in isotope has no effect on the potential surface, one may readily show that

$$\frac{\Gamma_\alpha}{\Gamma_\beta} = \frac{\omega'_{1\alpha}}{\omega'_{1\beta}} \tag{2.5.5}$$

where $\omega'_{1\alpha}$, $\omega'_{1\beta}$ are the imaginary frequencies at the saddle point. For uncoupled motion this reduces to (2.5.4) but in general one must write

$$\frac{\Gamma_\alpha}{\Gamma_\beta} - 1 = \Delta K\left(\left(\frac{m_\beta}{m_\alpha}\right)^{\frac{1}{2}} - 1\right) \tag{2.5.6}$$

where ΔK is usually called the kinetic energy factor. Calculations of ΔK have been performed for CoO and NiO (Harding 1986) and compared with the experimental determinations of Chen *et al* (1969) (see Figure 2.21). The experiments are performed by measuring f where it is known from the geometry of the problem and taking the ratio of the experiment to the geometrical result. The results are entirely satisfactory. Some earlier calculations (Huntingdon *et al* 1970) gave an anomalous result, but these were performed considering a very small region of crystal around the migrating ion.

The second point of importance about the isotope effect is that it can be used to gain understanding about the detailed dynamics of the migrating ion. The kinetic energy factor has been discussed using a variety of different theories of the diffusion process. We have discussed it here in the context of reaction rate theory. The reaction rate theory leads naturally to an interpretation in terms of the kinetic energy of the migrating particle. It suggests that ΔK is a measure of the fraction of the kinetic energy of the unstable mode that is associated with this particle. Flynn (1975) has discussed the isotope effect in terms of the geometry of the saddle-surface and showed that in certain circumstances the effect of return jumps can be significant. This was demonstrated for model systems by Jacucci and various coworkers (see Jacucci 1984). These authors have also pointed out an alternative way of interpreting ΔK. Their geometrical model interprets the ratio of the isotope jump rates as depending on the reciprocal effective mass along the normal to the saddle-surface and ΔK therefore as depending on the form of the reaction coordinate. A similar point has been made by Feit (1971). The reaction rate theory interpretation is a consequence of the particular choice of the reaction coordinate that it makes; a single normal mode corresponding to ion motion. In this case the more general interpretations reduce to the simple kinetic energy picture.

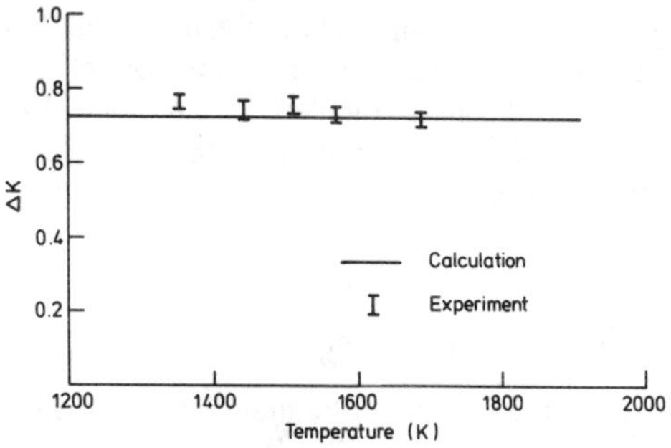

FIGURE 2.21 (a) KINETIC ENERGY FACTOR FOR ^{55}Co AND ^{60}Co IN COLBALT OXIDE

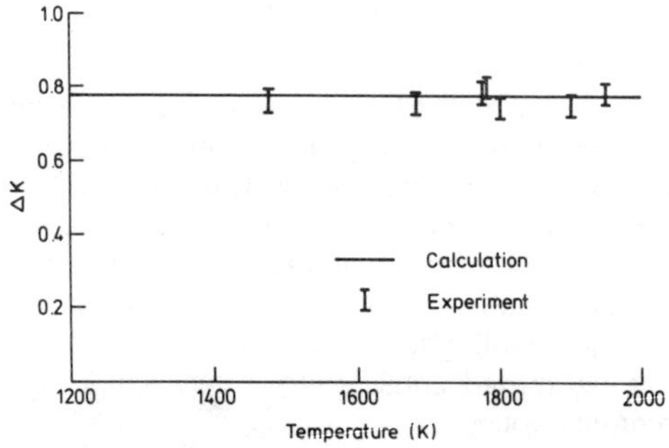

FIGURE 2.21 (b) KINETIC ENERGY FACTOR FOR ^{57}Ni AND ^{66}Ni IN NICKEL OXIDE

2.5.4 Defect motion at high concentrations. The calculations discussed so far all assume the dilute limit. We have discussed briefly at the beginning of Section 2.4 one problem we must face in going beyond this; the increasing number of increasingly complex defect paths to consider and calculate. We must do two things. First, we must calculate the equilibrium concentrations of the various defects. This matter was discussed in Section 2.3. If we can solve this, we face a new difficulty in the breakdown of the random jump approximation. If defect-defect interactions are important, we can no longer assume that individual hopping events are isolated. There will no longer be one activation enthalpy for a jump but a whole spectrum of enthalpies depending on the local environment. Allnatt and Lidiard (1987) give a general discussion using linear response theory. A possible computational approach to the problem has been suggested by Murray *et al* (1986). Here the idea is to use static lattice simulations to obtain the energies of a large variety of defect configurations in the ground state and at the saddle-point and then use Monte Carlo to solve the statistical mechanics. Short-range defect-defect interactions can be taken into account by using the lattice statics code to calculate the defect configurations as clusters. Figure 2.22 shows the results they obtain for the conductivity of yttrium-doped calcium fluoride. As can be seen, their calculation reproduces the main features of the experiments. They show that the decrease in conductivity at high dopant concentration is due to the trapping of the mobile vacancies by clusters of Y^{3+} ions. Beyond a certain level of dopant concentration, the effect of the extra vacancies produced is more than cancelled out by the increasing efficiency of the trapping. There is no obvious problem in extending this type of calculation to use the free energies of the various configurations rather than just the internal energies. Such a calculation would be of considerable interest.

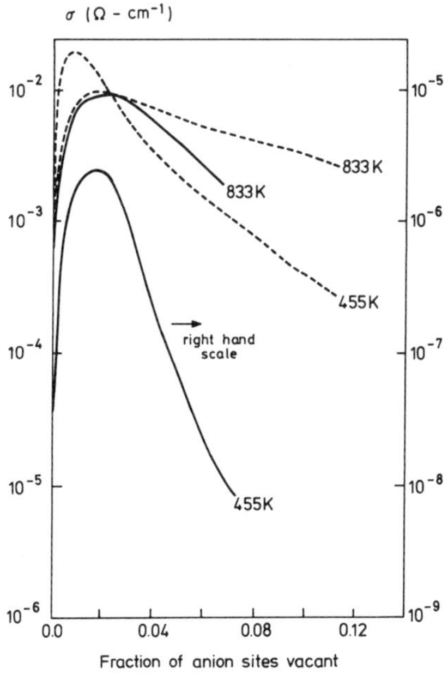

FIGURE 2.22 CONDUCTIVITY OF HEAVILY DOPED CALCIUM FLUORIDE (AFTER MURRAY et al 1986) THE FULL LINES ARE EXPERIMENT ; THE DASHED LINES CALCULATION

2.6 DIFFUSION AND THERMOTRANSPORT.

Up to this point we have considered diffusion as caused purely by gradients in the chemical or electrochemical potential. Yet diffusion is closely linked to the question of heat transport. First, there is the possibility of diffusion being caused by temperature gradients; the phenomenon of thermotransport. Second, there are important issues concerning heat transport even in the case of isothermal diffusion. A hopping event occurs because of a thermal fluctuation. For a hop to occur, the lattice must not only bring the necessary energy together at one spot, it must also dissipate it rapidly afterwards. If it cannot do the first there is, of course, no hop. If it cannot do the second, the defect may either simply jump back, or go on moving in a series of correlated jumps. In neither case are we in the simple hopping régime . This is another way of expressing the assumptions about damping made by the reaction rate theory. In this Section we shall briefly review the phenomenon of thermotransport and its relation to the diffusion process, following the approach of Gillan (1977, 1983).

The essential features of thermotransport are exhibited by the simple case of impurity diffusion in a Bravais lattice. If we consider a temperature gradient in addition to a non-uniform distribution of particles, we may write the flux of particles $\mathbf{J}(\mathbf{r})$ as

$$\mathbf{J}(\mathbf{r}) = -Dn\left(\frac{1}{n}\nabla n + \frac{Q^{*\prime}}{kT^2}\nabla T\right) \tag{2.6.1}$$

where $N(\mathbf{r})$ is the concentration of particles, D the (isotropic) diffusion constant and $Q^{*\prime}$ the reduced heat of transport. Equation (2.6.1) expresses the convention that $Q^{*\prime}$ is positive if the impurities move down the temperature gradient.

If we maintain this gradient and allow the particles to come to a steady state then, provided there are no sources or sinks for the impurities, we have $\mathbf{J} = 0$, i.e.

$$\nabla \ln n = \frac{Q^{*\prime}}{k\,\nabla\left(\frac{1}{T}\right)} \tag{2.6.2}$$

If we consider the motion of intrinsic defects, there will be sources or sinks. The point defect concentration will be governed by the local temperature (provided that the temperature gradient is not too large). Thus in the case of vacancies in our Bravais lattice we have

$$\frac{1}{n}\nabla n = \frac{h_{vac}}{kT^2}\,\nabla T \tag{2.6.3}$$

where h_{vac} is the enthalpy of formation of a vacancy. Thus the flux $\mathbf{J}$ is given by

$$\mathbf{J} = -Dn(h_{vac} + Q^{*\prime})\frac{\nabla T}{kT^2} \qquad (2.6.4)$$

The quantity $(h_{vac} + Q^{*\prime})$ is called the heat of transport, Q^*. There are clearly two contributions to the heat flux, one coming from the formation enthalpy of the defect, the other being associated with the process of defect motion. The second is of interest to us here.

In ionic crystals, matters are further complicated by the charge on the defect. If we consider thermotransport in alkali halides, for example, we have flux equations for the anion and cation vacancies produced by the Schottky process.

$$\mathbf{J}_{cv} = -n_{cv}D_{cv}\left(\frac{1}{n_{cv}}\nabla n_{cv} - \frac{Z_{cv}|e|}{kT}\mathbf{E} + \frac{Q^{*\prime}_{cv}}{kT^2}\nabla T\right) \qquad (2.6.5)$$

$$\mathbf{J}_{av} = -n_{av}D_{av}\left(\frac{1}{n_{av}}\nabla n_{av} - \frac{Z_{av}|e|}{kT}\mathbf{E} + \frac{Q^{*\prime}_{av}}{kT^2}\nabla T\right) \qquad (2.6.6)$$

where $\mathbf{E}$ is the local electric field. This can be used as a convenient way to obtain the reduced heats of transport for the cation and anion vacancies. If we solve these two equations for the case where the current is zero, we obtain an expression for $\mathbf{E}$. We can therefore obtain $Q^{*\prime}_{cv}$ and $Q^{*\prime}_{av}$ from measurements of the potential difference across a specimen using a variety of dopants and temperatures. The use of this, the thermoelectric effect, is not quite as simple as this implies; there are problems with unwanted contact potentials. Details of how the measurements are done are given in a review by Allnatt and Chadwick (1967).

Our main concern, however, is with the microscopic interpretation of these phenomena. Early theories (see Allnatt and Chadwick (1967) for a review) made great use of the idea of a local temperature that could be defined from site to site and even at the saddle-point. From our discussion of diffusion theory it is clear that such a concept is not likely to be useful. A better starting point is a result obtained from irreversible thermodynamics (see Howard and Lidiard (1964) for a discussion and proof) namely

$$\mathbf{J}'_q = \sum_\alpha Q^{*\prime}_\alpha \mathbf{J}_\alpha \qquad (2.6.7)$$

where $\mathbf{J}'_q$ is the reduced heat flux in isothermal diffusion; $\mathbf{J}_\alpha$ are the fluxes of the species in the lattice and $Q^{*\prime}_\alpha$ the appropriate reduced heats of transport. This equation permits us to regard the reduced heat of transport as the ratio of the particle flux to the heat flux at constant temperature. Gillan (1983) has shown how it is possible to exploit this relationship by writing the heat flux in terms of the positions and velocities of the ions. It is then possible to use a molecular dynamics simulation to obtain the flow of heat associated with a single hopping event, $q(t)$. If many events are considered, we can construct the average over a large number of saddle-plane crossings, $\bar{q}(t)$. This may be shown to be related to the heat of transport Q^*, by

$$Q^* = -(1/a) \int_0^\infty dt\ \bar{q}(t) \tag{2.6.8}$$

where a is the jump distance. Since $\bar{q}(t)$ refers to the total heat flux, we must subtract the defect formation enthalpy to obtain the part of the heat flux referring to the jump process, i.e.

$$Q^{*\prime} = -(1/a) \int_0^\infty dt\ \bar{q}(t) - h_{vac} \tag{2.6.9}$$

in the case of simple vacancy diffusion. These results assume that each saddle-plane crossing is statistically independent so that each crossing corresponds to a definite hop. As we have seen, this is not always so. However, a correction for multiple crossings of the saddle- plane may be obtained by replacing a by an effective distance a_{eff} where

$$a_{eff} = a\langle S\rangle \tag{2.6.10}$$

where $\langle S\rangle$ is the transmission coefficient discussed in Section 2.4.

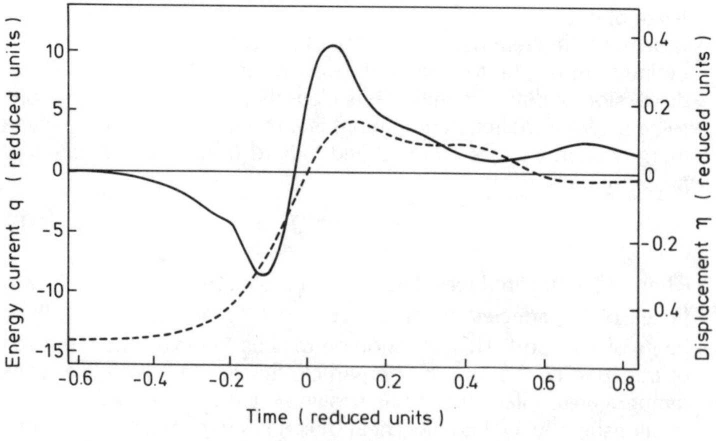

FIGURE 2.23 HEAT CURRENT $\bar{q}$ FOR VACANCY MOTION (AFTER GILLAN AND FINNIS 1978)

Gillan and Finnis (1978) have done simulations to obtain $\bar{q}$ for a model system of vacancy diffusion. The result is shown in Figure 2.23 Although this is a zero-temperature simulation, it shows the main features of the behaviour. The important time variation of $\bar{q}$ occurs over a very short period when the hopping particle is descending from the saddle-point to its equilibrium position. This suggests that the dissipation forces are very effective in removing the localisation of energy required to produce a jump. This is precisely what is needed by reaction rate theory. Preliminary calculations at finite temperature (Gillan 1983) show similar behaviour close to the saddle-plane, but at long times $\bar{q}$ decays exponentially due to anharmonic effects. This is the so-called phonon wind effect. We note finally that for metals there may be an electron wind effect as well. This arises from the fact that the electrons in metals are the principal carriers of heat and will drift down a temperature gradient. This electron wind can transfer momentum to the migrating particles. The driving force on the particles produced by an electric field

$$\mathbf{F} = |e|Z^{*}\mathbf{E} \tag{2.6.11}$$

where the effective charge number Z^{*} is the sum of the ionic charge and the electric wind term Z_{wind}. The size of this effect and thus values for Z_{wind} are, however, very difficult to calculate accurately (Huntingdon 1975).

5.7 SUMMARY AND CONCLUSIONS.

In this chapter we have discussed the calculation of the thermodynamic and transport properties of ionic crystals that depend on point defects. In the main we have confined ourselves to the dilute limit since that is where most of the work has been done. Where possible we have indicated what might be done when the defect concentration is large, say 1% or more.

There is still much to be done in calculating cases within the dilute limit. The free energies of defects have only been calculated for a few systems, as opposed to the large number of systems where the internal energies have been calculated and compared with experiment. The fuller calculations sometimes confirm the earlier ones (as for MgO) but sometimes they show up the inadequacies of the approximations made. Here the main need is for better potentials, better validated. In particular, some scheme for calculation, better than the electron-gas approximation but not ruinously expensive in computer time, is badly needed.

Much progress has been made in the detailed understanding of diffusion. We have shown how the reaction rate theory does provide a way of calculating defect hopping rates to respectable accuracy. Of course, the theory has its limitations. Because of its concentration on the saddle-point configuration, it is inappropriate for systems where quantum effects are important. Because it evades discussion of the details of the trajectory of the migrating ion it cannot consider such problems as the thermopower where these details are of the essence. Even within the limits of classical defect hopping it is still an approximation, as shown by the molecular dynamics simulations of Gillan *et al* (1987) and DeLorenzi and Jacucci (1986). When the hopping rate becomes very high, as in fast-ion conduction, the theory is not useful and we must turn to molecular dynamics. Examples of what can be accomplished here are given by the chapter written by Gillan in this volume.

The problems surrounding the detailed understanding of the diffusion process are not the only ones that should be addressed; the problems associated with transport in highly defective crystals are of great theoretical and practical importance yet, despite the effort put in, progress has been slow. We have discussed briefly the problems that must be faced, both in the thermodynamics and in transport. In the past, progress has tended to be confined to special cases; now there are signs of a more general approach. It is, however fair to say that progress in this area is now likely to be along the lines of better computer algorithms rather than better analytical theories.

References

Abroagye J. K. and Friauf R. J., 1975, *Phys. Rev. B* **11,** 1654.

Acuna L. A. and Jacobs P. W. M., 1980, *J. Phys. Chem. Solids,* **41,** 595.

Agrawal V. K. and Garg H. C., 1973, *Phys. Rev. B* **8,** 843.

Allan N. L., Mackrodt W. C. and Leslie M., 1987, *Adv. Ceramics* in press.

Allnatt A. R. and Chadwick A. V., 1967, *Chem. Rev.* **67,** 681.

Allnatt A. R. and Cohen M. H., 1964, *J. Chem. Phys.* **40,** 1860.

Allnatt A. R. and Loftus E., 1979, *J. Chem. Phys.* **71,** 5388.

Allnatt A. R. and Allnatt E. L., 1982, *J. Chem. Phys.* **76,** 5388.

Allnatt A. R. and Lidiard A. B., 1987, *Rep. Prog. Phys.* **50,** 373.

Almond D. and West A. R., 1986, *Solid State Ionics* **18 & 19,** 1105.

Atlas L. M., 1970, in *Extended Defects in Non-Metallic Solids,* eds. L. Eyring and M. O'Keeffe, (Amsterdam, North-Holland).

Austin I. G. and Mott N. F., 1969, *Adv. Phys.* **18,** 41.

Ball R. D. and Harding J. H., 1983, *A.E.R.E. Harwell Report No* M-3294.

Barker J., 1979, *J. Chem. Phys.* **70,** 2914.

Bendall P. J., Catlow C. R. A., Corish J. and Jacobs P. W. M., 1984, *J. Solid State Chem.* **51,** 159.

Benière F., 1976, *J. Physique Coll.* **37,** C7-261.

Bennett C. H., 1975, in *Diffusion in Solids: Recent Developments,* eds. J. J. Burton and A. S. Nowick (New York, Academic).

Bollmann W. and Henniger H., 1972, *Phys. Stat. Solidi A* **11,** 367.

Boureau G., 1981, *J. Phys. Chem. Sol.* **42,** 743.

Boureau G., 1985, *Z. Phys. Chem. N. F.* **143,** 89.

Catlow C. R. A., 1981, in *Nonstoichiometric Oxides,* ed. O. T. Sørensen, (New York, Academic Press)

Catlow C. R. A., 1986, in *Defects in Solids: Modern Techniques* eds. A. V. Chadwick and M. Terenzi (New York, Plenum).

Catlow C. R. A. (ed), 1987, *Adv. Ceramics,* **23** .

Catlow C. R. A., Corish J., Jacobs P. W. M. and Lidiard A. B., 1981, *J. Phys. C* **14,** L121.

Catlow C. R. A., Corish J., Harding J. H. and Jacobs P. W. M., 1987, *Phil. Mag. B* **55,** 481.

Catlow C. R. A., Mackrodt W. C., Norgett M. J. and Stoneham A. M., 1979, *Phil. Mag.* **40,** 161.

Catlow C. R. A. and Parker S. C., 1982, in *Computer Simulation of Solids,*
eds C. R. A. Catlow and W. C. Mackrodt, (New York, Springer-Verlag)

Catlow C. R. A. and Stoneham A. M., 1983, *J. Phys. C* **16,** 4321.

Chadwick A. V. and Corish J., 1987, in *Defects in Solids: Modern Techniques,* eds. A. V. Chadwick and M. Terenzi, (New York, Plenum).

Chandler D., 1978, *J. Chem. Phys.* **68,** 2959.

Chandler D. and Wolynes P. G., 1981, *J. Chem. Phys.* **74,** 7.

Chen W. K. and Peterson N. L., 1972, *J. Phys. Chem. Sol.* **33,** 881.

Chen W. K., Peterson N. L. and Reeves W. T., 1969, *Phys. Rev.* **186,** 887.

Cox A. and Sangster M. J. L., 1982, *J. Phys. C* **15,** 4473.

Da Fano A. and Jacucci G., 1977, *Phys. Rev. Lett.* **39,** 950.

DeLorenzi G. and Jacucci G., 1986, *Phys. Rev.* **33,** 1993.

DeLorenzi G., Jacucci G. and Flynn C. P., 1984, *Phys. Rev.* **30,** 5430.

Dick B. G. and Overhauser A. W., 1958, *Phys. Rev.* **112,** 90.

Dieckmann R., 1977, *Z. Phys. Chem. N. F.* **107,** 189.

Duffy D. M., Hoare J. and Tasker P. W., 1984, *J. Phys. C* **17** L195.

Feit M. D., 1971, *Phys. Rev. B* **3,** 1223.

Flynn C. P., 1968, *Phys. Rev.* **171,** 682.

Flynn C. P. and Stoneham A. M., 1970, *Phys. Rev. B* **1,** 3966.

Flynn C. P., 1972, *Point Defects and Diffusion,* (Oxford, Clarendon Press)

Flynn C. P., 1975, *Phys. Rev. Lett.* **35,** 1721.

Flynn C. P. and Jacucci G., 1982, *Phys. Rev.* **25,** 6225.

Flynn C. P., 1986, preprint.

Gillan M. J., 1977, *J. Phys. C* **10,** 1641.

Gillan M. J., 1981a, *Phil. Mag. B* **43,** 301.

Gillan M. J., 1983, in *Mass Transport in Solids,* eds. F Benière and C. R. A. Catlow, (New York, Plenum).

Gillan M. J., 1987, *Phys. Rev. Lett.* **58,** 563.

Gillan M. J. and Finnis M. W., 1978, *J. Phys. C* **11,** 4469.

Gillan M. J., Harding J. H. and Tarento R. J., 1987, *J. Phys. C* **20,** 2331.

Gillan M. J. and Jacobs P. W. M., 1983, *Phys. Rev. B* **28,** 759.

Glasstone S., Laidler K. J. and Eyring H., 1941, *The Theory of Rate Processes* (New York, McGraw-Hill).

Glyde H. R., 1967, *Rev. Mod. Phys.* **39,** 373.

Grimes R. W., Anderson A. B. and Heuer A. H., 1986, *J. Amer. Ceram. Soc.,* **68,** 619.

Gordon R. G. and Kim Y. S., 1972, *J. Chem. Phys.* **56,** 3122.

Harding J. H., 1985, *Phys. Rev. B* **32,** 6861.

Harding J. H., 1986, *J. Phys. C* **19,** L731.

Harding J. H. and Harker A. H., 1985, *Phil. Mag.* **51,** 119.

Harding J. H. and Sangster M. J. L., 1986, *J. Phys. C* **19,** 6153

Harding J. H. and Sangster M. J. L., 1987, *Cryst. Latt. Defects and Amorphous Solids* **15,** 1.

Harding J. H., Sangster M. J. L. and Stoneham A. M., 1987, *J. Phys. C* **20,** 5281.

Harding J. H. and Stoneham A. M., 1981, *Phil. Mag.* **B 43,** 705.

Harding J. H. and Stoneham A. M., 1982, *J. Phys. C* **15,** 4649.

Harding J. H. and Stoneham A. M., 1984, *J. Phys. C* **17,** 1179.

Harding J. H. and Tarento R. J., 1987, *Adv. Ceramics* in press.

Hönig J., 1984, in *Basic Properties of Binary Oxides,* eds. A. Dominguez-Rodriguez, J. Castaing and R. Marquez, (University of Seville)

Howard R. E., and Lidiard A. B., 1964, *Rep. Prog. Phys.* **27,** 161.

Hudson R. A., Farlow G. C. and Slifkin L. M., 1987, *Cryst. Latt. Defects and Amorphous Solids* **15,** 239.

Huntingdon H. B., Feit M. D. and Lortz D., 1970, *Cryst. Latt. Defects* **1,** 193.

Huntingdon H. B., 1975, in *Diffusion in Solids: Recent Developments,* eds. J. J. Burton and A. S. Nowick (New York, Academic).

Hutchings M. T. and Kjems J. private communication.

Jacucci G., 1984, in *Diffusion in Crystalline Solids* eds. G. E Murch and A. S. Nowick, (New York, Academic)

Jacucci G., Toller M., DeLorenzi G. and Flynn C P., 1984, *Phys. Rev.* **52,** 295.

Jacobs P. W. M. and Ong S. H., 1976, *J. Phys. (Paris) Coll.* **37,** C7-331.

Jacobs P. W. M., Nerenberg M. A. and Govindarajan J., 1982, in *Computer Simulation of Solids,* eds C. R. A. Catlow and W. C. Mackrodt, (New York, Springer-Verlag)

Kac M., 1943, *Ann. J. Math.* **65,** 609.

Kendrick J. and Mackrodt W. C., 1983, *Solid State Ionics* **8,** 2477.

Koehler T.R., 1975, in *Dynamical Properties of Solids,* Vol. **2,** eds. G. K. Horton and A. A. Maradudin, (North-Holland, Netherlands).

Kroger F. A., 1974, *The Chemistry of Imperfect Crystals: Volume* **3,** (Amsterdam, North-Holland)

Laskar A. L., 1984, *Mater. Sci. Forum* **1,** 59.

Le Claire A. D., 1970 in *Physical Chemistry – an advanced treatise, Vol X,* (New York, Academic Press).

Le Claire A. D., 1966, *Phil. Mag.* **14,** 1271.

Leslie M., 1982, *Daresbury Laboratory Report No* DL/SCI/TN31T.

Leslie M. and Gillan M. J., 1985, *J. Phys. C* **18,** 973.

Lidiard A. B., 1955, *Phil. Mag.* **46,** 1218.

Lidiard A. B., 1957, in *Handbuch der Physik* Vol **20,** 246 (Berlin, Springer).

Lidiard A. B., 1974, in *Crystals with the Fluorite Structure,* ed W. Hayes, (Clarendon Press, Oxford).

Mackrodt W. C., 1982, in *Computer Simulation of Solids,* eds C. R. A. Catlow and W. C. Mackrodt, (New York, Springer-Verlag)

Mackrodt W. C. and Stewart R. F., 1979, *J. Phys. C* **12,** 431.

Manes L., 1981, in *Nonstoichiometric Oxides,* ed. O. T. Sørensen, (New York, Academic Press).

Manning J. R., 1964, *Phys. Rev.* **136A,** 1758.

Maradudin A. A., Montroll E. W., Weiss G. H. and Ipatova I. P., 1971, *Theory of Lattice Dynamics in the Harmonic Approximation* (New York, Academic Press).

Mayer J. E., 1950, *J. Chem. Phys.* **18,** 1426.

Meyer W. and Neldel H., 1937, *Z. Phys.* **18,** 588.

Mitra S. K. and Allnatt A. R., 1979, *J. Phys. C* **12,** 2261.

Mott N. F. and Littleton M. J., 1938, *Trans. Farad. Soc.* **34,** 485.

Muhlhausen C. and Gordon R. G., 1981, *Phys. Rev. B* **23,** 900.

Mullen J. G., 1961, *Phys. Rev.* **121,** 1649.

Murch G. E., 1981, *Atomic Diffusion Theory in Highly Defective Solids,* (Switzerland, Trans-Tech).

Murch G. E., 1984, in *Diffusion in Crystalline Solids* eds. G. E. Murch and A. S. Nowick, (New York, Academic)

Murray A. D., Murch G. E. and Catlow C. R. A., 1986, *Solid State Ionics* **18 & 19,** 196.

Norgett M. J., 1974, *A.E.R.E. Harwell Report No.* **R.7650.**

Nowick A and Lee W. K., 1987, *Solid State Ionics* in press.

Oberschmidt J. and Lazarus D., 1980, *Phys. Rev. B* **21,** 5823.

Onsager L., 1927, *Phys. Z.* **28,** 286.

Petot-Ervas G., Matzke Hj., and Monty C. (eds), 1984, *Solid State Ionics* **12.**

Parinello M. and Rahman A., *J. Chem. Phys.* **80,** 860.

Pitts E., 1953, *Proc. Roy. Soc. Lond. Ser. A* **217,** 43.

Pyper N. C., 1986, *Phil. Trans. R. Soc. A* **320,** 107.

Pyper N. C., Marketos P. and Malli G. L., 1987, *J. Phys. C* **20,** 4711.

Rice S. A., 1958, *Phys. Rev.* **112,** 804.

Rothman S. J., 1984, in *Diffusion in Crystalline Solids,* eds. G. E. Murch and A. S. Nowick, (New York, Academic Press)

Sangster M. J. L. and Rowell D. K., 1982, *J. Phys. C* **15,** 5153.

Schmalzreid H., 1981, *Solid State Reactions, (2nd edition),* (Weinheim, Verlag Chemie).

Schoen A. H., 1959, *Phys. Rev. Lett.* **1,** 138.

Sempolinsky D. R. and Kingery W. D., 1980, *J. Amer. Ceram. Soc.* **63,** 11.

Sinha S. K., 1973 *Crit. Rev. Solid State Sci.* **3,** 273.

Slater N. B., 1959, *Theory of Unimolecular Reactions,* (Ithaca, Cornell U. P.)

Sørensen O. T. (ed), 1981, *Nonstoichiometric Oxides,* (New York, Academic Press)

Stoneham A. M. and Harding J. H., 1986, *Ann. Rev. Solid State Chem.* **37,** 53.

Stoneham, A. M., Tominson S. M., Catlow C. R. A. and Harding J. H., 1985, in *Physics of Disordered Solids,* eds (D. Adler, H. Fritzsche and D. R. Ovshinsky, (New York, Plenum).

Sykora G. P. and Mason T. O., 1987, *Adv. Ceramics* in press.

Tasker P. W. and Bullough T. J., 1981, *Phil. Mag. A* **43,** 313.

Teltow J., 1949, *Ann. Phys. Lpz.* **5,** 63,71.

Tharmalingam K. and Lidiard A. B., 1959, *Phil. Mag.* **4,** 899.

Thiemer O., 1958, *Phys. Rev.* **112,** 1857.

Toller M., Jacucci G., DeLorenzi G. and Flynn C. P., 1985, *Phys. Rev. B* **32** 2082.

Ure R. W., 1957, *J. Chem. Phys.* **26,** 1363.

Valleau J. P. and Whittington S. G., 1977, *Statistical Mechanics Part A,* ed B. J. Berne (New York, Plenum).

Vineyard G. H., 1957, *J. Phys. Chem. Solids* **3,** 121.

Uvarov N. F. and Hairetdinov E. F., 1986, *J. Solid State Chem.* **62,** 1.

Waldram J., 1985, *The theory of thermodynamics,* (Cambridge, Cambridge University Press)

Wedepohl P. T., 1967, *Proc. Phys. Soc.* **92,** 79.

Wert C. A. and Zener C., 1949, *Phys. Rev.* **76,** 1169.

Wigner E., 1938, *Trans. Farad. Soc.* **34,** 39.

Wuench B. J., 1983, in *Mass Transport in Solids,* eds. F Benière and C. R. A. Catlow, (New York, Plenum).

Chapter 3

DYNAMICAL SIMULATIONS OF SUPERIONIC CONDUCTORS

M.J. Gillan

Physics Department, University of Keele,
Keele, Staffordshire, ST5 5BG, U.K.

and

Theoretical Physics Division, Harwell Laboratory,
Didcot, Oxon OX11 0RA, U.K.

Superionic conductors are ionic solids in which one ion species is disordered and highly mobile. Molecular dynamics (MD) simulation gives a powerful way of studying models of superionic conductors. We explain the MD technique and describe how it is applied to ionic systems. We than show how simulation of fluorite superionic conductors has successfully reproduced a range of experimental measurements on quantities including the diffusion coefficient, the spatial distribution of disordered ions, the van Hove self-function and the dynamical structure factors. The simulations have been used to provide a unified interpretation of the observations. This shows that superionic conduction in fluorites occurs by the motion of vacancy and interstitial defects, both of which jump between regular lattice sites. the coherent quasielastic peak recently observed by neutron scattering is shown to arise from the motion of these defects.

1. Introduction

All crystalline materials show some degree of disorder at high temperatures. This disorder is of two kinds: firstly, there are vibrational displacements of the ions from their regular sites, and secondly, there are point defects. The presence of point defects – vacancies and interstitials – is responsible for the diffusion of ions that is observed in all ionic crystals. Vibrational motions are usually treated in the harmonic approximation, which assumes that the displacements are small. As one goes to higher temperatures, this approximation becomes less and less valid and anharmonic effects become important: the lattice parameter depends on temperature, the constant-pressure and constant-volume specific heats differ, and the vibrational modes become damped. A more sophisticated theoretical description is then called for – an example is the self-consistent phonon approximation described by Ball elsewhere in this book. The point-defect disorder is also easiest to discuss when the temperature is not too high. One can then use energy-minimization techniques to determine the energies of formation and migration of the defects, and harmonic theory to calculate the associated entropies. These highly successful methods are discussed in the article by Harding. Again, at very high temperatures one encounters difficulties, especially if the defect concentration becomes high and the defects begin to interact with each other.

The theoretical difficulties take an extreme form in the case of superionic conductors. These are ionic crystals showing a very large amount of disorder, in which the mobility of one of the ionic species is similar to that found in liquids (Hayes 1978, Salamon 1979, Bates and Farrington 1981, Perram 1983, Kleitz 1983). In such materials, many of the vibrational modes are often so strongly damped as to be experimentally unobservable. Sometimes the disorder is so great that the concept of point defects is no longer appropriate. Even when it is appropriate, the nature of the defects may not be obvious, and their mobility may be so high that a simple description is impossible. The theoretical approximations that work for normal ionic crystals have had very limited success for superionic conductors. The approach that has proved most powerful is that of molecular dynamics (MD) simulation. The idea here is to represent the material as a collection of ions, whose interactions are described by realistic model potentials. The ionic motions in the model system are computed by numerically integrating the equations of motion. Physical quantities are then calculated in terms of averages over the trajectories of the ions. The MD technique has now been applied to several important superionic conductors. In some cases, the numerical predictions are in close agreement with experimental results and the simulations have yielded important conclusions about the mechanisms operating in superionic conductors.

Our plan in this paper is to describe the MD simulation work that has been done on superionic conductors having the fluorite structure. The fluorite superionics form a particularly important class, because of the simplicity of their lattice structure, and because the interactions between the ions are in many cases rather well understood. They have been intensively studied experimentally, particularly by neutron scattering, which has proved to be one of the most powerful experimental techniques for studying superionic conductors (see the article by Hayes and Hutchings in this book). They have also been investigated by MD simulation more thoroughly than most other kinds of superionic conductor. As a result, our atomistic understanding of superionic conduction in the fluorites is probably more complete than for any other kind of superionic conductor. Simulation has contributed significantly to this understanding, as we hope to show here.

A substantial part of this paper will be devoted to the *technique* of MD simulation. We shall describe in some detail the mechanics of simulating superionic conductors, and the methods used for calculating physical quantities such as the diffusion coefficient. However, the paper is not mainly about technique. Our more important aim is to show how simulation can help to build up a coherent atomistic interpretation of superionic conduction in real materials. Here, the interplay between simulation and experiment is very important. Simulation is fallible, like any other approach, and needs to be tested against experiment. We shall therefore emphasize the close relation between experimental observables and quantities that can be computed by simulation. On the other hand, simulation yields essentially *complete* information about a model system, in a way that cannot be matched by experiment. It gives us access to things such as the detailed trajectories of the diffusing ions, which cannot be observed experimentally. This allows one to establish an interpretation of the observed phenomena with a certainty that would be impossible if one had only the experimental observations.

We shall begin (§2) by summarizing briefly some of the main properties of fluorite materials, and by describing the interaction models used in the simulations. We then discuss in some detail the technique of MD simulation (§3). The use of computer graphics to examine the behaviour of the simulated system is described in §4. In subsequent sections (§§5 – 10) we discuss the comparison between simulation and experiment for some key quantities: the diffusion coefficient, the incoherent quasielastic scattering cross-section, the spatial distribution of the ions, the electrical conductivity, and the coherent scattering cross-section. We then turn to the question of interpretation, showing how the diffusive motion can be understood in terms of point defects. We shall see in §§11 – 13 how the defect interpretation gives an explanation of the coherent scattering results. We conclude by summarizing the picture of superionic conduction that emerges from this work.

2. Fluorite Materials

Our main concern in this paper will be with the superionic behaviour of fluorite materials (for general reviews, see e.g. Hayes 1978, Catlow 1980, Hayes 1980, Lidiard 1980, Schoonman 1980, Chadwick 1983a,b). What we are talking about is indicated in Figure 1, where we show the electrical conductivity and the specific heat of $SrCl_2$ as a function of temperature. At low temperatures, $SrCl_2$, like other materials having the fluorite structure, is a perfectly normal ionic crystal. The concentration of point defects in this region is very low, and the electrical conductivity is very small. As the temperature is raised, the defect concentration

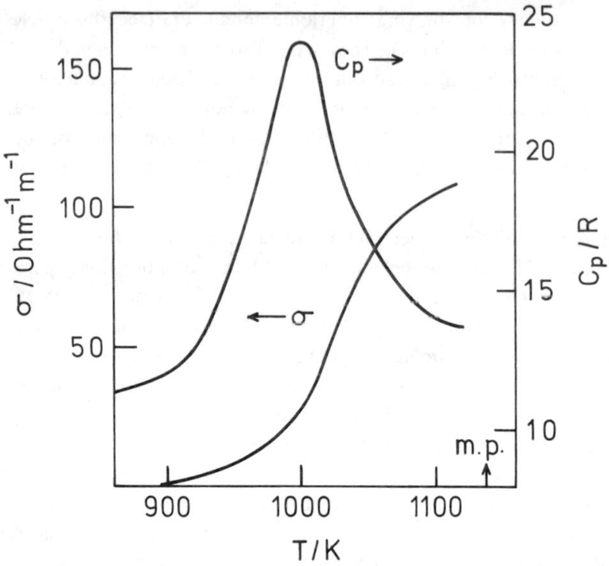

Figure 1. The electrical conductivity σ and molar specific heat C_p divided by the gas constant R for $SrCl_2$ as a function of temperature T. The conductivity data are from Carr *et al.* (1978), and the specific heat data from Schröter and Nölting (1980).

increases until, when it reaches a value of around 1 %, there is a smooth transition to the superionic state. The mobility of the ions and the electrical conductivity have now attained values typical of the molten state, and their rate of increase with temperature diminishes dramatically. The transition is marked by a peak in the specific heat, and by other thermodynamic anomalies. The transition temperature T_c is conveniently taken to be the temperature of the specific heat peak. Values of T_c for some fluorite compounds are given in Table 1. We note that T_c is usually about 4/5 of the melting temperature. It must be stressed that the transition is a continuous one. The lattice symmetry remains unchanged and there is no evidence for any discontinuity in the thermodynamic functions (Schröter and Nölting 1980).

Although we shall be primarily interested in the high-temperature superionic state, we want to summarize here some important facts about the low-temperature 'normal' behaviour, which will be relevant later. In the second part of this section, we discuss the problem of modelling the interionic forces in fluorite materials.

2.1 *Point defects in fluorites*

The perfect fluorite lattice is shown in Figure 2. Notice that the anions form a simple cubic lattice and the cations occupy the centres of alternate cubes; the remaining cube centres are vacant. Many compounds of the form MX_2 crystallize in this structure (Pauling 1960): typical examples are the halides CaF_2, $SrCl_2$ and PbF_2 and the oxides UO_2 and ThO_2. In addition, there are a number of compounds M_2X, such as K_2S and Li_2SO_4, which have the 'antifluorite' structure; this is the

	T_c (K)	T_m (K)
CaF_2	1430	1691
SrF_2	1400	1723
BaF_2	1275	1550
$SrCl_2$	1001	1146
PbF_2	712	1158
UO_2	2300	3120

Table 1. Temperature of diffuse transition to the superionic state T_c and melting point T_m for some important fluorite materials.

same as in Figure 2, except that the cations are now the black circles and the anions are the white circles. Whether or not a material MX_2 adopts the fluorite structure depends on the relative ionic radii. In order for the structure to be stable, the cation radius must not be appreciably smaller than that of the anion, for otherwise core repulsion between the anions will become important. A smaller cation radius will favour reduction of the cation coordination number. This is what happens in the case of, e.g. MgF_2 and $CaCl_2$, which adopt a structure in which the coordination number is six.

At temperatures well below the melting point, the ionic electrical conductivity in fluorites is due to a low concentration of point defects. These are overwhelmingly of the anion Frenkel type: anion vacancies and interstitials created by removing an anion from its regular site and putting it on an interstitial site. The location of the site for free interstitials is difficult to determine by direct experimental means. However, theoretical work on realistic models (Catlow and Norgett 1973, Catlow *et al.* 1977) leaves no doubt that the anion interstitial site is at the vacant cube centre (Figure 2). This is corroborated by endor and epr measurements on bound interstitial-impurity pairs (Baker *et al.* 1968, Baker 1974).

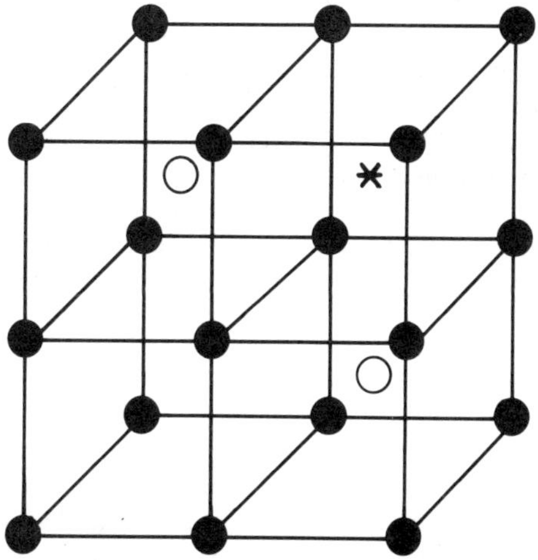

Figure 2. The fluorite lattice structure. Open and filled circles show cations and anions respectively; the asterisk shows the low-temperature interstitial site.

The concentration of vacancies and interstitials in thermal equilibrium is determined by their free energy of formation (Howard and Lidiard 1964, Corish and Jacobs 1973). Let g_F be the free energy needed to remove an anion from its regular site and replace it on a distant interstitial site. The equilibrium concentrations c_v^- and c_i^- of anion vacancies and interstitials are related by the mass-action formula

$$c_v^- c_i^- = \exp(-g_F/k_B T) \ . \tag{1}$$

It is usually convenient to think of g_F in the form $g_F = h_F - Ts_F$, where h_F and s_F are the enthalpy and entropy of formation. The concentration of cation defects can be expressed in a similar way. The energy needed to create cation interstitials is so large that they can be ignored. The concentration c_v^+ of cation vacancies is given by

$$c_v^+ (c_v^-)^2 = \exp(-g_S/k_B T) \ , \tag{2}$$

where g_S is the free energy to remove a cation and two anions from their regular sites and replace them on the crystal surface. The associated enthalpy h_S is much larger than h_F in fluorites. For example, in CaF_2 $h_F = 2.7$ eV, $h_S \simeq 6$ eV (Jacobs and Ong 1976, Catlow *et al.* 1977). Consequently, the concentration of cation vacancies is normally negligible compared with that of anion vacancies and interstitials. The basic reason why $h_S \gg h_F$ has to do with Coulombic energy. Since the cations have twice the charge of anions it is much more costly to displace them from their regular sites. Coulomb considerations also help to explain why the cube centre is a favourable interstitial site for anions: in spite of the presence of the eight neighbouring anions, the Madelung potential turns out to be positive at this site. It is worth pointing out that purely geometrical considerations, which might perhaps suggest that the cations would diffuse by passing to the vacant cube-centre positions, would be utterly misleading in this situation.

Calculations (Catlow and Norgett 1973, Catlow *et al.* 1977) show that the anion vacancy migrates in the way one would expect, by jumping between nearest-neighbour regular anion sites. Interstitial migration occurs by the so-called 'interstitialcy' mechanism: the interstitial ion passes from the interstitial site to one of the neighbouring regular sites as the occupant of that site simultaneously passes to another interstitial site. The rate of migration of the defects is governed by the (free) energy of migration, i.e. the (free) energy barrier that has to be surmounted as the defect jumps between sites. Thus the rate Γ_v at which a vacancy jumps from an anion regular site to any one of the 6 neighbouring anion sites is

$$\Gamma_v = \nu_0 \, e^{\Delta g_v/k_B T} \ , \tag{3}$$

where Δg_v is the migration free energy for vacancies and ν_0 is a constant having the dimension of frequency – for a discussion of the significance of ν_0, see the article by Harding. A similar expression holds for the jump rate of interstitials. In fluorites, the migration enthalpy for vacancies Δh_v is generally lower than that for interstitials Δh_i (Lidiard 1974, Catlow and Norgett 1973, Chadwick 1983a,b), so that at low temperatures vacancies are more mobile than interstitials. However, the migration entropies seem to be usually in the order $\Delta s_i > \Delta s_v$ (Chadwick

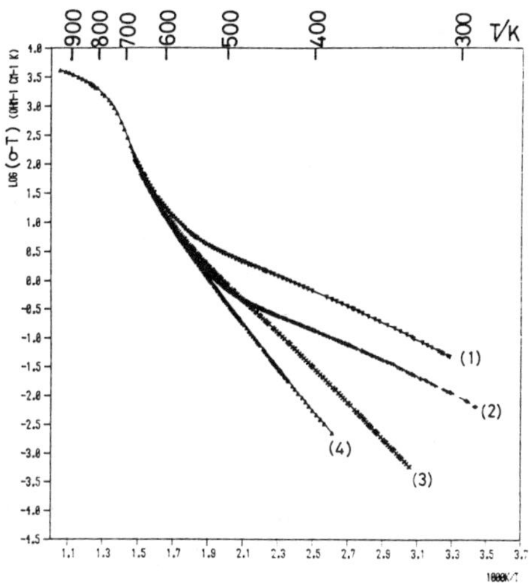

Figure 3. The ionic conductivity σ measured by Azimi *et al.* (1984) for doped and pure PbF_2 crystals. The four curves correspond to dopant levels: (1) 1000 ppm Na^+; (2) 100 ppm Na^+; (3) 100 ppm La^{3+}; (4) pure.

1983a,b), which means that the interstitial becomes more mobile than the vacancy if the temperature is high enough. In PbF_2, for example, the mobilities of the two defects become comparable at about 100 K below T_c (Chadwick 1983a,b).

The temperature dependence of the experimental ionic conductivity for pure and doped PbF_2 is shown on an Arrhenius plot in Figure 3. This shows that there is a dramatic reduction of the Arrhenius slope as the conductivity reaches liquid-like values. Analysis of experimental results such as these allows the determination of the enthalpies and entropies of formation and migration of the anion vacancies and interstitials. As an example, we reproduce in Table 2 the experimental defect parameters for PbF_2. The analysis that gives these values applies only in the 'normal' region, i.e. for temperatures up to within about 100 K below T_c. The substantial rise in Arrhenius slope just below T_c is not accounted for by the analysis, and is presumed to be due to the interaction between the defects, which becomes increasingly important as the superionic regime is approached.

h_F (eV)	1.04 – 1.14	s_F/k_B	3.7 – 5.1
Δh_v (eV)	0.22 – 0.23	$\Delta s_v/k_B$	0.8 – 1.5
Δh_i (eV)	0.47 – 0.65	$\Delta s_i/k_B$	3.5 – 5.2

Table 2. Experimental defect parameters for PbF_2. The quantities h_F, Δh_v and Δh_i are the anion Frenkel formation enthalpy and the enthalpies of migration of the anion vacancy and interstitial. The corresponding entropies s_F, Δs_v and Δs_i are divided by Boltsmann's constant k_B. The effective vibrational frequency used in the definition of the migration entropies is 10.1 THz. Data are from Azimi *et al.* (1984).

It is instructive to use the experimental defect parameters to estimate the concentration and the jump rates of the defects as one approaches T_c. As an example, we give results for PbF_2 at $T = 625$ K, which is where the description in terms of non-interacting low-temperature point defects first shows signs of breaking down. The parameters of Table 2 together with equation (1) give a defect concentration of about 10^{-3}. This value is consistent with the idea that the transition occurs when the concentration approaches the percent level, and defect interactions become important. Equation (3) gives for the jump rates at this temperature $\Gamma_v = 3.0\times10^{11}$, $\Gamma_i = 2.4\times10^{11}$ $\sec^{-1}$. Note that these are the rates of jumping of the defect to any *one* neighbouring site. To get the *total* rate of jumping away from an initial site, one has to multiply by 6 for the vacancy and by 12 for the interstitial. The mean times for which the defects reside on any site thus come to 0.6 psec and 0.4 psec for the vacancy and the interstitial. Note that these are very short times. The typical vibration period in PbF_2 is 0.2 psec. Thus even before we reach T_c, the defects migrate extremely rapidly, remaining on each site for only a few vibrational periods. Note also that the vacancy and interstitial jump rates are roughly the same at this temperature.

2.2 *Interaction models*

In order to do calculations of any kind, we must have some model for the interactions between the ions which make up the material. For some of the fluorite compounds, the bonding is close to the text-book ionic type: the ions have closed-shell configurations, the electronic band gap is large, and the ions interact predominantly through Coulombic and overlap-repulsion interactions (Pauling 1960, Hodby 1974). This should be a particularly good description for the alkaline-earth fluorides CaF_2, SrF_2 and BaF_2; it will be slightly less good for the chlorides, such as $SrCl_2$, and presumably even less good for a case like PbF_2, where the metal ion does not have a rare-gas configuration.

The modelling of ionic crystals has often been discussed in the literature (see e.g. Catlow *et al.* 1982). All the simulations to be described later are based on models in which the energy is a sum of pair-wise interactions between the ions. The potential $V_{\alpha\beta}(r)$ between ions of types α and β is taken to be

$$V_{\alpha\beta}(r) = z_\alpha z_\beta e^2/r + A_{\alpha\beta}\exp(-r/\rho_{\alpha\beta}) - C_{\alpha\beta}/r^6 \ , \tag{4}$$

where, in addition to the Coulomb and overlap contributions, represented by the first two terms, the van der Waals dispersion interaction is also included; the charge on ions of type α, in units of the protonic charge e, is denoted by z_α.

The parameters in these potentials are determined partly by fitting to selected experimental data for the low-temperature crystal and partly by appeal to quantum-mechanical calculation. All the materials we shall discuss are highly ionic, and the charges z_α are therefore assumed to have their nominal ionic values. Some of the remaining parameters can be eliminated by physical considerations. Because of their double charge, Coulomb repulsion ensures that the distance between cations is always large enough for their overlap repulsion and dispersion interactions to be neglected; we can therefore set $A_{++} = C_{++} = 0$. The dispersion interaction between cations and anions is also set equal to zero, on the grounds that unlike nearest neighbours are close enough for this contribution to be quenched (Catlow *et al.* 1982). The overlap interaction between anions is rather small in the perfect crystal, but it plays an important role in determining the anion defect energies. This is particularly true for the anion interstitial, since the distance to the anion neighbours is rather small. The parameters A_{--} and ρ_{--} are taken from quantum-mechanical calculations. The value of the anion-anion dispersion parameter C_{--} is taken in a rather *ad hoc* way from models that have been developed for defect calculations. The remaining two parameters A_{+-} and ρ_{+-} are adjusted to fit the lattice parameter a_0 and the anion Frenkel formation enthalpy (or in the case of $SrCl_2$, the static dielectric constant). The construction of the potentials for the various compounds of interest here is described in more detail in the original papers (Gillan and Dixon 1980b, Dixon and Gillan 1980b, Walker *et al.* 1982, Gillan 1986a).

The model we have described does not take explicit account of the electronic polarizability of the ions. In reality, the electronic charge distribution on each ion is displaced by the local electric field and by the repulsion of neighbouring ions. This displacement gives rise to electronic dipole moments, whose interaction contributes to the energy. The electronic polarizability can be important for some quantities. For example, it determines the high-frequency dielectric constant ε_∞. The unpolarizable, or 'rigid-ion' dscription that we use necessarily makes ε_∞ equal to unity, whereas its true value in ionic crystals is typically 2 (Ashcroft and Mermin 1976). The effects of polarizability are commonly taken into account by using the 'shell model', in which each ion is modelled by a core and a massless shell coupled by a harmonic spring (Dick and Overhauser 1958). Unfortunately, it is difficult to use the shell model in molecular dynamics simulation (Sangster and Dixon 1976). The main difficulty comes from the fact the shells have no mass. This means that at each instant in the simulation the force acting on every shell must vanish. This can only be ensured by a time-consuming relaxation of the shell positions at each simulation step. For this and other reasons, shell-model MD demands at least an

	Calc.	Expt.
ε_0	5.20	6.47
ε_∞	1.00	2.05
$c_{11}(10^{11}$ dyn. cm$^{-2})$	15.8	17.12
$c_{12}(10^{11}$ dyn. cm$^{-2})$	4.15	4.68
$c_{44}(10^{11}$ dyn. cm$^{-2})$	3.96	3.62
U (eV)	−27.10	−26.76
E_F (eV)	2.71	2.71
ΔE_v (eV)	0.19	0.42
ΔE_i (eV)	0.73	0.79

Table 3. Perfect-crystal and anion-defect quantities (ε_0, ε_∞ static and high-frequency dielectric constants, c_{ij} elastic constants, U cohesive energy per unit cell, relative to free ions, E_F Frenkel formation energy, ΔE_v, ΔE_i vacancy and interstitial migration energies, calculated from the potential model and and compared with the experimental results listed by Catlow and Norgett (1973) and by Jacobs and Ong (1976).

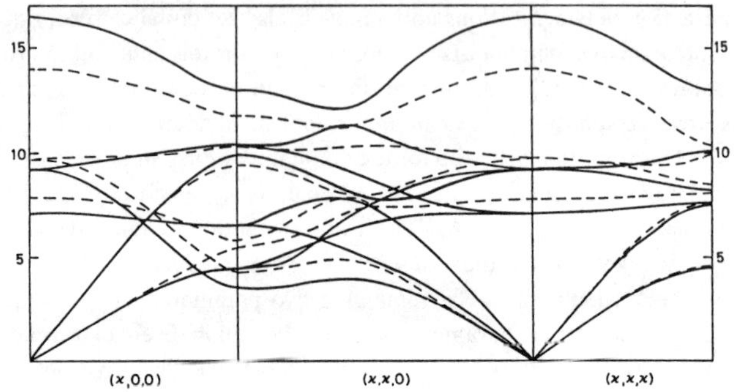

Figure 4. The harmonic phonon frequencies calculated from the rigid-ion potential used in simulations of CaF_2 (full lines), compared with the experimental frequencies of Elcombe and Pryor (1970) (broken lines).

order of magnitude more computer time than rigid-ion simulation (Sangster and Dixon 1976). MD simulation of ionic systems is therefore almost always done with rigid-ion potentials, even though a slight loss of realism is inevitable.

It must be stressed, though, that for many materials a rigid-ion model can give a very satisfactory description, if the potential parameters are suitably chosen. To illustrate this, we show in Table 3 the predictions of a number of important bulk and point-defect properties for CaF_2 obtained from the rigid-ion model used in the MD simulations, and from a good shell model. Except in the case of ε_∞, the two models reproduce the experimental values almost equally well. The usefulness of the rigid-ion model can also be seen by comparing the predicted phonon dispersion curves with experimental measurements (Figure 4). Since the high-frequency dielectric response is incorrectly given, some discrepancies are inevitable, but the overall agreement is very reasonable. There seems every reason to expect, then, that simulations based on the rigid-ion model will reproduce the properties of the real material, at least semi-quantitatively.

Calcium fluoride is, of course, a favourable case, since neither ion is very polarizable. The simulation results to be presented later indicate that the interaction model gives a very good description of this material in the superionic state. We shall also describe calculations on PbF_2, which we should expect to be more difficult, because Pb^{++} is not a closed-shell ion. It is in fact much more difficult to derive a rigid-ion model which satisfactorily reproduces the low-temperature bulk and defect properties of PbF_2, as has been discussed by Walker *et al.* (1982). Indeed, even a fully adequate shell-model description of PbF_2 turns out to be hard to obtain (Matar *et al.* 1984). Even in this case, though, the MD simulations appear to be very successful, as we shall see.

3. The Molecular Dynamics Method

The simulation of liquids and solids by the molecular dynamics (MD) approach has often been described in the literature (e.g. Sangster and Dixon 1976, Hansen and McDonald 1986, Allen and Tildesley 1987). In order to make this article self-contained, though, it will be useful to recall the general ideas underlying this kind of simulation, and to sketch how it is performed in practice.

3.1 *General ideas*

The ambition of MD is to construct a realistic working model of a material on the atomic scale. The model is based on a specification of the interionic potentials describing the forces between the ions (see §2.2). A small piece of the material is represented as a collection of ions contained in a box; it is usual to eliminate the boundaries of the box by applying periodic boundary conditions. Given the positions of the ions at any instant, the interionic forces, and hence the accelerations, can be calculated. The equation of motion is now integrated numerically to determine the dynamical evolution of the entire system. The simulation produces a set of numerical trajectories of the ions extending typically over a few tens of picoseconds. Finally, observable quantities, like diffusion coefficients, can be calculated from suitable averages taken over the duration of the simulation.

We will go into some of the details of how this done a little later, but we want to say something first about the meaning of MD simulation. We need to recall some ideas about the statistical mechanics of thermal equilibrium. Statistical mechanics tells us that the average value of any quantity in thermal equilibrium is equivalent to the time average of that quantity as the system evolves in time (Landau and Lifshitz 1980). This is a rough-and-ready statement of the general principle, but it will serve the present purpose. Let us take an example. Suppose we are interested in the average value of the total kinetic energy K of the system. If the temperature is T, then we know that the thermal average $\langle K \rangle$ is equal to $\frac{3}{2}Nk_B T$, where N is the total number of particles. Now consider the time average. Let us follow the system as it evolves in time at the constant total energy corresponding to the temperature T. We observe the time-varying kinetic energy $K(t)$ starting at some arbitrary time 0 for a duration $\mathcal{T}$. The time average is

$$\overline{K} = \frac{1}{\mathcal{T}} \int_0^{\mathcal{T}} dt \, K(t) \ . \tag{5}$$

The statement is that as $\mathcal{T}$ is made ever larger, the time average $\overline{K}$ tends ever closer to the exact thermal average $\langle K \rangle$. This is true whatever the starting time 0, and regardless of the configuration the system finds itself in at that time. We have considered the kinetic energy as an example, but the same principle holds for any dynamical variable.

Now the point is that MD simulation generates time-dependent trajectories of the system and allows us to calculate the time averages that we have discussed. Suppose then that we perform an MD simulation on our model system, the duration of the simulation being $\mathcal{T}$. As the particles follow the equation of motion, the total energy E is conserved. We calculate the time average $\overline{K}$ of the kinetic energy, and perhaps other time averages. Then, from what we have said, as the duration $\mathcal{T}$ is made ever longer, $\overline{K}$ and the other time averages will tend ever closer to their exact average values in thermal equilibrium. The thermodynamic state is specified by the total energy of the system and by its density – this is fixed, since we simulate a given number of particles in a box of given volume. In practice, one is normally interested in the temperature of a system rather than its energy. But the average $\overline{K}$ gives us the temperature, since $\overline{K} \to \langle K \rangle = \frac{3}{2}Nk_B T$.

The key idea, then, is that MD simulation gives us, by time averaging, the mean values of quantities in thermal equilibrium; as we shall see later, it also allows us to calculate dynamical quantities like diffusion coefficients. It is important to appreciate that although the trajectories of the particles are completely determined by the equation of motion, the calculation of mean quantities has a statistical character. The kinetic energy, for example, fluctuates with time in a rather random way about its mean value. These fluctuations are thermal noise – they have the same kind of randomness as is observed in Brownian motion or in the Johnson noise of electrical circuits. The calculation of a time average can thus be regarded as the sampling of a signal over a given time span. The time average tends closer to the exact thermal average with increasing $\mathcal{T}$ because the sampling becomes more and more extensive.

It follows from this that MD simulation gives results that are in a sense exact: they become exact as $\mathcal{T}$ tends to infinity. This means exact only for the chosen interaction model, of course: for a given model of the interionic potentials, we can obtain essentially exact values for diffusion coefficients etc. under chosen thermodynamic conditions. In this respect, simulation is quite different from calculations made with some approximate physical theory. In the harmonic theory of crystal vibrations, for example, one neglects from the outset anharmonic effects, which may often be important. MD simulation is not approximate in this way. It does not prejudge the physics of the system. Provided the interionic potentials are realistic, everything that happens in the real material will also happen in the simulation.

There is one qualification to be made to this idea of exactness, and that concerns the size of the simulated system. The number N of ions in a simulation is usually less than 10^3, which is much smaller than the number used in real experiments – usually $\sim 10^{23}$. It might be imagined from this that the properties of real and simulated systems would differ substantially. This is not the case. The point is that the properties of the simulated system become ever closer to those of the bulk as N increases. The N–dependence can be studied empirically by comparing results for different N, and it usually turns out that for most quantities the bulk limit is already reached when N is a few hundred. Nevertheless, there are circumstances when size effects are appreciable, and we shall see an example later.

Lastly, there is a rather minor point to be noted. The MD method assumes the validity of classical mechanics, and does not include quantum effects. Since we will be concerned here with temperatures well above the Debye temperature, the neglect of such effects is fully justified.

We now want to turn from the general principles of MD simulation to discuss some of the practical details.

3.2 *The equation of motion*

The equation of motion for the ions is

$$m_i d^2\mathbf{r}_i/dt^2 = \mathbf{F}_i \ , \tag{6}$$

where m_i is the mass of ion i, $\mathbf{r}_i$ its time-varying position, and $\mathbf{F}_i$ the force on it due to interaction with the other ions. In order to integrate this numerically, time must be treated as a discrete variable, so that we generate the positions of all the ions at a sequence of times separated by some chosen step Δt. The equation of motion has to be converted into an algorithm for determining the positions at time $t + \Delta t$, given the positions and velocities at time t. A number of methods are in use, of which the simplest is the so-called Verlet algorithm. (For a good recent review of algorithms, see Berendsen and van Gunsteren 1986). In the Verlet algorithm (Verlet 1967), one expands the positions $\mathbf{r}_i(t+\Delta t)$ as a Taylor series in Δt and truncates at quadratic order:

$$\mathbf{r}_i(t+\Delta t) \simeq \mathbf{r}_t(t) + \mathbf{v}_i(t)\Delta t + \tfrac{1}{2}\mathbf{F}_i(t)\Delta t^2/m_i \ . \tag{7}$$

The velocities $\mathbf{v}_i(t)$ are represented by

$$\mathbf{v}_i(t) \simeq (\mathbf{r}_i(t+\Delta t) - \mathbf{r}_i(t-\Delta t))/2\Delta t \ . \tag{8}$$

This algorithm is often recast so that the time evolution is expressed in terms of the positions alone:

$$\mathbf{r}_i(t+\Delta t) = 2\mathbf{r}_i(t) - \mathbf{r}_i(t-\Delta t) + \mathbf{F}_i(t)\Delta t^2/m_i \ . \tag{9}$$

In this form, the positions at the next time step are obtained from the present positions (from which the present forces $\mathbf{F}_i(t)$ are calculated) and the preceding positions. The algorithm is only an approximation to the true equation of motion, since it is correct only to order Δt^3 (for a discussion of the order to which it is correct, see Berendsen and van Gunsteren 1986). Other more accurate algorithms, correct to higher order in Δt, can also be used – an example is the Gear predictor-corrector method (Gear 1971); these, of course, demand more computation per time step. However, the Verlet algorithm (and any other sensible algorithm) becomes increasingly exact as Δt is reduced, in the sense that the trajectories calculated for a given time span t tend to the exact trajectories as $\Delta t \to 0$.

With the true equation of motion (6), the energy E of the system is exactly conserved. It is usual to calculate E at each time step in the course of a simulation. One can therefore check empirically that Δt is small enough by monitoring the constancy of E. It is generally enough to take Δt smaller than 1/20 of the typical period of motion. It is a waste of computer time to take it *much* smaller than this. On the other hand, if Δt is taken much larger, the evolution becomes not merely inaccurate, but unstable, and the energy suffers large divergent fluctuations.

3.3 *Initial conditions and equilibration*

The important thing in MD simulation is the generation of trajectories. We are free to start from whatever configuration we please. This is true because, as we stressed in §3.1, the time averages will always tend to the exact thermal-equilibrium averages as the duration of the simulation is increased. As a matter of convenience, some of the simulations to be described later were initiated by setting the ions on their regular sites and giving them random velocities. Others were initiated from the final configuration of a previous simulation. In either case, the temperature is monitored by calculating the average kinetic energy, and the velocities are rescaled so as to bring the system to the temperature of interest. In this early part of the simulation, the time averages are influenced by the special character of the initial conditions, which are not generally typical of the configurations found in the thermodynamic state of interest; if we rescale the velocities, the time averages are rendered meaningless by our interference with the system. After the system has arrived at the required thermodynamic state, it is usual to allow it to evolve for a certain further period (typically a few psec) to allow it to forget the special initial conditions and/or the velocity rescaling. Only after this is the accumulation of time averages begun, and the simulation is continued for long enough to make the statistical accuracy of these averages acceptable. The preparatory part of the simulation is often called 'equilibration'. This somewhat misleading term has, of course, nothing to do with *mechanical* equilibrium. The key idea is the loss of memory of the way in which the system has been prepared.

3.4 *Periodic boundary conditions*

There is no reason in principle why bulk matter should not be simulated by a system of ions contained in a box with rigid reflecting walls. The properties of such a system would tend to those of the bulk as its size was increased. However, we would expect surface effects to be important up to a few interparticle spacings from the walls. For a system of a few hundred ions, these effects would then extend over most of the system. Size effects are very much reduced if one imposes periodic boundary conditions. One imagines the N ions to be contained in a primary cell, which is often taken to be cubic, though this is not necessary. This primary system is periodically repeated. Each ion in the primary cell has images in the repeated cells: if the primary positions are $\mathbf{r}_{0i}$, then the image positions are

$$\mathbf{r}_{\mathbf{n}i} = \mathbf{r}_{0i} + n_1\mathbf{a}_1 + n_2\mathbf{a}_2 + n_3\mathbf{a}_3 \ , \qquad (10)$$

where $\mathbf{n}$ is the vector of integers (n_1, n_2, n_3) and $\mathbf{a}_s$ are the primitive repeat vectors. The simulated system thus forms a superlattice whose primitive unit cell contains the N ions. Which cell is regarded as 'primary' is of course arbitrary, since all the cells are images of one another. For book-keeping purposes, the configuration of the system is specified by the positions of the ions in one cell chosen as primary. From time to time, ions will cross the cell boundary. When an ion leaves the primary cell in this way, one of its images enters across the opposite face. At this point, the book-keeping coordinates of the ion are shifted by the appropriate translation vector, so as to keep them within the primary cell. This is merely a book-keeping transaction – no physical discontinuity is involved. Indeed, it must be stressed that the cell boundaries are purely fictitious, since the primary cell could be redefined by translating it in any way we please, though its orientation is fixed by the superlattice structure.

We pay for the elimination of surface effects by imposing an artificial periodicity on the system. Experience shows, however, that the effects of this are usually benign, and convergence towards the bulk limit is very much enhanced. There is an important consequence, though, when we are interested in wavevector-dependent properties, as we will see later.

3.5 *Calculation of the energy and forces*

In all the simulations to be described, the total potential energy U is expressed as a sum of pair potentials. Since the simulations employ periodic boundary conditions, the energy of interest is the energy per unit cell, which is half the sum of the interactions between each ion in a cell and all the other ions and their images. At this point, we must distinguish between ions of different types, and we shall denote the position of ion i of type α in cell $\mathbf{n}$ by $\mathbf{r}_{\mathbf{n}\alpha i}$; if it is unnecessary to distinguish between different images, we shall shorten the notation to $\mathbf{r}_{\alpha i}$. The energy per cell is

$$U = \tfrac{1}{2} \sum_{\mathbf{n}} \sum_{\alpha\beta} \sum_{i=1}^{N_\alpha} \sum_{j=1}^{N_\beta} V_{\alpha\beta}(|\mathbf{r}_{\mathbf{0}\alpha i} - \mathbf{r}_{\mathbf{n}\beta j}|) \ , \tag{11}$$

where we implicitly exclude $i = j$ if $\alpha = \beta$ and $\mathbf{n} = \mathbf{0}$; the number of ions of type α is N_α.

3.5.1 *Short-range part*

The interionic potentials $V_{\alpha\beta}(r)$ consist of the long-range Coulomb term and the short-range overlap and dispersion terms (see equation (4)). We separate the long-range and short-range parts by writing

$$V_{\alpha\beta} = z_\alpha z_\beta e^2/r + V^s_{\alpha\beta}(r) \ . \tag{12}$$

The short-range energy per cell is then

$$U^s = \tfrac{1}{2} \sum_{\mathbf{n}} \sum_{\alpha\beta} \sum_{ij} V^s_{\alpha\beta}(|\mathbf{r}_{\mathbf{0}\alpha i} - \mathbf{r}_{\mathbf{n}\beta j}|) \ . \tag{13}$$

Since $V^s_{\alpha\beta}(r)$ decays rapidly with distance, it can be set equal to zero beyond some cut-off distance r_c. Almost always, r_c can be taken to be less than half the box length L. Then for each pair i,j in equation (13), only a single $\mathbf{n}$ can give a non-zero contribution, and this $\mathbf{n}$ has $|n_s| \leqslant 1$ for all three s. The calculation of U^s thus requires only a sum over all distinct pairs $\alpha i,\beta j$; for each pair, one needs to determine which $\mathbf{n}$ (if any) makes $|\mathbf{r}_{\mathbf{0}\alpha i} - \mathbf{r}_{\mathbf{n}\beta j}| < r_c$.

The short-range force $\mathbf{F}^s_{\alpha i}$ acting on the ith ion of species α is

$$\mathbf{F}^s_{\alpha i} = -\nabla_{\alpha i} U^s = -\sum_{\mathbf{n}} \sum_{\beta} \sum_{j=1}^{N_\beta} \nabla_{\alpha i} V^s_{\alpha\beta}(|\mathbf{r}_{\mathbf{0}\alpha i} - \mathbf{r}_{\mathbf{n}\beta j}|) \ . \tag{14}$$

Once again, only a single $\mathbf{n}$ can give a non-zero contribution for each term βj, and all the short-range forces can be constructed through a double sum over all distinct pairs $\alpha i,\beta j$.

3.5.2 *Coulomb part*

The Coulomb energy per unit cell can be written in the form

$$U^c = \tfrac{1}{2} \sum_{\alpha} \sum_{i=1}^{N_\alpha} z_\alpha e \phi_{\alpha i} \ , \tag{15}$$

where $\phi_{\alpha i}$ is the electric potential at ion αi due to all the other ions and all their images (excluding, of course, ion αi itself in the primary cell, but including all its images in other cells). This electric potential is given by

$$\phi_{\alpha i} = e \sum_{\mathbf{n}} \sum_{\beta} z_\beta \sum_{j=1}^{N_\beta} \frac{1}{|\mathbf{r}_{\mathbf{0}\alpha i} - \mathbf{r}_{\mathbf{n}\beta j}|} \ . \tag{16}$$

It is a familiar fact that Coulomb sums of this kind are not uniquely defined, since the sum is only conditionally convergent. This problem has been much discussed in the recent literature (e.g. de Leeuw *et al.* 1980). It has been pointed out that the precise meaning to be attached to the sum depends on the boundary conditions applied to the bulk system one is attempting to represent. The conventional procedure for calculating $\phi_{\alpha i}$, which is what we shall describe, corresponds to having conducting boundaries (de Leeuw *et al.* 1980).

The strategy for calculating $\phi_{\alpha i}$ is originally due to Ewald (1921). We first consider the potential $\phi^{(1)}_{\alpha i}$ which would be observed at ion αi if the charge on every ion, instead of being concentrated at the position of the ion, were spread out into a Gaussian distribution centred at that position. We include in $\phi^{(1)}_{\alpha i}$ the potential due to the Gaussian distribution of the ion αi itself; let this latter potential be denoted by $\phi^{(2)}_{\alpha i}$. We then compute the difference potential $\phi^{(3)}_{\alpha i}$, which is the potential due to all the ions, other than αi, when they are point charges, *minus* the potential of these ions when they are Gaussian distributions. The quantity $\phi^{(1)}_{\alpha i} - \phi^{(2)}_{\alpha i} + \phi^{(3)}_{\alpha i}$ is the potential $\phi_{\alpha i}$ we seek. The potentials $\phi^{(1)}_{\alpha i}$ and $\phi^{(3)}_{\alpha i}$ can be expressed as rapidly convergent sums in reciprocal space and real space, respectively, as we now describe.

Consider first the contribution $\phi^{(1)}_{\alpha i}$. When all the ionic charges are spread out into Gaussians, the charge density distribution in the system is

$$\rho(\mathbf{r}) = e(\eta/\sqrt{\pi})^3 \sum_{\mathbf{n}} \sum_{\alpha} z_\alpha \sum_{i=1}^{N_\alpha} \exp(-\eta^2 |\mathbf{r} - \mathbf{r}_{\mathbf{n}\alpha i}|^2) \ . \tag{17}$$

Here, η is an inverse length specifying the width of the Gaussian, which at this stage is arbitrary. We want to determine the electric potential due to $\rho(\mathbf{r})$, which at some arbitrary point $\mathbf{r}$ we call $\phi^{(1)}(\mathbf{r})$; the potential at ion αi will then be $\phi^{(1)}_{\alpha i} = \phi^{(1)}(\mathbf{r}_{\alpha i})$. We have:

$$\phi^{(1)}(\mathbf{r}) = \int d\mathbf{r}' \ \rho(\mathbf{r}')/|\mathbf{r} - \mathbf{r}'| \ . \tag{18}$$

It is simplest to consider the relation between $\phi^{(1)}$ and ρ in Fourier space. Since both quantities have the periodicity of the repeating cells, they can be written:

$$\begin{aligned} \rho(\mathbf{r}) &= \sum_{\mathbf{G}} \hat{\rho}(\mathbf{G}) e^{-i\mathbf{G}\cdot\mathbf{r}} \\ \phi^{(1)}(\mathbf{r}) &= \sum_{\mathbf{G}} \hat{\phi}^{(1)}(\mathbf{G}) e^{-i\mathbf{G}\cdot\mathbf{r}} , \end{aligned} \tag{19}$$

where $\hat{\rho}(\mathbf{G})$ and $\hat{\phi}^{(1)}(\mathbf{G})$ are the Fourier components at the reciprocal lattice vectors of the superlattice: if we use a cubic repeating cell of length L, then $\mathbf{G} = (2\pi/L)\,(h_1, h_2, h_3)$, where h_s are integers. The Fourier components are given by

$$\hat{\rho}(\mathbf{G}) = \frac{1}{\Omega_0} \int_{\text{cell}} d\mathbf{r}\; e^{i\mathbf{G}\cdot\mathbf{r}} \rho(\mathbf{r}) , \tag{20}$$

and similarly for $\hat{\phi}^{(1)}(\mathbf{G})$; the integral goes over a single repeated cell, Ω_0 being the cell volume (i.e. L^3 for a cubic cell).

Now equation (18) for the electric potential takes the following form in Fourier space:

$$\hat{\phi}^{(1)}(\mathbf{G}) = \frac{4\pi}{G^2}\, \hat{\rho}(\mathbf{G}) . \tag{21}$$

From equations (17) and (20), one readily finds that

$$\hat{\rho}(\mathbf{G}) = \frac{e}{\Omega_0}\, S(\mathbf{G})\, e^{-G^2/4\eta^2} , \tag{22}$$

where

$$S(\mathbf{G}) = \sum_{\alpha} z_{\alpha} \sum_{i=1}^{N_\alpha} e^{i\mathbf{G}\cdot\mathbf{r}_{\alpha i}} . \tag{23}$$

On putting equation (21) into (19), we find

$$\phi_{\alpha i}^{(1)} = \phi^{(1)}(\mathbf{r}_{\alpha i}) = \frac{4\pi e}{\Omega_0} \sum_{\mathbf{G}\neq 0} S(\mathbf{G}) G^{-2} e^{-G^2/4\eta^2} e^{-i\mathbf{G}\cdot\mathbf{r}_{\alpha i}} . \tag{24}$$

The term $\mathbf{G} = \mathbf{0}$ is omitted, since $S(\mathbf{G}=\mathbf{0})$ vanishes because of the electrical neutrality of the system.

This gives us the potential due to all the Gaussians, including the one centred on αi itself. The potential $\phi_{\alpha i}^{(2)}$ due to this particular Gaussian is readily shown to be

$$\phi_{\alpha i}^{(2)} = 2ez_{\alpha}\eta/\sqrt{\pi} . \tag{25}$$

Now consider the difference potential $\phi_{\alpha i}^{(3)}$. The electric potential at distance r from the centre of a Gaussian charge distribution of total charge $z_\alpha e$ is $(z_\alpha e/r)\,\mathrm{erf}(\eta r)$, where erf is the error function. The difference between the potential due to a point charge and a Gaussian distribution is therefore $(z_\alpha e/r)\,\mathrm{erfc}(\eta r)$, where erfc is the complementary error function (erfc = 1 − erf). Hence $\phi_{\alpha i}^{(3)}$ is given by

$$\phi_{\alpha i}^{(3)} = e \sum_{\mathbf{n}} \sum_{\beta} z_\beta \sum_{j=1}^{N_\beta} \frac{\operatorname{erfc}(\eta|\mathbf{r}_{\mathbf{0}\alpha i} - \mathbf{r}_{\mathbf{n}\beta j}|)}{|\mathbf{r}_{\mathbf{0}\alpha i} - \mathbf{r}_{\mathbf{n}\beta j}|} \ . \tag{26}$$

Since

$$\phi_{\alpha i} = \phi_{\alpha i}^{(1)} - \phi_{\alpha i}^{(2)} + \phi_{\alpha i}^{(3)} \ , \tag{27}$$

we obtain from equations (24-26) the following expression for the Coulomb energy of the system:

$$\begin{aligned} U^c = {} & \tfrac{1}{2} \sum_{\mathbf{n}} \sum_{\alpha\beta} z_\alpha z_\beta e^2 \sum_{ij} \frac{\operatorname{erfc}(\eta|\mathbf{r}_{\mathbf{0}\alpha i} - \mathbf{r}_{\mathbf{n}\beta j}|)}{|\mathbf{r}_{\mathbf{0}\alpha a} - \mathbf{r}_{\mathbf{n}\beta j}|} \\ & + \frac{2\pi e^2}{\Omega_0} \sum_{\mathbf{G} \neq \mathbf{0}} |S(\mathbf{G})|^2 G^{-2} e^{-G^2/4\eta^2} - \frac{e^2 \eta}{\sqrt{\pi}} \sum_\alpha N_\alpha z_\alpha^2 \ . \end{aligned} \tag{28}$$

Both the real-space and the reciprocal-space summations are absolutely convergent: the first, because of the erfc, which falls off as a Gaussian for large distances, and the second, because of the Gaussian $\exp(-G^2/4\eta^2)$.

It will be recalled that the parameter η is arbitrary: the expression for U^c, if evaluated exactly, is independent of η. The value of η determines the balance of computational effort between the two summations: if η is small, the real-space convergence will be slow and the reciprocal-space convergence rapid, while if η is large, the convergence rates will be reversed. What is commonly done is to choose η so that $\operatorname{erfc}(\eta r)/r$ can be neglected for $r > \frac{1}{2}L$. In this case, the 'interaction' $z_\alpha z_\beta e^2 \operatorname{erfc}(\eta r)/r$ can be handled in exactly the same way as the true short-range potential. A satisfactory value of η is $5.6/L$, which is what we generally use in the calculations to be described later.

An expression for the Coulomb force $\mathbf{F}_{\alpha i}^c$ on ion αi can be obtained directly from equation (28) by using the relation

$$\mathbf{F}_{\alpha i}^c = -\nabla_{\alpha i} U_c \ . \tag{29}$$

This is likewise a sum of real-space and reciprocal-space contributions.

3.6 *Storing trajectories*

Some time averages, for example that of the kinetic energy, are routinely calculated in the course of the simulation. Others cannot be handled in this way. In order to calculate these, the positions of all the ions at every time step must be stored, so that they can be processed later. The quantity of information is large. If, to take a typical example, we have a run of 10^4 steps for 300 particles, the number of coordinates to be stored is 9×10^6. The usual procedure is to write out these coordinates at each step onto a mass-storage medium, like magnetic tape.

4. Trajectory-plotting

The raw output from a simulation is an enormous undigested mass of numbers representing the trajectories of all the ions. It is often helpful to obtain an overall impression of the behaviour of the system by plotting the trajectories using computer graphics. In superionic conductors, this allows one to see directly the diffusion of the mobile ions and to understand some of the characteristics of the diffusive motion.

We show in Figure 5 two trajectory plots for simulated CaF_2, one at a low temperature and the other at just below the melting point. What we have done here is to plot the path followed by every ion in the system over a short interval of time, corresponding to about two vibrational periods. The plots represent a projection onto the [100] plane, so that the trajectories of different ions are superposed. The cation trajectories, for example, come in superposed pairs. The grid has been drawn to guide the eye. These pictures show at once that at the low temperature the system is a well-behaved regular crystal, with every ion vibrating about its own lattice site. One notices that the anion amplitudes are much larger than those of the cations. This is a characteristic feature of fluorite materials, which has been known for many years from diffraction measurements (Willis 1963a,b, 1965, Dawson *et al.* 1967, Cooper 1970). The second plot shows the dramatic events that appear at high temperature. The anions are considerably disordered, and a number of them are seen in transit between neighbouring lattice sites. The cations meanwhile continue to vibrate about their regular sites, though with larger amplitudes than before. This plot incidentally illustrates the periodicity of the simulated system: one of the diffusing ions passes through the top face of the box as its image enters through the bottom.

A more complete impression of the behaviour at high temperature is given by plotting the trajectories over a longer period of time. The exposure time of the plot shown in Figure 6 is 3 psec, which corresponds to about 30 vibrational periods. The busy activity of the anions is very striking. The time lapse here is long enough to give an idea of the *average* behaviour. For example, one gets a rough picture of the spatial distribution of the ions: the anions, though certainly disordered, still spend almost all their time in the neighbourhood of their regular sites. One also sees that much of the diffusion is occurring by direct jumps between these sites.

The diffusion process can be understood in more detail by following the motion of a chosen anion over a long period of time. Two plots of this kind are shown in Figure 7. One sees more clearly now that most of the time is spent in vibration about the regular sites. When diffusive jumps occur, they are usually between neighbouring sites in a $\langle 100 \rangle$ direction – though this is not quite clear from these plots, because of the projection.

It is of great interest in superionic conductors to understand something about the correlated motion of different ions. This is not straightforward to examine graphically, since one first needs to identify the significant events. It can be done, though, and Figure 8 shows a dramatic example of correlated hopping in CaF_2 at high temperature. We have identified a sequence of hops involving 8 different ions,

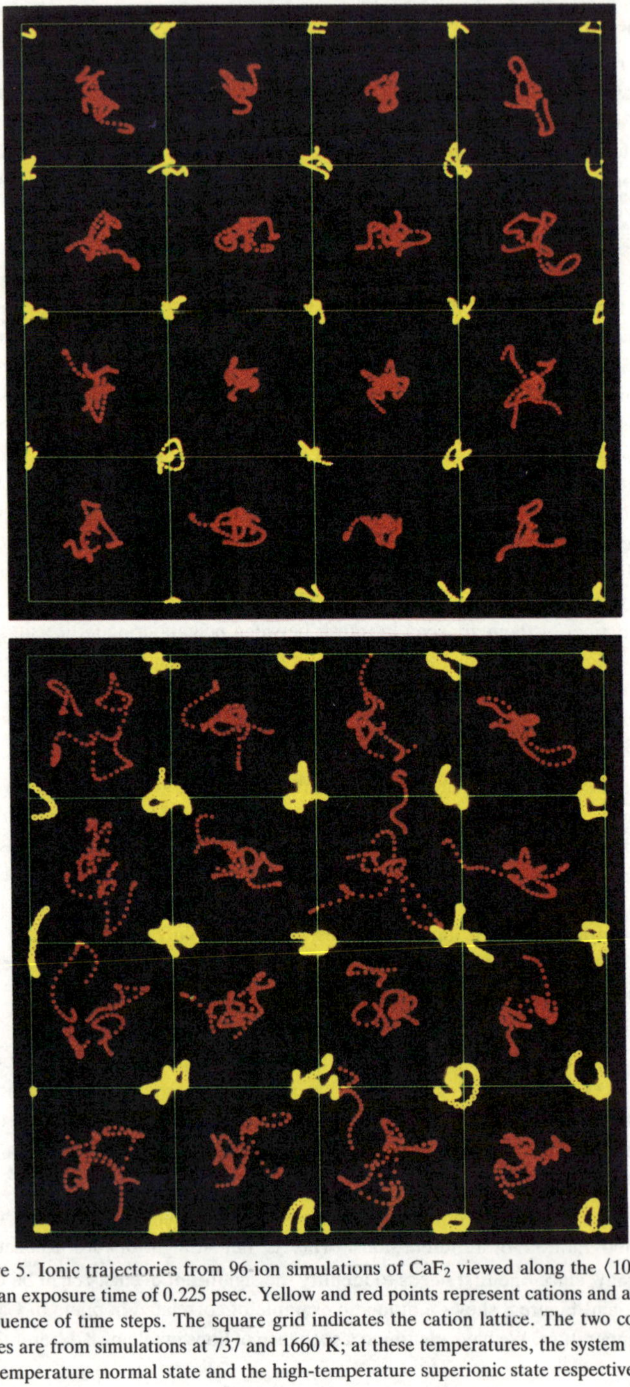

Figure 5. Ionic trajectories from 96 ion simulations of CaF_2 viewed along the $\langle 100 \rangle$ axis, with an exposure time of 0.225 psec. Yellow and red points represent cations and anions at a sequence of time steps. The square grid indicates the cation lattice. The two computer images are from simulations at 737 and 1660 K; at these temperatures, the system is in the low-temperature normal state and the high-temperature superionic state respectively.

Figure 6. Ionic trajectories from a 96-ion simulation of CaF_2 viewed along the $\langle 100 \rangle$ axis, with an exposure time of 3 psec. Colours have the same meaning as in Figure 5. The mean temperature in the simulation is 1657 K.

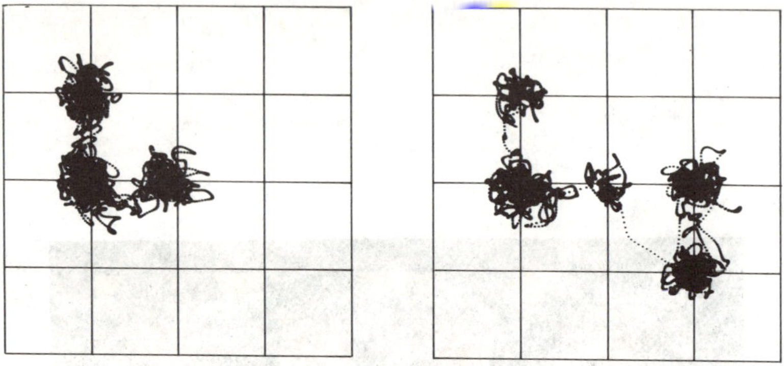

Figure 7. Trajectories followed by two randomly chosen anions in a simulation of PbF_2 at 991 K over a time of 39.6 psec. The square grid shows the lattice of regular anion sites.

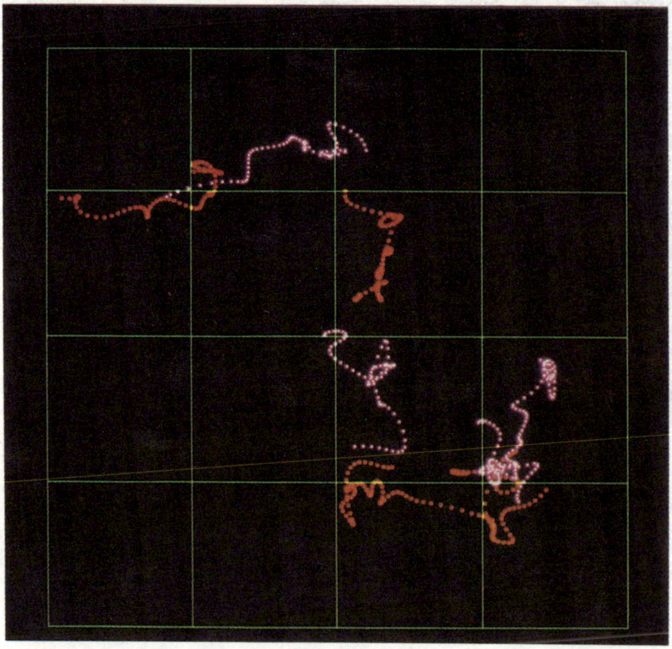

Figure 8. Correlated trajectories of jumping anions in simulated CaF_2 at 1657 K over a time interval of 0.6 psec. The computer image shows the trajectories of eight different anions, chosen because they are known from the defect analysis (§ 11) to make inter-site jumps during the given time interval. The anions are distinguished by different colours for clarity of presentation.

and plotted the trajectories of these ions alone over a time of 0.6 psec. In order to distinguish the ions, we show them in different colours.

We end this section on a cautionary note. Computer graphics gives a very valuable intuitive view of what is happening in the system. But because it tends to be unsystematic, it is somewhat akin to travellers' tales. It suggests what might be characteristic of the system, but there is always the danger of being misled by untypical events. This must be true because of the statistical nature of the system. Almost anything can happen, so long as it is allowed by the equation of motion – but what is interesting is what happens on the average. Computer graphics should therefore be treated as an adjunct to the physical analysis of the simulation. This means the calculation of physical quantities by statistical averaging, which is needed not only to arrive at quantitative statements, but also to make contact with experiment. How this works in practice will become clear in the following sections.

5. The Diffusion Coefficient

One of the most important quantities for a superionic conductor is the tracer diffusion coefficient of the mobile species. This can be calculated from the time-dependent positions generated in a simulation. Suppose that at some arbitrary time t_0 a chosen ion is found to be at position $\mathbf{r}(t_0)$. It wanders away from this position, and after a time t arrives at a second position $\mathbf{r}(t_0+t)$. The mean square value of $\mathbf{r}(t_0+t) - \mathbf{r}(t_0)$ for a given time difference t will be denoted by $\langle \Delta r_\alpha(t)^2 \rangle$ in the case of ions of species α; we refer to it as the mean square displacement (m.s.d.). It is an important quantity, because for diffusing particles it increases linearly in t at long times, and the constant of proportionality is 6 times the diffusion coefficient (Hansen and McDonald 1986):

$$\langle \Delta r_\alpha(t)^2 \rangle \to B_\alpha + 6D_\alpha |t| \ , \tag{30}$$

where D_α is the tracer diffusion coefficient for species α. For ions bound to fixed sites, $\langle \Delta r_\alpha(t)^2 \rangle$ goes to the constant B_α, which in this case is equal to twice the vibrational mean square displacement.

We calculate the m.s.d. by systematically averaging the values of $|\mathbf{r}(t_0+t) - \mathbf{r}(t_0)|^2$ for given t over time 'origins' t_0 and over the ions of each species. Suppose we wish to compute $\langle \Delta r_\alpha(t)^2 \rangle$ by averaging over a section of simulation containing P steps $t_1,...t_P$. Then if the time difference t corresponds to s steps ($t = s\Delta t$), all the steps $t_1,...t_{P-s}$ can be used as time origins, so that the average is constructed as

$$\langle \Delta r_\alpha(t)^2 \rangle = \frac{1}{(P-s)N_\alpha} \sum_{u=1}^{P-s} \sum_{i=1}^{N_\alpha} |\mathbf{r}_{\alpha i}(t_{u+s}) - \mathbf{r}_{\alpha i}(t_u)|^2 \ . \tag{31}$$

Here and in the following, it will usually be unnecessary to distinguish between the different periodic images of each ion, and we use the shortened notation $\mathbf{r}_{\alpha i}$ for the positions. When applying this formula in practice, one must, of course, keep track of crossings of the cell boundaries.

Provided the necessary storage capacity is available, it is convenient to calculate the m.s.d. in the course of the simulation. If the maximum required t corresponds to S steps, and we want to use every step as a time origin, this means having available at each step the positions of all the ions for the preceding S steps. In fact, we lose little if instead of making every step a time origin, we take time origins at intervals of $n\Delta t$, where n is a smallish integer (say 5). This allows us to go to large enough S without demanding exorbitant amounts of storage.

The diffusion coefficients can equivalently be obtained by calculating the velocity autocorrelation function (v.a.f.) $z_\alpha(t)$ (Hansen and McDonald). This is defined, for ions of species α, by

$$z_\alpha(t) = \langle \mathbf{v}_{\alpha i}(t_0+t)\cdot\mathbf{v}_{\alpha i}(t_0)\rangle \ , \tag{32}$$

for any ion of this species. There is a close connection between $z_\alpha(t)$ and the m.s.d: it follows directly from the definition of $\langle \Delta r_\alpha(t)^2\rangle$ that

$$z_\alpha(t) = \tfrac{1}{2}\,\frac{d^2}{dt^2}\,\langle \Delta r_\alpha(t)^2\rangle \ . \tag{33}$$

From this and equation (30), it follows that

$$D_\alpha = \tfrac{1}{3}\lim_{t\to\infty}\int_0^t dt'\ z_\alpha(t') \ . \tag{34}$$

We can therefore obtain D_α from the v.a.f. by calculating the time integral of the latter for large t. In practice, the v.a.f. is calculated, just like the m.s.d., by averaging the product of velocities in equation (32) over time origins t_0 and over the ions of each species. In our own calculations, we generally prefer to derive the D_α *via* equation (30). The relation (34) will, however, be important when we come to discuss the electrical conductivity.

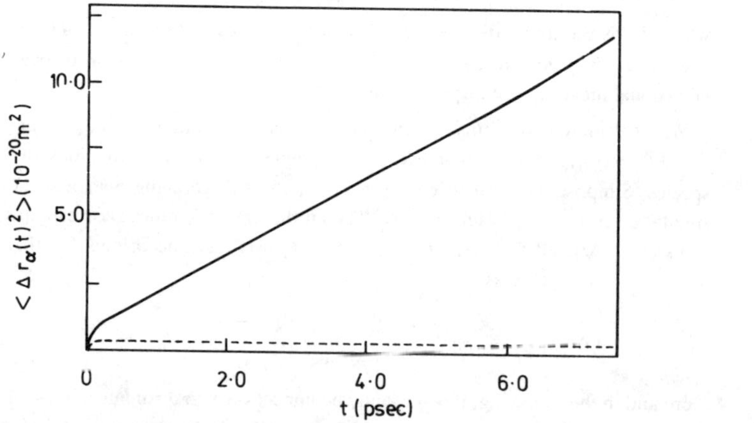

Figure 9. The mean square displacement $\langle \Delta r_\alpha(t)^2\rangle$ (see equation (31)) for cations (broken line) and anions (full line) from a simulation of CaF_2 at 1529 K. The simulation was performed on a system of 324 ions, and had a duration of 35 psec.

As an illustration, we show in Figure 9 the m.s.d. for cations and anions calculated in a long simulation of CaF_2 at 1609 K (Gillan 1986a), which is some 200 K above the experimental transition temperature T_c. The very different behaviour of the two ionic species is clear. For the cations, the m.s.d. attains a constant value after a few oscillation periods. This shows that the cation diffusion coefficient D_+ is completely negligible, which is what we expect. The anion m.s.d. rapidly settles down to a linear dependence on time. The slope gives us the anion tracer diffusion coefficient D_- *via* equation (30). The value we obtain is 3.4×10^{-5} cm^2 sec^{-1}, which is a typically liquid-like value, as we expect for CaF_2 above T_c.

It might possibly seem surprising that the linear regime of $\langle \Delta r_-(t)^2 \rangle$ is attained so quickly. After all, D_- is a *macroscopic* quantity, characterizing diffusion over long distances, and one might think that linear behaviour would only become established after a time long enough for the ions to diffuse several lattice spacings. The simulation results show that this is not so. We shall see later that the anion diffusion occurs essentially by a random walk on the regular anion sublattice. This makes our result less surprising. At any instant, some of the ions will be preparing to jump, or in the process of doing so, and the diffusive behaviour will become well characterized in a time comparable with the mean flight time – the time spent in transit between the sites. As we shall see, this is only a few vibrational periods.

Recently, we have calculated D_- in a series of long simulations of CaF_2 over a wide range of temperatures (Gillan 1986). Unfortunately, accurate measurements of D_- for CaF_2 at high temperature are not available. However, we can deduce a rough estimate of D_- from the electrical conductivity measurements of Derrington *et al.* (1975) by using the Nernst-Einstein relation. This expresses D_- as

$$D_- = H_R k_B T \sigma_0 / \rho_- z_-^2 e^2 \ , \qquad (35)$$

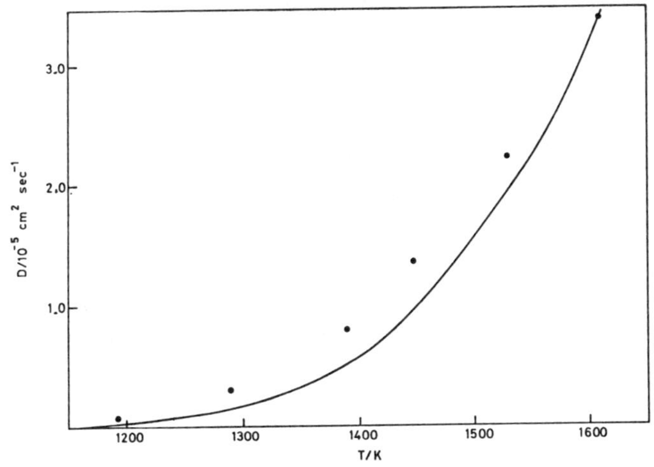

Figure 10. Anion diffusion coefficient D_- calculated from simulations of CaF_2 (solid circles) compared with values deduced *via* the Nernst-Einstein relation from conductivity measurements of Derrington *et al.* (1975). From Gillan (1986a).

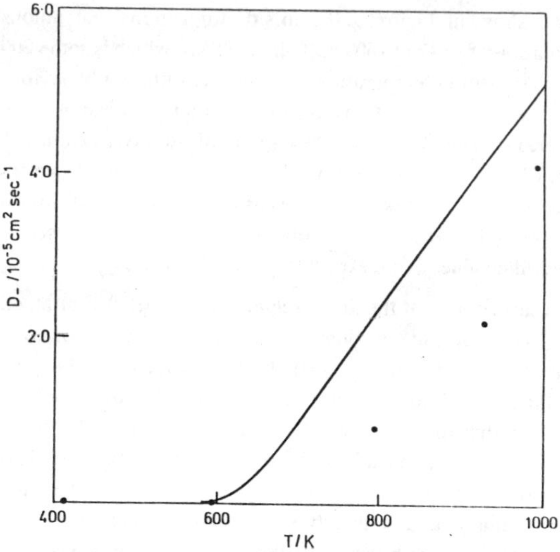

Figure 11. Anion diffusion coefficient D_- calculated from simulations of PbF_2 (solid circles), compared with experimental n.m.r. data of Gordon and Strange (1978). From Walker *et al.* (1982).

where σ_0 is the electrical conductivity, $\bar{\rho}_-$ is the bulk number density of anions, and H_R is the so-called Haven ratio (Compaan and Haven 1956a,b, Lidiard 1957). The latter would be equal to unity if there were no correlation between the diffusion of different ions (see §9.1). A comparison of the simulated values of D_- with 'experimental' ones obtained by setting $H_R = 1$ in equation (35) is presented in Figure 10. The good agreement confirms that the simulated system is a fairly faithful representation of the real material. The agreement is, in fact, not quite so significant as it appears, for two reasons: firstly, because the experimental results themselves are not of high accuracy, as noted by the experimentalists themselves; and secondly, because H_R is not equal to unity. Our simulation results for σ_0 described later suggest that H_R is more like 0.5. Nevertheless, the comparison shown in the figure leaves no doubt that the simulations are semi-quantitatively reliable.

We have reported a similar comparison for PbF_2 (Walker *et al.* 1982), where the experimental situation is more satisfactory. The fairly close agreement (Figure 11) is particularly gratifying in this case, in view of the difficulty of constructing satisfactory potentials for PbF_2 alluded to in §2.2.

6. The Diffusion Process

The calculation of the diffusion coefficient tells us how fast the ions are diffusing, but it tells us nothing about the manner in which they are diffusing. We should like to know whether diffusion occurs by well-defined jumps between sites, and about the directions of such jumps. One way of doing this is by examining trajectories, as we have discussed in §4. The results we presented there strongly suggested that the anions diffuse mainly by jumping between regular sites and there was some indication that most of the jumps are between neighbouring sites in the cube-axis directions. But this approach is too subjective: there is too much scope for being misled by untypical events. We want to discuss now a more objective way of describing the diffusion process.

6.1 *The van Hove self-correlation function*

A convenient tool for this purpose is provided by the van Hove self-correlation function $G^s_\alpha(\mathbf{r},t)$ (Hansen and McDonald 1986). Instead of considering the mean square displacement of ions, we now ask for the probability that the displacement occurring in time t lies in the volume element $d\mathbf{r}$ centred on $\mathbf{r}$. We call this $G^s_\alpha(\mathbf{r},t)d\mathbf{r}$ for ions of species α. The definition of $G^s_\alpha(\mathbf{r},t)$ is thus:

$$G^s_\alpha(\mathbf{r},t) = \langle \delta(\mathbf{r} - \mathbf{r}_{\alpha i}(t_0+t) + \mathbf{r}_{\alpha i}(t_0)) \rangle \tag{36}$$

for any ion of this species. As in the definition of the m.s.d., the time origin t_0 is arbitrary. From this definition, we see that the m.s.d. is just the spatial second moment of $G^s_\alpha(\mathbf{r},t)$:

$$\langle \Delta r_\alpha(t)^2 \rangle = \int d\mathbf{r}\, r^2\, G^s_\alpha(\mathbf{r},t) \ . \tag{37}$$

Since $G^s_\alpha(\mathbf{r},t)$ describes the distribution of vector displacements, it contains enough information to tell us something about jump lengths and directions.

In practice, it turns out to be more convenient to work not with G^s_α itself, but with its spatial Fourier transform, which we call $F^s_\alpha(\mathbf{q},t)$:

$$\begin{aligned} F^s_\alpha(\mathbf{q},t) &= \int d\mathbf{r}\, e^{i\mathbf{q}\cdot\mathbf{r}}\, G^s_\alpha(\mathbf{r},t) \\ &= \langle \exp[i\mathbf{q}\cdot(\mathbf{r}_{\alpha i}(t+t_0) - \mathbf{r}_{\alpha i}(t_0))] \rangle \ , \end{aligned} \tag{38}$$

where the second equation follows by substitution of equation (36).

One reason why this quantity is more useful is that it is directly related to the cross-section measured in inelastic neutron scattering. For a detailed discussion of neutron scattering from superionic conductors, we refer the reader to the article by Hayes and Hutchings in this book. We recall briefly (see e.g. Lovesey 1984) that the scattering cross-section consists in general of two parts. The *coherent* part comes from the interference of scattered waves from different nuclei, and contains information about correlations between different ions. The other part is a sum of *incoherent* scattering from the individual ions; it contains information only about correlations of each ion with itself. The incoherent part arises because the neutron scattering amplitude is not the same for all nuclei of a given ionic species: partly because it depends on the relative spin state of neutron and nucleus, and partly because different isotopes may be present. The function $F^s_\alpha(\mathbf{q},t)$ describes the correlation of an ion with itself at different times and is directly related to the double-differential cross-section for incoherent scattering, which is given by

$$(\partial^2\sigma/\partial\Omega\partial E)_{\rm inc} = \frac{k_f}{\hbar k_i}\sum_\alpha N_\alpha(\overline{b_\alpha^2} - \overline{b}_\alpha^2)S_\alpha^s(\mathbf{q},\omega) \ , \tag{39}$$

where k_i and k_f are the initial and final neutron wavevectors, $\overline{b}_\alpha$ and $\overline{b_\alpha^2}$ are the mean and mean square scattering lengths and N_α is the number of ions of species α in the specimen; the incoherent structure factor $S_\alpha^s(\mathbf{q},\omega)$ is given by

$$S_\alpha^s(\mathbf{q},\omega) = \frac{1}{2\pi}\int_{-\infty}^{\infty} dt\, e^{i\omega t}\, F_\alpha^s(\mathbf{q},t) \ , \tag{40}$$

with $\hbar\mathbf{q}$ and $\hbar\omega$ the momentum and energy transfers.

We now want to consider in more detail how $F_\alpha^s(\mathbf{q},t)$ gives us information about the diffusion process. A useful way of thinking about this is provided by an analysis given many years ago by Chudley and Elliott (1961). The essential ideas are simple enough to summarize here. Suppose we ignore all vibrational motion, and assume that diffusion occurs by instantaneous jumps of the anions between nearest-neighbour sites on the simple-cubic lattice of regular anion sites. The separations $\mathbf{r}_i(t_0+t) - \mathbf{r}_i(t_0)$ must then be translation vectors $\mathbf{R}$ of the lattice, and $G_-^s(\mathbf{r},t)$ consists of δ-functions at these separations. The weight of the δ-function at $\mathbf{R}$, which we call $p_\mathbf{R}(t)$ is just the probability that in a time period t an anion has arrived at a site separated by $\mathbf{R}$ from its site at the beginning of the period. We can write down an equation of motion for $p_\mathbf{R}(t)$ if we assume that successive jumps of the chosen anion are uncorrelated with each other:

$$dp_\mathbf{R}/dt = \gamma\,[\sum_\boldsymbol{\delta} p_{\mathbf{R}+\boldsymbol{\delta}} - zp_\mathbf{R}] \ , \tag{41}$$

where γ is the rate of hopping from a given site to any one of its neighbours, $\boldsymbol{\delta}$ are the separation vectors of the neighbours, and z is the coordination number of the lattice (6 in the present case). This equation just expresses the rate of change of $p_\mathbf{R}$ due to hopping from the neighbouring sites at $\mathbf{R}+\boldsymbol{\delta}$ and due to hopping to these sites.

Now $F_-^s(\mathbf{q},t)$, being the Fourier transform of $G_-^s(\mathbf{r},t)$, is given by

$$F_-^s(\mathbf{q},t) = \sum_\mathbf{R} e^{i\mathbf{q}\cdot\mathbf{R}}\, p_\mathbf{R}(t) \ , \tag{42}$$

and its equation of motion follows directly from equation (41):

$$dF_-^s(\mathbf{q},t)/dt = -\Gamma(\mathbf{q})\, F_-^s(\mathbf{q},t) \ , \tag{43}$$

where

$$\Gamma(\mathbf{q}) = \gamma\sum_\boldsymbol{\delta}(1 - \cos\mathbf{q}\cdot\boldsymbol{\delta}) \ . \tag{44}$$

From equation (42), we have

$$F_\alpha^s(\mathbf{q},t{=}0) = 1 \ , \tag{45}$$

so that the solution to equation (43) is

$$F_-^s(\mathbf{q},t) = \exp(-\Gamma(\mathbf{q})|t|) \ . \tag{46}$$

The consequence of the lattice hopping model is thus that $F_-^s(\mathbf{q},t)$ decays exponentially. The $\mathbf{q}$–dependence of the decay rate $\Gamma(\mathbf{q})$ is determined by the directions and magnitudes of the jump vectors $\boldsymbol{\delta}$.

This result is easily generalized to the case where jumps occur not only to nearest-neighbour sites, but also to more distant sites (Rowe *et al.* 1971). If only first- and second-neighbour jumps are included, $F_-^s(\mathbf{q},t)$ is still given by equation (43), but $\Gamma(\mathbf{q})$ is now a sum of the terms for the two kinds of jump (Jacucci and Rahman 1978):

$$\Gamma(\mathbf{q}) = \gamma_1 \sum_{\boldsymbol{\delta}_1} (1 - \cos \mathbf{q}\cdot\boldsymbol{\delta}_1) + \gamma_2 \sum_{\boldsymbol{\delta}_2} (1 - \cos \mathbf{q}\cdot\boldsymbol{\delta}_2) \ , \tag{47}$$

where γ_j is the hopping rate from a site to any one of the jth-nearest neighbours and $\boldsymbol{\delta}_j$ are the corresponding jump vectors. The effect of lattice vibrations can also be included. This modifies the form of $F_-^s(\mathbf{q},t)$ at short times, but its long-time behaviour is still given by equation (46), but multiplied by a vibrational Debye-Waller factor (Chudley and Elliott 1961).

One now asks whether this inter-site jumping does in fact describe the diffusion process in fluorites. This question was examined using simulation by Jacucci and Rahman (1978) for a model of CaF_2 in the superionic regime. Their results for $F_-^s(\mathbf{q},t)$ for wavevectors $\mathbf{q}$ in the $\langle 100 \rangle$ direction are shown on a logarithmic plot in Figure 12. The linearity in t for long times reveals the exponential behaviour predicted by equation (46). An important consequence of the jump model is that $\Gamma(\mathbf{q})$ should be the same for wavevectors that differ by a reciprocal lattice vector of the jump-site lattice. Jacucci and Rahman find that this relation is very well satisfied. They then go on to test whether the $\mathbf{q}$–dependence of the simulated $\Gamma(\mathbf{q})$ can be satisfactorily described by the Chudley-Elliott formula (47) including first- and second-neighbour jumps (i.e. jumps in the $\langle 100 \rangle$ and $\langle 110 \rangle$ directions). They find that the formula gives a very good fit to the results. The fractions of jumps that are to first and second neighbours are 0.79 and 0.21 respectively.

Exactly the same procedure can be carried through experimentally, as discussed in the article by Hayes and Hutchings in this volume. If we insert equation (46) into (40), we have

$$S_\alpha^s(\mathbf{q},\omega) = \pi^{-1}\Gamma(\mathbf{q})/(\Gamma(\mathbf{q})^2 + \omega^2) \ , \tag{48}$$

which, by equation (39), indicates that there is a quasielastic (i.e. zero-frequency) peak in the incoherent neutron-scattering cross-section, whose frequency width gives us the rate $\Gamma(\mathbf{q})$. In order for this to be measurable, the anion nucleus must have a sizeable incoherent cross-section, which unfortunately rules out fluoride compounds. However, Dickens *et al.* (1983) have succeeded in making measurements of $\Gamma(\mathbf{q})$ for $SrCl_2$ in the superionic regime. As expected from the results of Jacucci and Rahman (1978), they find that a very satisfactory fit to the $\mathbf{q}$–dependence of $\Gamma(\mathbf{q})$ can be achieved with the Chudley-Elliott model for jumps

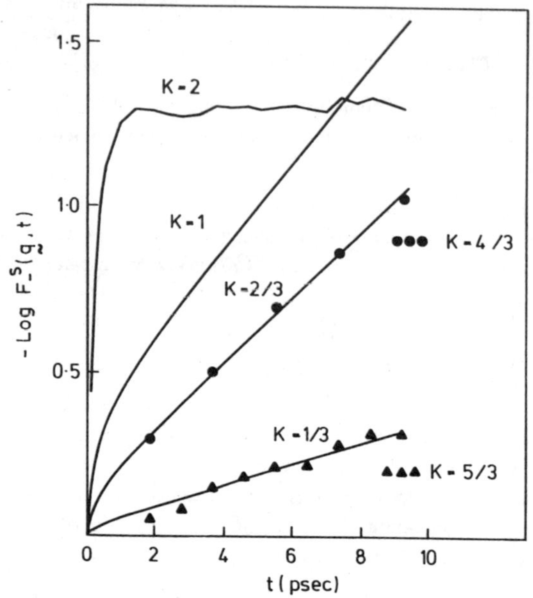

Figure 12. The anion incoherent scattering function $F_{-}^{s}(\mathbf{q},t)$ calculated from a simulation of superionic CaF_2 at 1590 K (Jacucci and Rahman 1978). The curves show results for wavevectors $\mathbf{q} = (2\pi/a_0)\,(\kappa,0,0)$ in the $\langle 100\rangle$ direction. Solid curves show results for $\kappa =$ 1/3, 2/3, 1 and 2; circles and triangles show results for κ = 4/3 and 5/3. The plots give $F_{-}^{s}(\mathbf{q},t)$ on a logarithmic scale, so that their asymptotic linearity shows the exponential decay of the function.

on the simple-cubic lattice of regular anion sites. They deduce that a fraction 0.73 of jumps are between nearest neighbours and 0.27 between second neighbours. In comparing with simulation, it should be remembered that the latter is for a model of CaF_2 rather than $SrCl_2$. Nevertheless, the close agreement regarding the fractions of jump types is striking, and gives important support for the validity of the simulations. As we stress below, the evidence is that the diffusion process has essentially the same characteristics in all fluorite materials.

6.2 *The hopping analysis*

There is another approach to the diffusion process that is illuminating, and will turn out to play an important role in the later discussion. The idea here is to analyze the trajectories themselves so as to identify the hopping events. The method described here is due to Dixon and Gillan (1978); a very similar scheme was developed independently by Jacucci and Rahman (1978).

The procedure rests on the fact that, since the cations do not diffuse, the position of the lattice of cation sites is well defined; it can be determined, for example, by averaging the positions of the cations over many vibrational periods (Gillan and Dixon 1980). From this, the position of the simple cubic lattice of anion regular sites follows. We now suppose that the anion motion consists of vibration about these sites and jumps between them. The correctness of this supposition, already suggested by trajectory plotting (Figure 7), and corroborated by the results for the van Hove function, is demonstrated by the present analysis. We define inter-site jump events using a simple geometrical construction. On each anion site we centre a sphere of radius λd, where d is the distance between neighbouring sites and λ is some number $\leq \frac{1}{2}$, so that the spheres do not overlap. A jump occurs when an anion leaves one sphere and enters another. In choosing a suitable value for λ, we note that if it is taken too large, say $\lambda = \frac{1}{2}$, then there will be large vibrational excursions which are counted as a jump immediately followed by a return jump. On the other hand, if it is too small, then an anion may execute many oscillations about a site without ever entering the sphere on that site. In practice, we generally take $\lambda = \frac{1}{3}$.

Using this definition, we now analyze the ion trajectories generated by simulation to construct a catalogue of all the jump events. Each event is characterized by the initial and final sites and the times of leaving the first sphere and entering the second. It is then straightforward to count the numbers of jumps between first-neighbour, second-neighbour sites, etc. We have carried through this procedure in our simulations of CaF_2, $SrCl_2$ and PbF_2 (Dixon and Gillan 1978, 1980a, Walker *et al.* 1982). The results for CaF_2 are typical, and particularly relevant because the potential model used was the same as that of Jacucci and Rahman, and the temperature was almost the same as in their simulation. We found jumps only between first–, second- and third-neighbour sites, the proportions being 0.89 : 0.08 : 0.03. The fairly close agreement with the values that emerge from their analysis of the van Hove function confirms that the same physical process is being described in the two approaches. The present hopping analysis shows more clearly, though, that hopping takes place between the regular sites.

The Chudley-Elliott analysis assumes that the jumps are instantaneous. More correctly, it requires that the flight time τ_f between sites be much smaller than the time τ_r for which the ions reside on the sites. These two times can be estimated directly from the catalogue of hops. They depend on temperature, of course, and we should expect that τ_r^{-1} would increase more rapidly with temperature than τ_f^{-1}, since τ_r^{-1} will be roughly proportional to the diffusion coefficient. The condition $\tau_r \gg \tau_f$ will therefore become less well satisfied at high temperature, and it is most important to test it in this regime. Representative results are given by Gillan and Dixon (1980b), who find for simulated $SrCl_2$ at 1522 K: $\tau_f = 0.61$ psec, $\tau_r = 7.1$ psec. This confirms the validity of the assumption of instantaneous jumps.

Very recently, this same hopping analysis has been used by Brass (1987) to study the *distribution* of transit times in simulations of CaF_2, $SrCl_2$ and PbF_2, using very

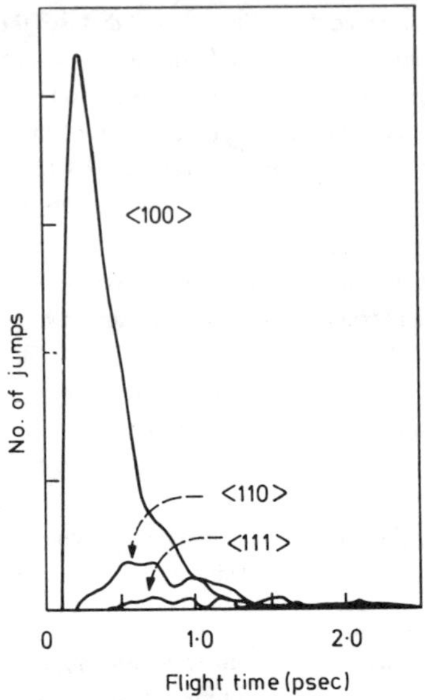

Figure 13. Distribution of anion flight times from a simulation of superionic CaF_2 at 1495 K (Brass 1987). The flight time is defined as the time taken for an anion to pass from within $a_0/6$ of one regular site to within $a_0/6$ of another regular site. Separate distributions are shown for jumps in the $\langle 100 \rangle$, $\langle 110 \rangle$ and $\langle 111 \rangle$ directions.

large systems of 1210 cations and 2420 anions. We reproduce in Figure 13 his results for this distribution in CaF_2 at 1495 K. The peak of the distribution occurs at the very short time of 0.3 psec.

The hopping catalogue also gives an alternative method of calculating the tracer diffusion coefficient D_- (Gillan and Dixon 1980). From the different numbers of jumps of different kinds, we obtain the single-ion jump rates. If we assume that different jumps of a given ion are uncorrelated, we can express D_- as

$$D_- = \tfrac{1}{6}(z_1 \gamma_1 \delta_1^2 + z_2 \gamma_2 \delta_2^2 + \ldots) , \tag{49}$$

where z_j is the number of jth neighbours and δ_j is the corresponding jump distance. This method has been studied in detail by Brass (1987), who finds that it systematically overestimates D_- by about 10 % compared with the value obtained from the m.s.d., this being, of course, the correct value. This difference indicates that the assumption of uncorrelated jumps is not fully justified. This is no surprise, since the correlation of successive jumps is a familiar phenomenon in defect-assisted diffusion.

To summarize this section: The neutron-scattering measurements and the two methods of analyzing the simulations point to a clear and simple picture of the diffusion process. Diffusion occurs by discrete well-defined jumps between the regular anion sites. The flight time between sites is very short compared with the time spent in vibration about each site. Most jumps (~ 80 %) are in the ⟨100⟩ direction between nearest-neighbour sites, but there is a sizeable fraction (~ 20 %) in the ⟨110⟩ direction between second neighbours. The total fraction of third-neighbour and other jumps is very small.

7. Spatial Distribution of the Ions

We have concluded from the hopping analysis that even in the superionic state the anions spend almost all their time vibrating about their regular sites. There is an alternative way of studying this question, and that is by calculating the spatial distribution of ions over the unit cell. Through this calculation, we also make contact with diffraction measurements.

We are interested here in the probability of finding ions of each species in different regions of the unit cell, or equivalently the proportion of their time spent by the ions in different regions. We can study this using the densities $\rho_\alpha(\mathbf{r})$ defined by

$$\rho_\alpha(\mathbf{r}) = \langle \sum_{i=1}^{N_\alpha} \delta(\mathbf{r} - \mathbf{r}_{\alpha i}) \rangle \quad , \tag{50}$$

where the sum goes over all N_α ions of species α. The probability of finding some ion of type α in the volume element $d\mathbf{r}$ centred on $\mathbf{r}$ is then $\rho_\alpha(\mathbf{r})d\mathbf{r}$. Because of the averaging implied in equation (50), $\rho_\alpha(\mathbf{r})$ is a periodic function of $\mathbf{r}$ having the periodicity of the crystal lattice: on the average, any crystal unit cell is the same as any other.

In practice, $\rho_\alpha(\mathbf{r})$ is calculated as a three-dimensional histogram. We cover the simulation cell with a fine cubic mesh. We then go through the record of MD positions step by step and count the total number of times that ions are found in each of the mesh elements. Appropriately normalized, this histogram gives a discrete representation of $\rho_\alpha(\mathbf{r})$. In detail, the calculation is somewhat more complicated than this. The statistical quality of the results is greatly improved if one makes full use of the lattice symmetry, but the technicalities of this need not concern us here.

As a preliminary illustration of the results, we show in Figure 14 contour plots of the anion density on a [110] plane; these come from simulations of PbF_2 (Walker *et al.* 1982) at two temperatures, one well below and one well above the transition temperature 711 K. At the low temperature, both cations and anions are of course strongly localized on their regular sites. The spread of the anion distribution is much greater, as suggested by the trajectory plots, and is also markedly anisotropic. Both these effects have been well known experimentally for many years. The anisotropy is expected because the anion sites have tetrahedral symmetry. It is striking that the high temperature plot is qualitatively similar. The anion distribution is still strongly concentrated on the regular site, though it is broader than at the lower temperature. The onset of superionic conduction appears not to lead to any dramatic qualitative change in the distribution of anions.

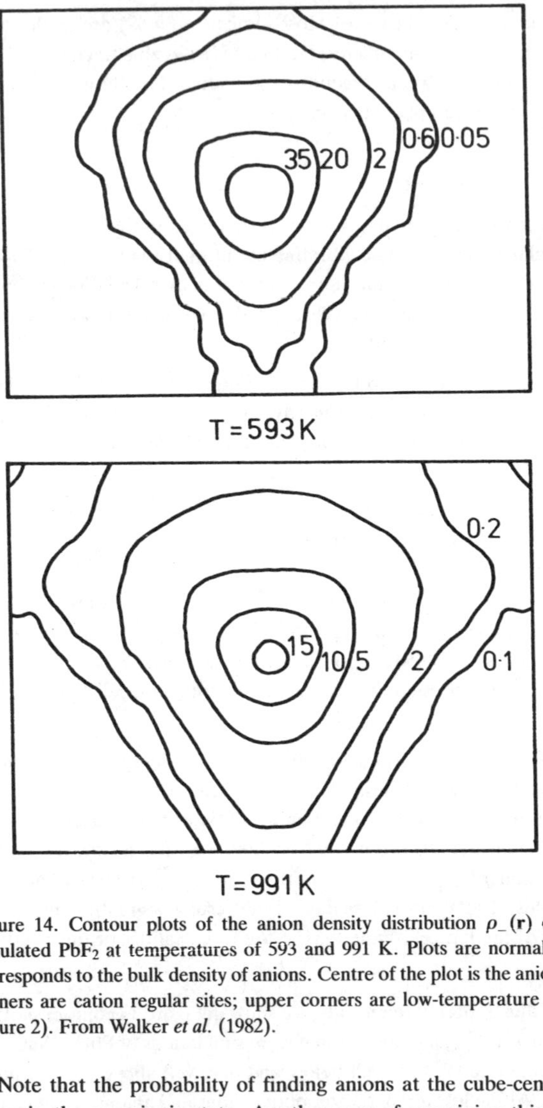

Figure 14. Contour plots of the anion density distribution $\rho_-(\mathbf{r})$ on a (110) plane in simulated PbF_2 at temperatures of 593 and 991 K. Plots are normalized so that $\rho_- = 1$ corresponds to the bulk density of anions. Centre of the plot is the anion regular site; lower corners are cation regular sites; upper corners are low-temperature interstitial sites (see Figure 2). From Walker *et al.* (1982).

Note that the probability of finding anions at the cube-centre site is very low even in the superionic state. Another way of examining this is to calculate the radial distribution of anions with respect to this site. Let the probability of finding an anion at a distance between r and $r + dr$ of a chosen cube-centre site be written as $4\pi r^2 \bar{\rho}_- g_1(r)dr$, where $\bar{\rho}_-$ is the bulk number density of anions. Calculations of this have been made for simulated CaF_2, using both a rigid-ion model and a shell model (Dixon and Gillan 1980b); the results are shown in Figure 15. For the rigid-ion model, two sizes of system were used. The results from the different simulations are in close agreement, and show very clearly that the distribution

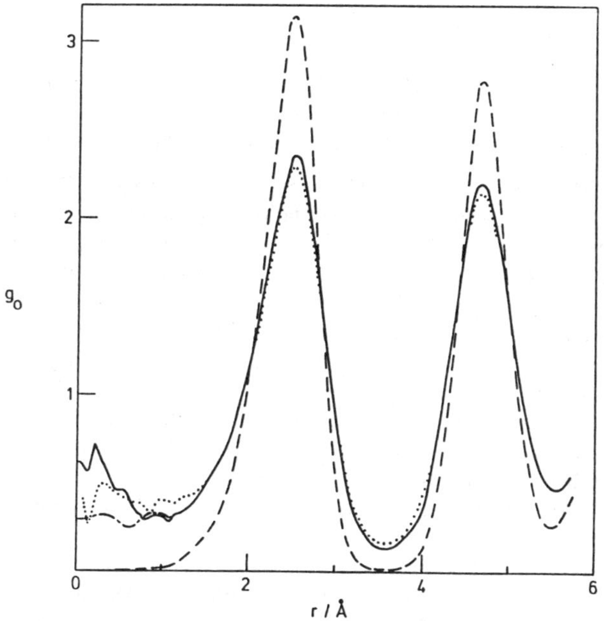

Figure 15. Radial distribution function of anions with respect to the cube-centre site (see text). Chain, solid, dotted and broken curves show results from the following simulations: 96-particle rigid-ion at 1651 K; 324-particle rigid-ion at 1585 K; 96-particle shell-model at 1639 K; 324-particle rigid-ion at 1277 K. Chain and dotted curves are omitted where they coincide with the solid curve. From Dixon and Gillan (1980b).

around the cube-centre is very flat and has a small value. This is an important point, because the question of cube-centre occupation has been controversial. We shall return to this shortly.

The densities are important, because they are related to the intensities measured in X-ray or neutron diffraction experiments. The diffraction intensity $I(\mathbf{G})$ at reciprocal lattice vector $\mathbf{G}$ is given, apart from an overall scale factor, by

$$I(\mathbf{G}) = |\sum_{\alpha} f_{\alpha}(\mathbf{G}) F_{\alpha}(\mathbf{G})|^2 \ , \tag{51}$$

where $f_{\alpha}(\mathbf{G})$ are the (wave-vector-dependent) form factors in the case of X-rays, or the (wave-vector-independent) scattering lengths in the case of neutrons. The $F_{\alpha}(\mathbf{G})$ are the elastic scattering amplitudes, which are the Fourier transforms of the densities $\rho_{\alpha}(\mathbf{r})$:

$$F_{\alpha}(\mathbf{G}) = \int_{\text{cell}} d\mathbf{r}\, e^{i\mathbf{G}\cdot\mathbf{r}}\, \rho_{\alpha}(\mathbf{r}) \ , \tag{52}$$

where the integration goes over a unit cell of the crystal. If we substitute equation (50) into (52), we find that

$$F_\alpha(\mathbf{G}) = N_{\text{cell}}^{-1} \langle \sum_{i=1}^{N_\alpha} \exp(i\mathbf{G}\cdot\mathbf{r}_{\alpha i}) \rangle \ , \tag{53}$$

where N_{cell} is the number of unit cells in the simulated system. We can use equation (53) to calculate the $F_\alpha(\mathbf{G})$ by simply averaging the quantity in angular brackets over the time steps of the simulation for the **G** of interest.

The most direct way of comparing simulation with experiment would be in terms of the $I(\mathbf{G})$. However, this would be rather uninformative, since the $I(\mathbf{G})$ have little intuitive meaning. It is therefore preferable to compare the $\rho_\alpha(\mathbf{r})$. Unfortunately, the inversion of diffraction data to obtain the $\rho_\alpha(\mathbf{r})$ is not straightforward because (i) the $F_\alpha(\mathbf{G})$ enter $I(\mathbf{G})$ in the form of a weighted sum; (ii) measurements cannot always be made for an adequate set of **G**; (iii) experimental corrections may not be fully under control. Some of the problems are discussed by Hayes and Hutchings. In spite of this, a detailed and painstaking analysis of diffraction data for superionic PbF_2 has been presented by Dickens *et al.* (1982), which appears to give enough information to yield the $\rho_\alpha(\mathbf{r})$ themselves. The procedure is to set up a parameterized model for the $\rho_\alpha(\mathbf{r})$ and perform a least-squares fit to the measured $I(\mathbf{G})$. The cation density $\rho_+(\mathbf{r})$ is modelled by a gaussian distribution on the cation site, representing harmonic vibrations of the cations. To represent the anion density, it is assumed that a certain fraction of the anions are displaced from their regular sites onto interstitial sites. The fraction remaining on regular sites is modelled by an isotropic gaussian multiplied by an anisotropic factor having tetrahedral symmetry. We have already referred to this anisotropy above. The interstitial contribution is modelled by a simple gaussian, and the position of the interstitial sites is also allowed to vary. A very satisfactory fit to the diffraction data is achieved for a wide range of temperatures. The model parameters can then be used to reconstruct the $\rho_\alpha(\mathbf{r})$.

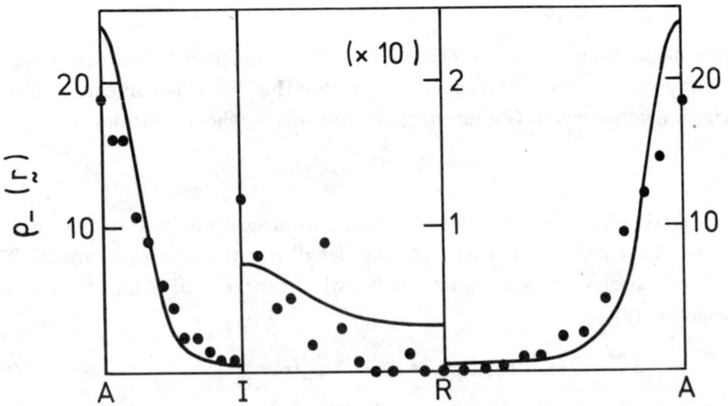

Figure 16. Comparison of anion density $\rho_-(\mathbf{r})$ for PbF_2 from simulation at 991 K (dots) and from diffraction measurements of Dickens *et al.* (1982) at 973 K (solid line). Points in the unit cell are: A = regular anion site, I = mid-point of cube edge, R = cube centre. Note 10-fold magnification of scale between I and R. From Walker *et al.* (1982).

A comparison between the experimental and simulated anion distribution for superionic PbF_2 has been presented by Walker *et al.* (1982). This comparison is reproduced in Figure 16. Given the difficulty of constructing a satisfactory potential model for PbF_2 discussed in §2.2, the agreement is very gratifying. These results emphasize once again that the anions are strongly localized on their regular sites and that the probability of finding them near the cube-centre position is small. (see also Shapiro and Reidinger 1979, Dickens *et al.* 1979, Koto *et al.* 1980). On the other hand, the probability is quite significant on the path connecting neighbouring anion sites. The trajectory plots of §4 suggest that this is largely associated with ions in transit between sites, though there is also a contribution from vibrational excursions. Since we have seen that diffusion occurs mainly by jumps between nearest-neighbour sites, this is what we should expect.

In this connection, it should be stressed that although the experimental fitting model employs interstitial sites, the success of the procedure does not necessarily mean that the sites have any physical significance (Walker *et al.* 1982). The fitting model makes it seem that the experimental density in the mid-site region is associated with an interstitial site. According to the simulations, this density

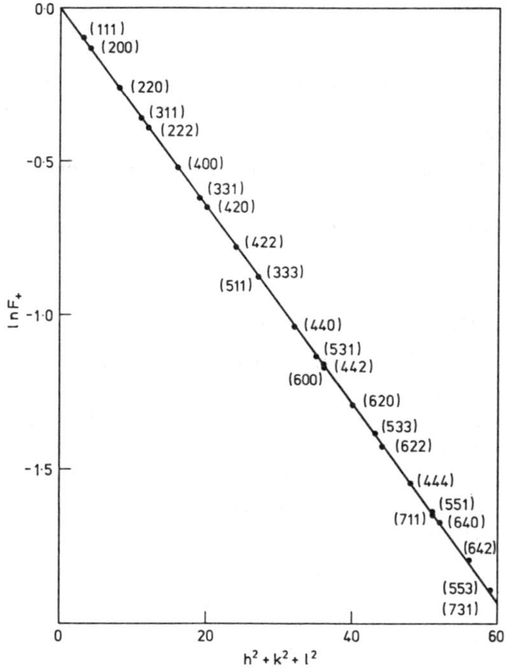

Figure 17. Cation elastic scattering amplitude $F_+(\mathbf{G})$ from simulation of PbF_2 at 991 K. The graph shows ln $F_+(\mathbf{G})$ plotted against $h^2 + k^2 + l^2$, where $\mathbf{G} = (2\pi/a_0)\,(h,k,l)$, each point being labelled by the triplet (h,k,l). The straight line is a guide to the eye. From Walker *et al.* (1982).

originates physically from ions jumping between regular sites. We suggest that the fitting model should be seen merely as a successful technical device for analyzing the diffraction measurements. A similar point has been made by Bachmann and Schulz (1983), who have studied alternative ways of analyzing the diffraction data.

One of the assumptions that enters the fitting model is that the cation distribution is gaussian and isotropic. Since the cation contribution to the diffraction intensities is large, and since the parts of interest in the anion distribution give only a small contribution, this assumption plays a rather important role. We can test it directly by simulation by calculating the cation amplitude $F_+(\mathbf{G})$ from equation (53). The gaussian assumption for $\rho_+(\mathbf{r})$ is exactly equivalent to saying that

$$F_+(\mathbf{G}) \sim \exp(-\tfrac{1}{6}\langle u_c^2 \rangle G^2) \ , \qquad (54)$$

where $\langle u_c^2 \rangle$ is the vibrational mean square displacement of the cations. We show in Figure 17 the results of Walker *et al.* (1982) for $F_+(\mathbf{G})$ in simulated PbF_2 at 991 K. This confirms that the gaussian assumption for the cations is valid to very high accuracy.

The main conclusion of this section is that the anion distribution is strongly concentrated on the anion regular sites. This is exactly what we should expect from the discussion of the previous section, where it was shown that the ions spend almost all their time vibrating about the regular sites. The anion distribution around the regular site has a strong tetrahedral anisotropy, as has long been known from diffraction work below T_c (Dawson *et al.* 1967, Cooper 1970). The most significant additional feature is the appearance of density along the path joining nearest-neighbour sites. This originates mainly from anions in transit between sites. The fairly close agreement between simulation and experiment for the anion distribution in superionic PbF_2 provides valuable support for the validity of the simulations.

8. Radial Distribution Functions

The quantities we have discussed up to now are single-particle quantities, in the sense that they could in principle be calculated from a sufficiently long trajectory of one particle of each species. They are influenced, perhaps strongly, by interparticle correlations, but they do not themselves describe these correlations. We now turn to an explicit consideration of the correlations between the ions. We start by discussing the radial distribution functions (r.d.f.) $g_{\alpha\beta}(r)$, which describe the probability of finding ions of type β at distance r from ions of type α.

We choose a particular ion of type α, and ask for the mean number of particles of type β at distances between r and $r + dr$. We write this as $4\pi r^2 \bar{\rho}_\beta g_{\alpha\beta}(r) dr$, which gives the definition of the r.d.f. $g_{\alpha\beta}(r)$. In practice, the r.d.f. are calculated in discrete form, by dividing the range of r into intervals and counting the numbers of interparticle distances which fall in each interval in the course of the simulation. The maximum value of r for this purpose is usually taken to be half the length of the simulation box.

We show in Figure 18 simulation results for the three $g_{\alpha\beta}(r)$ in CaF_2 at temperatures well below and well above T_c. At both temperatures, the cation subsystem is highly correlated: the peaks are tall and narrow, as one would expect for ions tightly bound to regular lattice sites. There is also a strong correlation between cations and anions. The most interesting r.d.f. is that for the anions. Even at the low temperature, the correlations between anions are markedly weaker than the other correlations. When we go to the high temperature, they become even weaker. The height of the first peak, for example, is somewhat lower than is found in simple liquids, like argon, at the triple point.

Large numbers of experimental measurements of the r.d.f. in liquids by neutron and X-ray scattering have been reported. In recent years, there has been a series of neutron-scattering experiments on molten salts, in which isotope substitution has been used to deduce the three separate $g_{\alpha\beta}(r)$ (Enderby and Neilson 1981). There

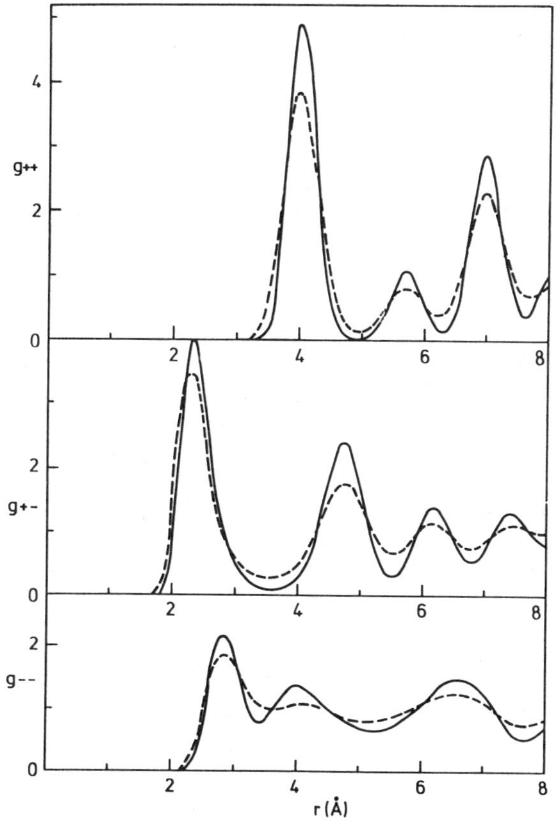

Figure 18. Radial distribution functions $g_{\alpha\beta}(r)$ from simulations of CaF_2 at 1193 K (solid lines) and 1611 K (broken lines).

seems no reason why the same thing could not be done for superionic $SrCl_2$, where the appropriate isotopes are available. This would allow a further useful way of comparing simulation with experiment.

9. Electrical Conductivity

We now want to consider the collective dynamical properties of superionic conductors. By a 'collective' property, we mean one that involves the correlated motion of ions. The frequency-dependent electrical conductivity is a quantity of this kind, since it involves the response of all the particles to an external perturbation applied to the system as a whole.

The conductivity $\sigma(\omega)$ is defined in the usual way as the coefficient of proportionality between the electric current density $\mathbf{j}$ and a weak uniform external electric field $\mathbf{E}(t) = \mathbf{E}_0 \exp(-i\omega t)$ of frequency ω:

$$\mathbf{j}(t) = \sigma(\omega)\mathbf{E}_0 \exp(-i\omega t) \ . \tag{55}$$

For cubic systems, like the fluorites, $\sigma(\omega)$ is a scalar. It is in general complex. Equivalently, we can work with the frequency-dependent dielectric function $\varepsilon(\omega)$, the relation between the two being

$$\varepsilon(\omega) = 1 + 4\pi i \sigma(\omega)/\omega \ . \tag{56}$$

We shall be particularly interested in the d.c. conductivity $\sigma_0 \equiv \sigma(0)$, and it will also be interesting to examine the static dielectric constant:

$$\varepsilon_0 = 1 - 4\pi \lim_{\omega \to 0} \mathcal{I}m\ \sigma(\omega)/\omega \ . \tag{57}$$

There are two approaches to the calculation of $\sigma(\omega)$. The most obvious is to apply an external electric field to the simulated system and to calculate the response of the electric current. An alternative method makes use of the fluctuation-dissipation theorem (Green-Kubo theory), which shows that $\sigma(\omega)$ can be expressed in terms of the spontaneous fluctuations of electric current in thermal equilibrium; this method does not involve the application of any external perturbation. For the fluorite superionics, we have found the second approach to be more useful, and we shall describe it first.

9.1 *The Green-Kubo method*

The Green-Kubo formula for $\sigma(\omega)$ is (Kubo 1957, 1966)

$$\sigma(\omega) = \frac{1}{3\Omega k_B T} \int_0^\infty dt\, e^{i\omega t} \, \langle \mathbf{J}_c(t) \cdot \mathbf{J}_c(0) \rangle \ , \tag{58}$$

where $\mathbf{J}_c$ is the fluctuating electric current associated with the motion of the ions:

$$\mathbf{J}_c = e \sum_\alpha z_\alpha \sum_{i=1}^{N_\alpha} \mathbf{v}_{\alpha i} \ , \tag{59}$$

with $\mathbf{v}_{\alpha i}$ the velocity of ion αi, and Ω the volume of the system. In order to see the physical meaning of this, it is instructive to think what happens for a perfect harmonic crystal. In this case, the electric current fluctuations correspond exactly to optic-mode vibrations at zero wavevector. With the periodic boundary conditions used in simulation, there is no long-range electric field, so that $\mathbf{J}_c$ will oscillate at the *transverse* optic frequency ω_T (a separate discussion is needed to show this). Therefore $\mathcal{Re}\ \sigma(\omega)$ will consist solely of δ-functions at $\omega = \pm\omega_T$:

$$\mathcal{Re}\ \sigma(\omega) = \tfrac{1}{8}\omega_p^2[\delta(\omega-\omega_T) + \delta(\omega+\omega_T)] \ , \tag{60}$$

where ω_p is the plasma frequency, which is defined by

$$\omega_p^2 = 4\pi e^2 \sum_\alpha \bar{\rho}_\alpha z_\alpha^2/m_\alpha \ . \tag{61}$$

Here we have used the fact that

$$\langle J_c^2 \rangle = e^2 \sum_\alpha z_\alpha^2 \sum_{i=1}^{N_\alpha} v_{\alpha i}^2 = 3\Omega k_B T \omega_p^2/4\pi \ , \tag{62}$$

since the velocities of different ions at any instant are uncorrelated. From this we deduce from equation (57) that the static dielectric constant is

$$\varepsilon_0 = 1 + \omega_p^2/\omega_T^2 \ , \tag{63}$$

which is a familiar formula for harmonic ionic crystals.

The practical calculation of the correlation function is straightforward: one has only to calculate $\mathbf{J}_c$ at each time step, form the products $\mathbf{J}_c(t_1+t)\cdot\mathbf{J}_c(t_1)$ and average over time origins t_1, just as for the m.s.d. An important difference is that in the present case one does not benefit from a separate average over particles, so that considerably longer simulations are needed to achieve good statistical accuracy. The Laplace transform needed to get $\sigma(\omega)$ presents no numerical problems, provided it is carefully performed.

We show in Figure 19 previously unpublished results for $\sigma(\omega)$ obtained in a simulation of CaF_2 at 1611 K. Instead of the δ-function peak we would get for a harmonic crystal, we find an extremely broad peak, which is however roughly at the harmonic TO frequency calculated from the interionic potentials. The d.c. conductivity σ_0 is found to be 3.3×10^{12} sec^{-1}. According to equation (57), the static dielectric constant can be obtained from the slope of $\mathcal{Im}\ \sigma(\omega)$ at $\omega = 0$; we find the result $\varepsilon_0 = 9.7$. Similar calculations at $T = 1390$ K give $\sigma_0 = 1.0 \times 10^{12}$ sec^{-1}, $\varepsilon_0 = 10.5$. The calculated value of ε_0 for the perfect crystal at zero temperature with the lattice parameter used in the high-temperature simulations is $\varepsilon_0 = 8.2$. We conclude that the existence of the large conductivity has a relatively weak effect on ε_0.

We alluded in §5 to the Nernst-Einstein relation between the electrical conductivity and the tracer diffusion coefficient. This has a rather simple interpretation from the present point of view. If we substitute equation (59) into (58) and ignore correlations between different ions, we get:

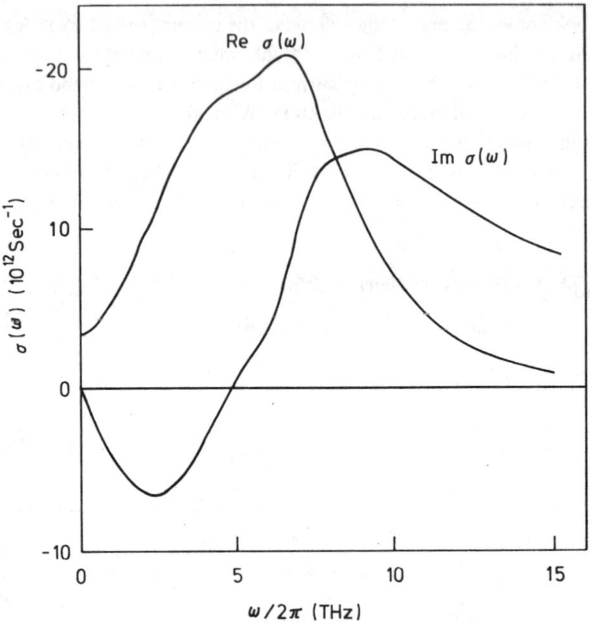

Figure 19. Simulation results for the real and imaginary parts of the frequency-dependent electrical conductivity $\sigma(\omega)$ of CaF_2 at 1611 K. The system contained 324 ions and the duration of the simulation was 300 psec.

$$\sigma_0 = \frac{e^2}{3\Omega k_B T}[N_+ z_+^2 \int_0^\infty dt\ \langle \mathbf{v}_{+i}(t)\cdot\mathbf{v}_{+i}(0)\rangle + N_- z_-^2 \int_0^\infty dt\ \langle \mathbf{v}_{-i}(t)\cdot\mathbf{v}_{-i}(0)\rangle]$$

$$= \frac{e^2}{\Omega k_B T}[N_+ z_+^2 D_+ + N_- z_-^2 D_-] \tag{64}$$

$$= e^2 z_-^2 \bar{\rho}_- D_- / k_B T\ ,$$

since the tracer diffusion coefficient of the cations is zero. This is just the Nernst-Einstein relation with the Haven ratio H_R set equal to unity (see equation (35)). This means that H_R is a measure of the contribution of interionic correlations to the conductivity.

From the values of D_- described in §5 and the two values of σ_0 given above, we deduce that H_R in the simulated system is 0.40 at 1390 K and 0.45 at 1611 K. Experimental values of H_R have likewise been obtained by comparing σ_0 and D_- in the case of PbF_2 (Carr *et al.* 1978). These indicate that H_R has a value between 0.7 and 1.0 in the superionic region. It is not quite clear whether our simulation value for CaF_2 above T_c is irreconcilable with the experimental value for PbF_2 in this region, given the fairly large error bars on the latter; but it does seem to us that the simulation result at 1390 K is probably giving too low a value, since below T_c we should expect H_R to be somewhere between the values 0.65 and 0.74

appropriate to the vacancy and interstitialcy mechanisms for fluorites. (As usual, we assume that CaF_2 and PbF_2 should behave in the same way.) It also seems likely that the problem lies with σ_0 rather than D_-, since the latter shows reasonable agreement with experiment.

We believe the difficulty arises from a size effect in the simulation, which affects σ_0 but not D_-, and which becomes serious only at lower temperatures. In Appendix 2, we describe the physical mechanism we believe to be responsible for this effect.

9.2 *The external field method*

The external field method is simple in principle, but a little complicated in practice. If we apply a constant electric field **E**, the equation of motion for the ions becomes

$$m_\alpha d^2 \mathbf{r}_{\alpha i}/dt^2 = \mathbf{F}_{\alpha i} + \mathbf{E} e z_\alpha \ , \tag{65}$$

where $\mathbf{F}_{\alpha i}$ is the interionic force. If we now calculate the mean value of the electric current $\mathbf{J}_c$, we can deduce the d.c. conductivity. This general approach to the calculation of transport coefficients is known as non-equilibrium molecular dynamics. A review of such simulation methods has been given by Evans and Morriss (1984). The first complication that has to be faced is that a steady field causes Joule heating, which will cause a continual upward drift of the temperature. This problem can be overcome by introducing a thermostat. A convenient way of doing this is to modify the equation of motion so that the total energy remains constant (Evans and Morriss 1984, Gillan and Holloway 1985, Gillan 1987), which can be achieved by writing

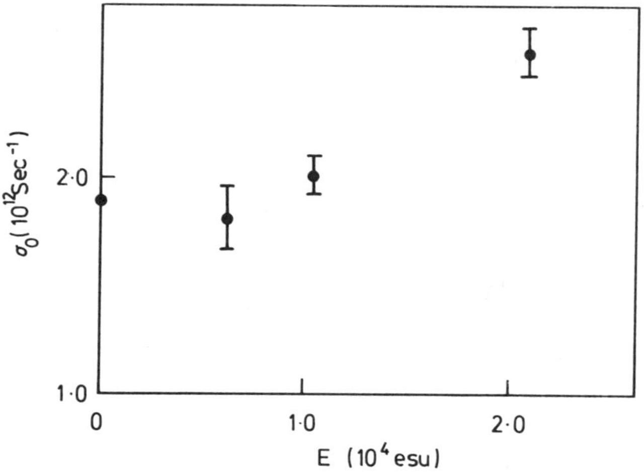

Figure 20. The apparent d.c. conductivity $\langle J_c \rangle / \Omega E$ from 96-ion non-equilibrium simulations of CaF_2 at 1543 K (dots with error bars). The point at $E = 0$ represents the result of a Green-Kubo calculation of σ_0 on the same system at the same temperature.

$$m_\alpha d^2\mathbf{r}_{\alpha i}/dt^2 = \mathbf{F}_{\alpha i} + \mathbf{E}ez_\alpha - m_\alpha\gamma\mathbf{v}_{\alpha i} \, , \tag{66}$$

where γ is a time-varying friction coefficient, whose value at any instant is given by

$$\gamma = \mathbf{E}\cdot\mathbf{J}_c/2K \, , \tag{67}$$

with $\mathbf{J}_c$ given by equation (59) and K the total kinetic energy of the system. A modification of the Verlet algorithm is needed to deal with this equation of motion, but there are no problems here.

The second complication is that the mean value $\langle \mathbf{J}_c \rangle$ will be a linear function of $\mathbf{E}$ (Ohm's law) only if E is small enough. The calculations must therefore be repeated for different values of E to check that we are in the linear regime.

Calculations of this kind have so far only been done for a 96-ion simulation of CaF_2. We show in Figure 20 the apparent value of σ_0 (i.e. $\langle J_c \rangle/\Omega E$) for different values of E at 1543 K. To provide a comparison, Green-Kubo calculations have been done with the same number of particles and at the same temperature; the result of these calculations is shown on the same Figure. The close agreement between the results confirms the correctness of the two approaches and thus, indirectly, the real existence of the discrepancy between the simulated and experimental values for H_R.

10. Dynamical Correlations between Ions

In §6.1, we discussed the van Hove self-correlation function, which describes the correlation between the positions of a *given* particle at different times. We showed how it allows us to characterize the diffusion process, and we noted that it could be be measured by neutron scattering. Here we are interested in the correlations *between* particles at different times, which are likewise described by an experimentally accessible quantity.

Let us start by considering the joint probability density of finding an ion of type β at position $\mathbf{r}'$ at some instant *and* of finding a particle of type α at position $\mathbf{r}$ a time t later. This quantity, denoted by $G_{\alpha\beta}(\mathbf{r},\mathbf{r}';t)$, is called the van Hove correlation function (Hansen and McDonald 1986). It is defined by

$$G_{\alpha\beta}(\mathbf{r},\mathbf{r}';t) = \langle \sum_{i=1}^{N_\alpha} \delta(\mathbf{r} - \mathbf{r}_{\alpha i}(t_0+t)) \sum_{j=1}^{N_\beta} \delta(\mathbf{r}' - \mathbf{r}_{\beta j}(t_0)) \rangle \, . \tag{68}$$

The time origin t_0 is, as usual, arbitrary. The function $G_{\alpha\beta}(\mathbf{r},\mathbf{r}';t)$ is like the self-correlation function $G_\alpha^s(\mathbf{r},t)$, except that it is generalized to describe the correlations of ions with each other, instead of with themselves.

Just as with the self-correlation functions, it is more helpful to work in Fourier space. We define the so-called intermediate scattering functions $F_{\alpha\beta}(\mathbf{q},t)$ (Hansen and McDonald 1986) as

$$\begin{aligned} F_{\alpha\beta}(\mathbf{q},t) &= (N_\alpha N_\beta)^{-\frac{1}{2}} \int d\mathbf{r}d\mathbf{r}' \, e^{i\mathbf{q}\cdot(\mathbf{r}-\mathbf{r}')} \, G_{\alpha\beta}(\mathbf{r},\mathbf{r}';t) \\ &= (N_\alpha N_\beta)^{-\frac{1}{2}} \, \langle \sum_{i=1}^{N_\alpha} \sum_{j=1}^{N_\beta} e^{i\mathbf{q}\cdot(\mathbf{r}_{\alpha i}(t)-\mathbf{r}_{\beta j}(0))} \rangle \, , \end{aligned} \tag{69}$$

where we have set $t_1 = 0$. The function $F_{\alpha\beta}(\mathbf{q},t)$ has a fairly simple physical interpretation. Let us define the dynamical variable

$$\hat{\rho}_\alpha(\mathbf{q}) = N_\alpha^{-\frac{1}{2}} \sum_{i=1}^{N_\alpha} e^{i\mathbf{q}\cdot\mathbf{r}_{\alpha i}} \ . \tag{70}$$

This represents the Fourier component at wavevector $\mathbf{q}$ of the density of ions of type α, as one can see by noting that the real-space density of these ions for a given configuration $\{\mathbf{r}_{\alpha i}\}$ is

$$\rho_\alpha(\mathbf{r}) = \sum_{i=1}^{N_\alpha} \delta(\mathbf{r} - \mathbf{r}_{\alpha i}) \ , \tag{71}$$

i.e. a δ-function at the position of each ion. On taking the Fourier transform of this, we verify that

$$\hat{\rho}_\alpha(\mathbf{q}) = N_\alpha^{-\frac{1}{2}} \int d\mathbf{r}\, e^{i\mathbf{q}\cdot\mathbf{r}}\, \rho_\alpha(\mathbf{r}) \ . \tag{72}$$

We can write

$$F_{\alpha\beta}(\mathbf{q},t) = \langle \hat{\rho}_\alpha(\mathbf{q},t)\hat{\rho}_\beta{}^*(\mathbf{q},0)\rangle \ , \tag{73}$$

so that it characterizes the fluctuations of density at wavevector $\mathbf{q}$.

To understand this more clearly, consider what would happen for a perfect harmonic solid. In this case, the density fluctuations are associated with vibrational displacements from the regular sites. If the displacements are small, then it can readily be shown that $\hat{\rho}_\alpha(\mathbf{q})$ is a sum of normal mode (i.e. phonon) coordinates associated with wavevector $\mathbf{q}$ (Lovesey 1984). Then $\hat{\rho}_\alpha(\mathbf{q},t)$ is a sum of harmonically oscillating terms, one for each phonon branch at that $\mathbf{q}$. Hence, $F_{\alpha\beta}(\mathbf{q},t)$ is a sum of harmonic oscillations.

Like its self-correlation counterpart, the time Fourier transform of $F_{\alpha\beta}(\mathbf{q},t)$ is important because it is accessible to experiment. The double-differential cross-section for coherent inelastic neutron scattering is given by (Lovesey 1984)

$$(\partial^2\sigma/\partial\Omega\partial E)_{\text{coh}} = \frac{k_f}{\hbar k_i} \sum_{\alpha\beta} (N_\alpha N_\beta)^{\frac{1}{2}} \bar{b}_\alpha \bar{b}_\beta S_{\alpha\beta}(\mathbf{q},\omega) \ , \tag{74}$$

where the symbols have the same meaning as in equation (39), and the 'dynamical structure factors' $S_{\alpha\beta}(\mathbf{q},\omega)$ are defined by

$$S_{\alpha\beta}(\mathbf{q},\omega) = \frac{1}{2\pi} \int_{-\infty}^{\infty} dt\, e^{i\omega t}\, F_{\alpha\beta}(\mathbf{q},t) \ . \tag{75}$$

In the perfect harmonic solid, the harmonic oscillations of the $F_{\alpha\beta}(\mathbf{q},t)$ give rise to δ-function peaks in the cross-section as a function of ω for each $\mathbf{q}$. This is, indeed, the usual way of measuring phonon dispersion relations.

An extensive series of inelastic neutron-scattering measurements has been made on several fluorites by Hayes, Hutchings and co-workers (Dickens *et al.* 1978, Hutchings *et al.* 1984; see also their article in the present book). At temperatures well below T_c, only phonon peaks are visible in the cross-section. As the temperature is raised through T_c, a new peak emerges, centred on zero ω. This 'quasielastic' peak signals the existence of non-oscillatory slowly-decaying fluctuations of the densities $\hat{\rho}_\alpha(\mathbf{q})$. The experimental quasielastic peak has several important characteristics: (i) its intensity varies strongly with temperature, increasing rapidly as the temperature passes through T_c; (ii) the intensity also depends strongly on $\mathbf{q}$, the main weight being concentrated in an island just beyond the reciprocal lattice vector at (2,0,0) and in weaker features at a similar radius in $\mathbf{q}$–space; (iii) the energy (frequency) width increases quite rapidly with temperature, but is typically 1 meV, corresponding to a decay time of order 1 psec (considerably longer than a typical phonon period, which is about 0.1 psec. Recently, a set of MD simulations of CaF_2 has been undertaken (Gillan 1986a,b), with the purpose of aiding the interpretation of these results.

The MD calculation of the $S_{\alpha\beta}(\mathbf{q},\omega)$ is straightforward. The densities $\hat{\rho}_\alpha(\mathbf{q})$ are calculated at each time step from equation (70) for the $\mathbf{q}$ of interest, and the correlation functions $F_{\alpha\beta}(\mathbf{q},t)$ are then constructed for some chosen range of t values by averaging in the usual way over time origins (equation (73)). Finally, the Fourier transform of equation (75) is performed numerically to obtain the $S_{\alpha\beta}(\mathbf{q},\omega)$. Technical details are given in the original paper (Gillan 1986a). The periodic boundary conditions restrict the allowed values of $\mathbf{q}$, which must be commensurate with the periodicity: for the usual cubic simulation cell, $\mathbf{q}$ must have the form $(2\pi/L)\ (n_1,n_2,n_3)$, where the n_i are integers.

The simulations were made over a wide range of temperatures, going from 500 to 1611 K (T_c for CaF_2 being at $\sim$ 1420 K). Figure 21 shows results for the anion-anion quantity $S_{--}(\mathbf{q},\omega)$ as a function of ω at $\mathbf{q} = (2\pi/a_0)\ (7/3,0,0)$, which is close to the wavevector where the experimental quasielastic intensity is largest. At the lowest temperature, where we expect the system to be fully ordered and nearly harmonic, the spectrum consists solely of phonon peaks at 3 and 15 THz. These are the longitudinal acoustic and optic modes – it can be shown by symmetry that other modes cannot be seen in $S_{\alpha\beta}(\mathbf{q},\omega)$ for $\mathbf{q} \parallel \langle 100 \rangle$. As the temperature approaches T_c, a new peak centered on $\omega = 0$ appears, its intensity growing rapidly with T until at the highest temperature it is the dominant feature. It will become clear that this corresponds to the quasielastic peak seen in the experimental cross-section. When one examines not only the anion-anion function, but also the cation-cation and cation-anion functions one finds that the quasielastic peak appears in all three. This can be seen in Figure 22, where we show the three $S_{\alpha\beta}(\mathbf{q},\omega)$ for four different $\mathbf{q}$ at $T = 1529$ K. It is particularly striking that at the smallest q the three functions are essentially identical, except for a scaling factor. This might seem strange, since it is clear that the quasielastic peak is closely connected with the superionic conduction, and we know that diffusion occurs only on the anion sublattice. We shall see later the reason for this.

The experiments yield the intensity and frequency width of the quasielastic peak as a function of temperature and wavevector. They do not separate the contributions from the three $S_{\alpha\beta}(\mathbf{q},\omega)$, but give only a weighted average. For the intensity, the experiments give absolute values of the quantity

$$S_m^{cp}(\mathbf{q}) = (\bar{b}_+^2/2\bar{b}_-^2)S_{++}^{cp}(\mathbf{q}) + (\bar{b}_+/2^{\frac{1}{2}}\bar{b}_-)S_{+-}^{cp}(\mathbf{q}) + S_{--}^{cp}(\mathbf{q})\ , \qquad (76)$$

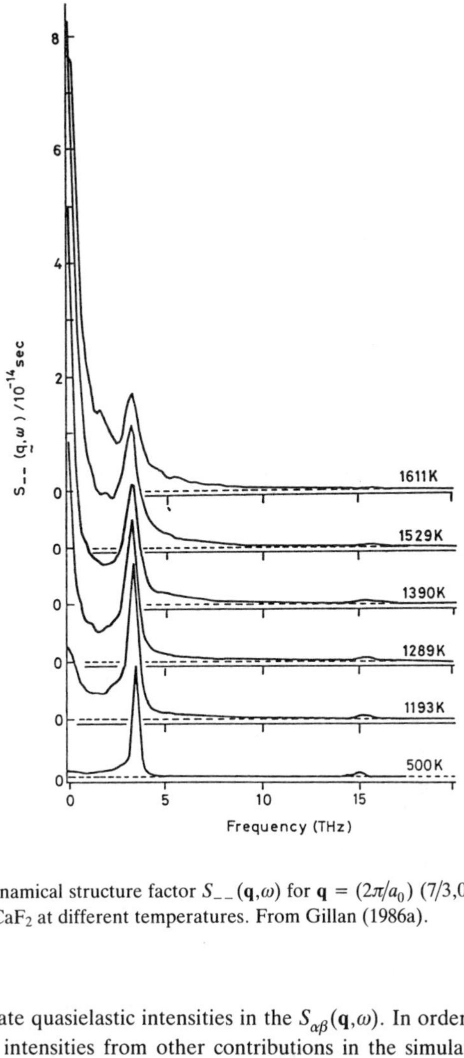

Figure 21. The anion-anion dynamical structure factor $S_{--}(\mathbf{q},\omega)$ for $\mathbf{q} = (2\pi/a_0)\ (7/3,0,0)$ obtained from simulations of CaF_2 at different temperatures. From Gillan (1986a).

where $S^{cp}_{\alpha\beta}(\mathbf{q})$ are the separate quasielastic intensities in the $S_{\alpha\beta}(\mathbf{q},\omega)$. In order to separate these quasielastic intensities from other contributions in the simulated $S_{\alpha\beta}(\mathbf{q},\omega)$, a fitting procedure is necessary; this is described in the original paper (Gillan 1986a; see also Gillan and Dixon 1980a, Gillan and Dixon 1981). The net result is that we can compare absolute values of $S^{cp}_{m}(\mathbf{q})$ from simulation and experiment. We show in Figure 23 this comparison as a function of temperature for a wavevector where the intensity is large. The experimental error bars are large, and the statistical accuracy of the calculations is not high. We believe that the differences are probably within the combined errors. The comparison as a function of $\mathbf{q}$ (in the $\langle 100 \rangle$ direction) is shown in Figure 24. Once again, the agreement is very satisfactory. The strong variation of intensity with $\mathbf{q}$ is well reproduced in the

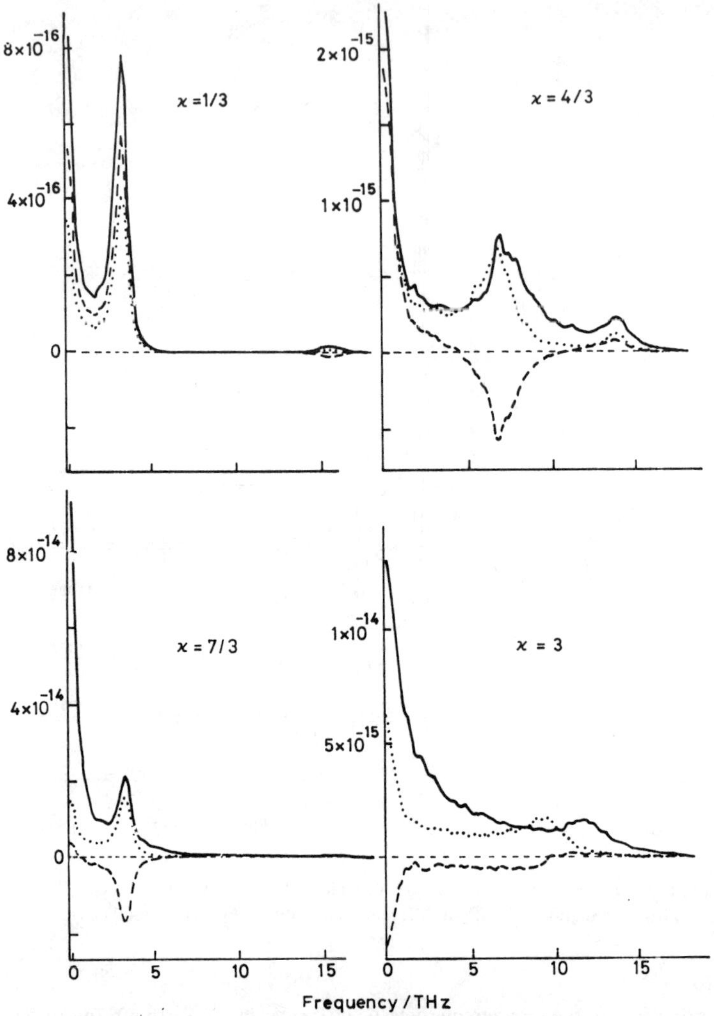

Figure 22. The simulated dynamical structure factors $S_{\alpha\beta}(\mathbf{q},\omega)$ of CaF_2 at 1529 K for wavevectors $(2\pi/a_0)$ $(\kappa,0,0)$ in the $\langle 100 \rangle$ direction: dotted line, cation-cation; broken line, cation-anion; full line, anion-anion. From Gillan (1986a).

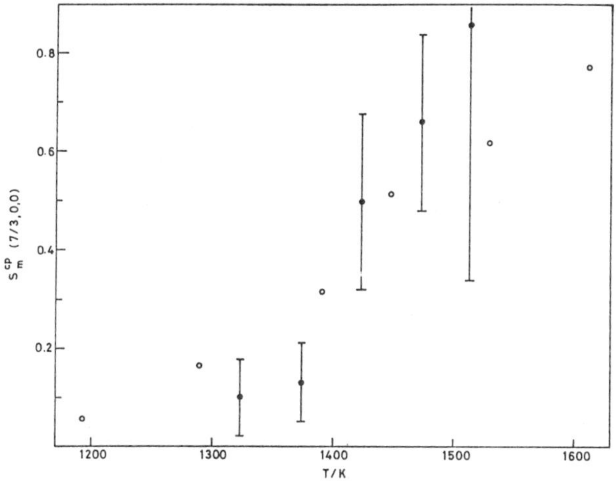

Figure 23. The simulated quasielastic intensity $S_m^{cp}(\mathbf{q})$ for CaF_2 at wavevector $\mathbf{q} = (2\pi/a_0)$ (7/3,0,0) (open circles), compared with experimental results (Hutchings *et al.* 1984) extrapolated to the same wavevector (full circles) over a range of temperatures; bars show the quoted experimental errors for wavevector (2.3,0,0). From Gillan (1986a).

simulations. Because of the restriction on the allowed **q** in the simulation, the resolution in **q**–space is rather poor, but this does not prevent us making an adequate comparison with experiment.

Experimental results for the width have been reported only for a rather restricted range of **q**. There is only a single available **q** in the simulations lying in this range (or almost in it). The comparison between the simulated and

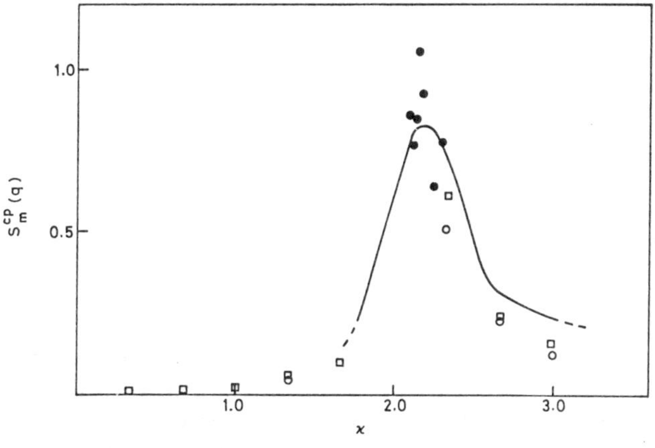

Figure 24. The simulated quasielastic intensity $S_m^{cp}(\mathbf{q})$ for CaF_2 at temperatures 1447 and 1529 K (open circle and square), compared with experimental results of Hutchings *et al.* (1984) at 1473 K (full circle and curve). From Gillan 1986a.

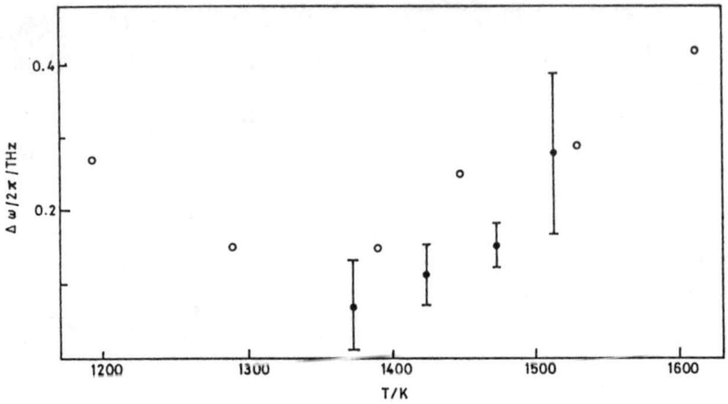

Figure 25. The simulated quasielastic frequency width for CaF_2 at $\mathbf{q} = (2\pi/a_0)\ (7/3,0,0)$ (open circle), compared with the experimental results of Hutchings *et al.* (1984) at (2.3,0,0) (full circles) for a range of temperatures; bars show the quoted experimental errors. From Gillan 1986a.

experimental widths as a function T for this $\mathbf{q}$ is shown in Figure 25. In fact, since the anion-anion intensity is dominant at this $\mathbf{q}$, the simulation results shown here are actually the quasielastic width of $S_{--}(\mathbf{q},\omega)$. Above T_c, the comparison is fairly satisfactory. At lower temperatures, it seems likely that something is wrong with the simulated widths. Although experimental results are not available for T much below T_c for CaF_2 itself, measurements on PbF_2 indicate that the width decreases steadily as one goes to lower temperatures (Hutchings *et al.* 1984). This does not happen in the simulations. This recalls the trouble experienced with the simulated electrical conductivity, particularly when one notes that both quantities have dimensions of frequency. We show in Appendix 2 that the two problems may both be explained by a physical mechanism associated with the size of the simulated system.

The general conclusion of this section is that the simulations reproduce rather satisfactorily the experimental observations on the quasielastic peak. The $\mathbf{q}$-dependence of the intensity, which contains information about the spatial character of interionic correlations, is in good agreement with experiment, as is its dependence on temperature. The frequence width is also in satisfactory agreement, except at temperatures below T_c.

11. Defects

Up to now, we have discussed mainly the empirical comparison of simulation and experiment. Overall, this leaves little to be desired: the results for the diffusion coefficient, the van Hove self function, the spatial distribution and the dynamical structure factors all correspond rather closely to what is observed. We now want to show how the simulations can be analyzed to reveal the mechanisms underlying the phenomena. We shall show that the key to a deeper understanding of superionic conduction in fluorites is provided by an analysis in terms of point defects.

In the low-temperature normal state, the point-defect description has been established for many decades, and there is a detailed quantitative understanding of the properties of the defects. It is not immediately obvious that the same ideas can be applied to the superionic state. One might imagine that the system is so disordered that the whole idea of discrete, localized defects loses its meaning. The results we have already discussed indicate that this is not so. We know from §6 that the anions diffuse by well defined jumps between the regular anion sites and from §§6 and 7 that they spend most of their time on these sites. This strongly suggests that the diffusion of the ions is due to the jumping of defects. What is the nature of these defects? Since only the anions diffuse, the defects must be anion vacancies and interstitials, present in equal numbers. There are two possible ways of interpreting the fact that diffusion occurs by jumps between the regular sites. The first is that diffusion is mainly due to vacancies; the interstitials reside on other sites and are relatively immobile. This seems reasonable, but turns out to be incompatible with the simulation results. The second interpretation is that diffusion is due to vacancies and interstitials, which both jump between the regular sites. This seems at first sight less reasonable, but turns out in fact to be correct.

The analysis that leads to this conclusion has been described in a number of papers (Dixon and Gillan 1978, Gillan and Dixon 1980b, Gillan 1986b); a related method has been developed by Jacucci and Rahman (1978). This analysis operates with the jump catalogue described in §6.2 – the list of all the jump events occurring in a given span of time. We recall that each event is characterized by the initial and final sites and by the associated departure and arrival times. For present purposes, we shall not be interested in the flight time and we shall *define* the instant of the jump to be the median of the departure and arrival times. Now – again as a definition – we shall say that, at the instant of any jump, the occupancy of the final site increases by unity and that of the initial site decreases by unity. This definition determines the occupancy of every site at any instant, except for an additive constant; in other words, if we know the occupancy of every site at some particular time, then the definition determines the occupancies at all other times. In order to specify the occupancies absolutely, we define them to be equal to unity for any configuration in which there is one anion inside each of the site-spheres (see §6.2). As a final piece of nomenclature, we shall say that an occupancy of 0 at a site corresponds to a vacancy, 1 corresponds to no defect, 2 to an interstitial, 3 to two interstitials ... at that site. The impossibility of an occupancy less than zero follows from the definitions. Occupancies of 3 or more are not excluded, but we expect them to be rare.

There are two important points to be made about this scheme. The first is that, at this stage of the discussion, it consists merely of a set of definitions – a formal book-keeping system which allows us to re-express the contents of the hopping catalogue in another way. It is formal in the sense that it could be applied to any system whatsoever: a liquid, for example, or even a perfect gas. It is completely neutral as to the existence or otherwise of possible sites that may play a physical role in addition to the regular sites. For the purposes of the scheme, the labels 'vacancy' and 'interstitial' have the formal meanings given to them by the definitions. But of course the scheme is not an idle game. We shall show in the following section that the 'vacancies' and 'interstitials' are directly responsible for the quasielastic peak.

The second important point concerns the role played by the site spheres. These are introduced so that we can identify the jumps between sites. The defects are defined in terms of these jumps. One can imagine other ways of proceeding. For example, one could identify a vacancy with an empty site sphere and an interstitial with an ion in the space between spheres. We stress that such a method would be entirely different from the one described above. We believe it would be unsatisfactory. The reason for this can be seen from the trajectory plot of Figure 6. This shows that the anion vibrational amplitudes in the superionic state are very large. Vibrational excursions of up to half the anion nearest-neighbour separation are not uncommon. This means that there is no natural way of deciding from the instantaneous position of an ion whether or not it should be associated with a defect. The defect analysis we have described does not work with instantaneous positions, but with *trajectories* – more specifically with those trajectories that connect different site spheres. In order to clarify further the meaning of our defect scheme, we illustrate in Appendix 1 its application to a simple one-dimensional model.

We now want to illustrate the kind of result produced by the defect analysis. There is a graphical method that is useful for qualitative purposes (Dixon and Gillan 1978, Gillan and Dixon 1980b). Let us choose to represent time increasing downwards on a vertical scale. Each intersite jump is represented by a dot at the time of its occurrence. We also represent the defects on the diagram. Each defect exists on a particular site for a given time: a time during which the occupancy of the site remains constant. We represent the defect on the site by a line extending over its period of residence at that site. In order to distinguish vacancy and interstitial lines, we mark the former by an upward-pointing arrow and the latter by a downward-pointing arrow. The appearance and disappearance of a defect at a site are associated with jumps involving that site. Therefore each line connects two dots.

In Figure 26, we show an example of this graphical representation of the defect analysis, taken from a short section of a simulation of $SrCl_2$ in the superionic state. Note that the horizontal extension of the diagram has no physical significance. We observe first that the configuration of the incoming and outgoing lines at a dot can be of four kinds: up/up, down/down, up/down and down/up. These signify respectively the intersite jump of a vacancy and of an interstitial and the creation and the annihilation of a vacancy-interstitial pair. We can therefore identify each ion jump with one of these four kinds of event. An analysis of the full simulation of which Figure 26 represents a small section shows that the proportions of jumps of these four kinds are 0.28 : 0.34 : 0.19 : 0.19. This shows two things: firstly that a substantial contribution to diffusion comes from creation/annihilation events; secondly that there is a roughly equal contribution from vacancy and interstitial jumps – this second point is important, and we shall return to it later.

The defect analysis yields a value for the concentration of vacancies and interstitials; the numbers of the two defects are of course the same by construction, since they are created and annihilated in pairs. The analysis gives us the site occupancies and therefore the numbers of vacancies and interstitials at each time step. In terms of the graph, these numbers are equal to the numbers of upward and

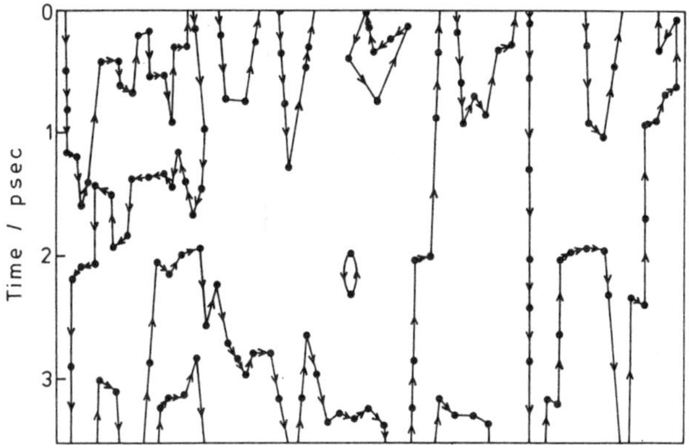

Figure 26. Hopping diagram for a short section of a simulation of $SrCl_2$ at 1522 K. Dots represent hopping events, the time at which the event occurs being shown by its position on the vertical time scale (time increasing downwards). Directed lines indicate the presence of vacancy (upward line) or interstitial (downward line) on the site involved in each jump event. From Dixon and Gillan (1980b).

downward lines cut when we draw a horizontal line at a particular time. The mean concentration is straightforwardly calculated by averaging over the length of the run. The results reported for our recent simulations of CaF_2 (Gillan 1986b) are typical. At temperatures of 1390 and 1611 K, the numbers of vacancies per regular anion site are found to be 0.011 and 0.031 respectively.

Now we want to return to the idea that is sometimes put forward, that diffusion is due almost entirely to vacancies, while the interstitials are immobilized on additional sites (se e.g. Mościński and Jacobs 1985, Allnatt *et al.* 1987). Can this idea be reconciled with the form of the hopping diagram shown in Figure 26? To see that it cannot, it will be helpful to consider what form this diagram would take in the low-temperature regime well below T_c, where we know that diffusion *is* mainly due to vacancies and that the less mobile interstitials reside on the cube centre sites (2.1). Note first that these physical vacancies correspond precisely to the 'vacancies' defined by our defect analysis. Their presence at a site is signalled by the zero occupancy of that site: an ion has made a jump from that site before another ion jumps to it. The physical interstitials will also be identified as 'interstitials' in the defect analysis. Suppose that at some time t_1 an ion passes from a regular site 1 onto a cube-centre interstitial position, where it resides for some period before passing again, at time t_2, to a regular site 2. As far as the defect analysis is concerned, between times t_1 and t_2 the ion was in flight from site 1 to site 2; the jump of the ion concerned occurred at the median time $\frac{1}{2}(t_1 + t_2)$. Before this median time, the interstitial was attributed to regular site 1, and after it to regular site 2. Thus the analysis, because it identifies defects through the hopping of the ions, assigns the interstitial to the regular sites involved in the hops. This description is not perhaps the most natural one in the low-temperature region. Nevertheless – and this is the crucial point – the physical vacancies are identified as

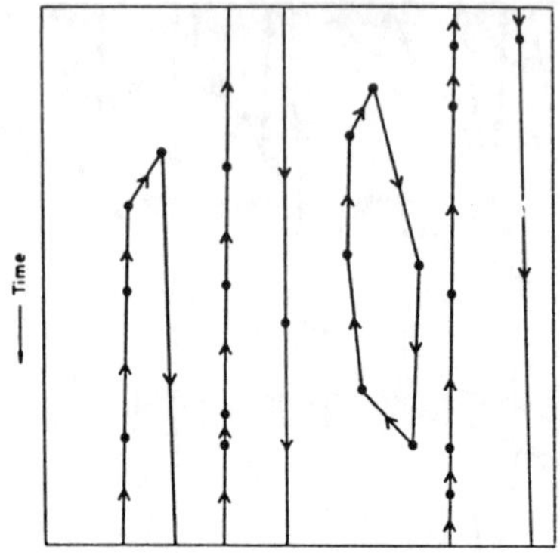

Figure 27. Hypothetical hopping diagram for a system in which diffusion is vacancy dominated.

'vacancies' and the physical interstitials as 'interstitials'. Therefore at low temperature, when the vacancies are more mobile, the hopping diagram will have the form shown in Figure 27. (Since the defect concentration in this region is very low, we must imagine the analysis applied to a very large system – otherwise we shall see nothing.) The point is that almost all the hopping events (dots) will be on up-going (vacancy) lines. The relative immobility of the interstitials shows itself by the sparsity of dots on the down-going (interstitial) lines. As the temperature is raised, experiment indicates that the mobilities of the two defects become comparable, as we have stressed in § 2.1. The dots will then become more equally distributed between vacancy and interstitial lines, and the hopping diagram will take on a form more like what we actually find in the superionic state (Figure 26).

The same arguments apply in the superionic state, where we already know that the cube-centre site is not significantly occupied. If we wanted to say that there were interstitials immobilized on some other site, the hopping diagram would still have to look like Figure 27 rather than Figure 26.

The argument can also be presented in another, perhaps simpler way. We know from §6.2 that essentially all the hopping events are associated with diffusion. If this were not so, our method of obtaining the diffusion coefficient from the rate of occurrence of these events (equation (49)) would not give the correct result. It follows that if most of the diffusion were to be attributed to vacancy motion, most of the dots in the hopping diagram would have to lie on up-going (vacancy) lines. This is far from true, as we have seen in Figure 26.

The conclusion is that the simulation results are incompatible with the idea that diffusion is mainly due to vacancies and that the interstitials are comparatively immobile.

12. Defects and the Quasielastic Peak

The defects discussed in the previous section are observable physical objects. It has recently been shown that they are directly responsible for the central peak in the coherent dynamical structure factor (Gillan 1986b). The full analysis that shows this is rather lengthy, and we shall only try to outline the argument here. The main physical idea can be stated quite simply (Gillan and Wolf 1985): a defect causes a disturbance of the density; as defects are created, diffuse, and are annihilated, the associated density disturbance fluctuates; the central peak is a manifestation of these fluctuations.

12.1 *Density fluctuations due to defects.*

Let us develop this idea further, by considering the density disturbance caused by a single defect in the absence of lattice vibrations (Gillan and Wolf 1985). Suppose the defect is a vacancy, which is at a regular anion site at position $\mathbf{R}_1$. The vacancy will cause a distortion of the surrounding lattice. Let the regular-lattice position of ion i of type α be called $\mathbf{R}_{\alpha i}$ and let its displacement from this in the presence of the vacancy be $\mathbf{u}_{\alpha i}$. From equation (70), the density at wavevector $\mathbf{q}$ is

$$\hat{\rho}_\alpha(\mathbf{q}) = N_\alpha^{-\frac{1}{2}} \sum_i{}' \exp[i\mathbf{q}\cdot(\mathbf{R}_{\alpha i} + \mathbf{u}_{\alpha i})] \ , \tag{77}$$

where the sum goes over all sites occupied by ions of type α; the vacancy site is of course omitted for α = anion. It is convenient to subtract from this the quantity $N_\alpha^{-\frac{1}{2}} \Sigma_i \exp(i\mathbf{q}\cdot\mathbf{R}_{\alpha i})$, where the sum goes over *all* sites, which is the value of $\hat{\rho}_\alpha(\mathbf{q})$ for the perfect non-defective lattice. Since this vanishes if $\mathbf{q}$ is not a reciprocal lattice vector, we can write:

$$\hat{\rho}_\alpha(\mathbf{q}) = N_\alpha^{-\frac{1}{2}} \{\sum_i{}' \exp[i\mathbf{q}\cdot(\mathbf{R}_{\alpha i}+\mathbf{u}_{\alpha i})] - \sum_i \exp[i\mathbf{q}\cdot\mathbf{R}_{\alpha i}]\} \ . \tag{78}$$

A little rearrangement then gives:

$$\hat{\rho}_\alpha(\mathbf{q}) = N_\alpha^{-\frac{1}{2}} e^{i\mathbf{q}\cdot\mathbf{R}_1} f_\alpha(\mathbf{q}) \ , \tag{79}$$

where

$$f_\alpha(\mathbf{q}) = \sum_i{}' e^{i\mathbf{q}\cdot(\mathbf{R}_{\alpha i}-\mathbf{R}_1)} [e^{i\mathbf{q}\cdot\mathbf{u}_{\alpha i}} - 1] - \delta_{\alpha -} \ . \tag{80}$$

Note that because of translational symmetry, $\mathbf{u}_{\alpha i}$ depends only on the relative positions $\mathbf{R}_{\alpha i} - \mathbf{R}_1$, so that $f_\alpha(\mathbf{q})$ does not depend on $\mathbf{R}_1$. Clearly $f_\alpha(\mathbf{q})$ plays the role of a form factor for the vacancy and its surrounding distortion. The last term in equation (80) is associated with the absence of the ion at the vacant site; the other term comes from the surrounding distortion. The same argument can be carried through for an interstitial at position $\mathbf{R}_1$. We get the same formula (79) for the density $\hat{\rho}_\alpha(\mathbf{q})$; the expression for the form factor is slightly different. In order to distinguish the two, we shall denote the vacancy form factor by $f_\alpha^v(\mathbf{q})$ and the interstitial form factor by $f_\alpha^i(\mathbf{q})$.

We are actually interested not in an isolated defect, but in a finite concentration of defects. If this concentration is not too large, the lattice distortion for one defect will not be greatly modified by the presence of other defects, and we can assume that their contributions to the $\hat{\rho}_\alpha(\mathbf{q})$ add linearly. Suppose we have vacancies at positions $\mathbf{R}_l^v$ and interstitials at positions $\mathbf{R}_l^i$. Then from equation (79), the density disturbance associated with the defects is

$$\hat{\rho}_\alpha(\mathbf{q}) = N_\alpha^{-\frac{1}{2}} [\sum_l e^{i\mathbf{q}\cdot\mathbf{R}_l^v} f_\alpha^v(\mathbf{q}) + \sum_l e^{i\mathbf{q}\cdot\mathbf{R}_l^i} f_\alpha^i(\mathbf{q})] \ . \tag{81}$$

It is convenient to write this in a more compact form. The Fourier component of the ion densities at wavevector $\mathbf{q}$ is defined by equation (70). In the same way, we can define the Fourier component of the vacancy and interstitial densities by

$$P_a(\mathbf{q}) = \bar{n}^{-\frac{1}{2}} \sum_l e^{i\mathbf{q}\cdot\mathbf{R}_l^a} \ , \tag{82}$$

where $\bar{n}$ is the mean number of vacancies (= number of interstitials) in the system, and a takes the values v (vacancy) or i (interstitial). Then equation (81) can be rewritten as

$$\hat{\rho}_\alpha(\mathbf{q}) = (\bar{n}/N_\alpha)^{\frac{1}{2}} \sum_a f_\alpha^a(\mathbf{q}) P_a(\mathbf{q}) \ . \tag{83}$$

Our derivation of this formula has implicitly assumed that the defects are fixed on certain sites. It can be taken to apply also when the defects are hopping between sites and being created and destroyed, if we can assume that the lattice distortion follows each defect instantaneously as it hops, and appears or disappears instantaneously when the defect is created or destroyed. This is a reasonable assumption in the present context, since the time-scale of the defect dynamics is considerably longer than that of the lattice vibrations.

As the defects are created, diffuse and are annihiliated, their densities $P_a(\mathbf{q})$ fluctuate. The contribution of the defects to the dynamical structure factors $S_{\alpha\beta}(\mathbf{q},\omega)$ of the ions can be expressed in terms of these fluctuations. If we put equation (83) into equations (73) and (75), we get this defect contribution, which we denote $S_{\alpha\beta}^{\mathrm{def}}(\mathbf{q},\omega)$:

$$S_{\alpha\beta}^{\mathrm{def}}(\mathbf{q},\omega) = \bar{n}(N_\alpha N_\beta)^{-\frac{1}{2}} \sum_{ab} f_\alpha^a(\mathbf{q}) f_\beta^b(\mathbf{q})^* \hat{\sigma}_{ab}(\mathbf{q},\omega) \ , \tag{84}$$

where

$$\hat{\sigma}_{ab}(\mathbf{q},\omega) = \frac{1}{2\pi} \int_{-\infty}^{\infty} dt \, e^{i\omega t} \langle P_a(\mathbf{q},t) P_b(\mathbf{q},0)^* \rangle \ . \tag{85}$$

The quantity $\hat{\sigma}_{ab}(\mathbf{q},\omega)$ has the form of a dynamical structure factor for the defects. Now the dynamics of the defects is non-oscillatory, so we expect $\hat{\sigma}_{ab}(\mathbf{q},\omega)$ to consist solely of a peak at zero frequency (we shall see in a moment that this is true). We therefore have every reason to expect the defects to give rise to a quasielastic peak in the dynamical structure factors. The idea is that this is exactly the peak discussed in §10.

We have noted that the intensity of the observed peak depends strongly on wavevector. The intensity of the peak predicted by equation (84) is

$$S^{\rm def}_{\alpha\beta}(\mathbf{q}) \equiv \int_{-\infty}^{\infty} d\omega \; S^{\rm def}_{\alpha\beta}(\mathbf{q},\omega)$$
$$= \bar{n}(N_\alpha N_\beta)^{-\frac{1}{2}} \sum_{ab} f^a_\alpha(\mathbf{q}) f^b_\beta(\mathbf{q})^* \hat{\sigma}_{ab}(\mathbf{q}) \quad , \tag{86}$$

where

$$\hat{\sigma}_{ab}(\mathbf{q}) = \langle P_a(\mathbf{q}) P_b(\mathbf{q})^* \rangle \quad . \tag{87}$$

This shows that the wavevector dependence of the intensity could be associated with either (i) the $f^a_\alpha(\mathbf{q})$, i.e. the distortion surrounding the defects; or (ii) the $\hat{\sigma}_{ab}(\mathbf{q})$, i.e. the correlations between the defects. We shall see that the distortion is the main source of $\mathbf{q}$–dependence.

12.2 *Testing the defect hypothesis*

If the observed quasielastic peak in the $S_{\alpha\beta}(\mathbf{q},\omega)$ does originate from the motion of defects, then the defect dynamical structure factors $\hat{\sigma}_{ab}(\mathbf{q},\omega)$ must show a quasielastic peak, and the frequency width of this peak must agree with the observed width. This is straightforward to test. The defect analysis described in §11 yields the positions of the vacancies and interstitials at each time step. From these, we can calculate the defect densities $P_a(\mathbf{q},t)$ and hence the correlation functions $\langle P_a(\mathbf{q},t) P_b(\mathbf{q},0)^* \rangle$, by the usual averaging over time origins. The Fourier transform required to get the $\hat{\sigma}_{ab}(\mathbf{q},\omega)$ is then performed numerically, exactly as in the calculation of the $S_{\alpha\beta}(\mathbf{q},\omega)$.

Figure 28 shows the simulation results for the three $\hat{\sigma}_{ab}(\mathbf{q},\omega)$ at four different wavevectors; these come from a simulation of CaF_2 at 1611 K. As we expect, the spectra consist solely of a central peak. We can now compare the width of this peak with the width of the quasielastic peak in $S_{\alpha\beta}(\mathbf{q},\omega)$ observed in the simulations. This comparison for the simulations of CaF_2 at two different temperatures is shown in Table 4. We remarked earlier on the fact that the three widths for the $S_{\alpha\beta}(\mathbf{q},\omega)$ were essentially the same. We note here that the same is roughly true also of the $\hat{\sigma}_{ab}(\mathbf{q},\omega)$. We are therefore comparing, for each $\mathbf{q}$, the 'ion width' with the 'defect width'. The results of the Table show that these two widths are close to each other and vary in the same way with $\mathbf{q}$. This provides strong evidence that there is a close connection between the observed quasielastic peak and the motion of the defects.

MD simulation allows us to probe this connection more deeply. Suppose we were to observe, in the simulation, one of the time-dependent densities $\hat{\rho}_\alpha(\mathbf{q},t)$ responsible for the quasielastic peak. We should see a quantity that fluctuates about zero in a rather unpredictable way. The quasielastic peak is the spectrum of these fluctuations at low frequency. What we are saying in equation (83) is that the low frequency fluctuations of $\hat{\rho}_\alpha(\mathbf{q},t)$ are a linear combination of the fluctuations of the two defect densities $P_a(\mathbf{q},t)$ at the same wavevector; the coefficients are the defect form factors. Since we can calculate both the $\hat{\rho}_\alpha(\mathbf{q},t)$ and the $P_a(\mathbf{q},t)$ at every time step, we can verify directly that this is the case. In order to do this, we must note two points. Firstly, the $\hat{\rho}_\alpha(\mathbf{q},t)$ contain fluctuations both in the low-frequency region of interest and in the high-frequency region of lattice vibrations. The presence of the high-frequency components would confuse the comparison we

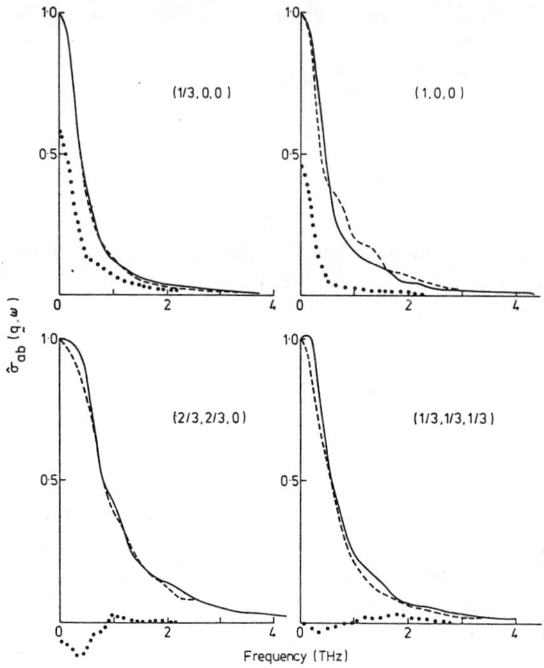

Figure 28. Simulated dynamical structure factors $\hat{\sigma}_{ab}(\mathbf{q},\omega)$ for vacancy and interstitial defects in CaF_2 at 1611 K:____ vacancy-vacancy; — interstitial-interstitial; ... vacancy-interstitial. The graph shows the normalized quantities $\hat{\sigma}_{ab}(\mathbf{q},\omega)/[\hat{\sigma}_{aa}(\mathbf{q},\omega=0)\hat{\sigma}_{bb}(\mathbf{q},\omega=0)]^{\frac{1}{2}}$ for wavevectors $\mathbf{q} = (2\pi/a_0)$ (1/3,0,0), (1,0,0), (2/3,2/3,0) and (1/3,1/3,1/3). From Gillan (1986b).

want to make. We can avoid this problem by filtering the signal so as to remove the high-frequency components. The second point is that we do not know in advance the form factors. If only a single defect density were involved, this would not be a problem, since we are only seeking to verify that the $\hat{\rho}_\alpha(\mathbf{q},t)$ are *proportional* to the $P_a(\mathbf{q},t)$. But since we have the densities for both vacancies and interstitials, we must have some way of deciding on the appropriate linear combination to use in equation (83). There is a systematic way of doing this, which we shall point out a little later.

We have made the comparison for a section of a simulation of CaF_2 in the superionic state. Figure 29 shows this for the three equivalent wavevectors $a_0\mathbf{q}/2\pi$ = (7/3,0,0), (0,7/3,0) and (0,0,7/3), where, we recall, the observed quasielastic intensity has its largest value. The two traces represent (i) the anion density $\hat{\rho}_-(\mathbf{q},t)$ filtered to remove high frequencies; and (ii) a linear combination of the vacancy and interstitial densities $P_v(\mathbf{q},t)$ and $P_i(\mathbf{q},t)$. The close correlation between the two quantities is immediately obvious. This demonstrates that in order to understand the observed quasielastic peak it suffices to understand the behaviour of the defects.

The direct comparison of the time dependence of the $\hat{\rho}_\alpha(\mathbf{q},t)$ with that of the $P_a(\mathbf{q},t)$ is a rather simple-minded way of proceeding. However, it does emphasize the key idea of *correlation* between the two kinds of density. We can show more systematically that the quasielastic peak is due to defect motions by examining the cross correlations between the two densities. In order to explain how this works, we need to recall something about the correlation of variables. Suppose we have two variables A and B. The degree of correlation between them is gauged by the quantity $\mu \equiv \langle AB^* \rangle / \langle |A|^2 \rangle^{\frac{1}{2}} \langle |B|^2 \rangle^{\frac{1}{2}}$. If $|\mu| = 1$, the variables are perfectly correlated: to all intents and purposes they are the same variable, except perhaps for a multiplicative factor. This can be extended to treat correlations at different frequencies. Let us define the cross-correlation spectrum of A and B as

$$\langle AB^* \rangle_\omega = \frac{1}{2\pi} \int_{-\infty}^{\infty} dt \, e^{i\omega t} \langle A(t) B^*(0) \rangle \; . \tag{88}$$

Here we have followed exactly the definition of the dynamical structure factors: if A and B are ion densities $\hat{\rho}_\alpha(\mathbf{q})$, then the $\langle AB^* \rangle_\omega$ are the $S_{\alpha\beta}(\mathbf{q},\omega)$; if they are the defect densities $P_a(\mathbf{q})$, then the $\langle AB^* \rangle_\omega$ are the $\hat{\sigma}_{ab}(\mathbf{q},\omega)$. On the other hand, if A is an ion density and B is a defect density, then $\langle AB^* \rangle_\omega$ is the cross-correlation spectrum of the two densities. The degree of cross-correlation at frequency ω is then $\mu(\omega) \equiv \langle AB^* \rangle_\omega / \langle |A|^2 \rangle_\omega^{\frac{1}{2}} \langle |B|^2 \rangle_\omega^{\frac{1}{2}}$: if $|\mu(\omega)| = 1$, the variables are perfectly correlated at this frequency.

$(\kappa_1, \kappa_2, \kappa_3)$	$\Delta\nu_{vv}$	$\Delta\nu_{ii}$	$\Delta\nu_{vi}$	$\Delta\nu_{++}$	$\Delta\nu_{--}$	$\Delta\nu_{+-}$
$(\frac{1}{3}00)$	0.36	0.36	0.24	0.38	0.39	0.38
	0.24	0.19	0.29	0.25	0.24	0.25
$(\frac{2}{3}00)$	0.58	0.64	0.64	0.56	0.56	0.56
	0.34	0.46	0.46	0.43	0.40	0.41
(100)	0.39	0.39	0.19	0.42	0.42	0.41
	0.42	0.52	0.57	0.55	0.50	0.50
$(\frac{1}{3}\frac{1}{3}0)$	0.58	0.48	0.57	0.90	0.84	0.87
	0.39	0.29	–			
$(\frac{2}{3}\frac{2}{3}0)$	0.77	0.77	–	0.92	0.91	0.93
	0.57	0.56	–			
(110)	1.0	0.78	–	–	1.4	–
	0.71	0.63	–			
$(\frac{1}{3}\frac{1}{3}\frac{1}{3})$	0.60	0.56	–	1.2	1.1	1.2
	0.54	0.37	–			

Table 4. Frequency widths of the defect dynamical structure factors $\hat{\sigma}_{ab}(\mathbf{q},\omega)$ compared with those of the ion dynamical structure factors $S_{\alpha\beta}(\mathbf{q},\omega)$. Values are for $\Delta\nu \equiv \Delta\omega/2\pi$ in THz. Upper and lower values for each wavevector are results from the simulations at 1611 and 1390 K respectively. A dash indicates that the width could not be reliably estimated.

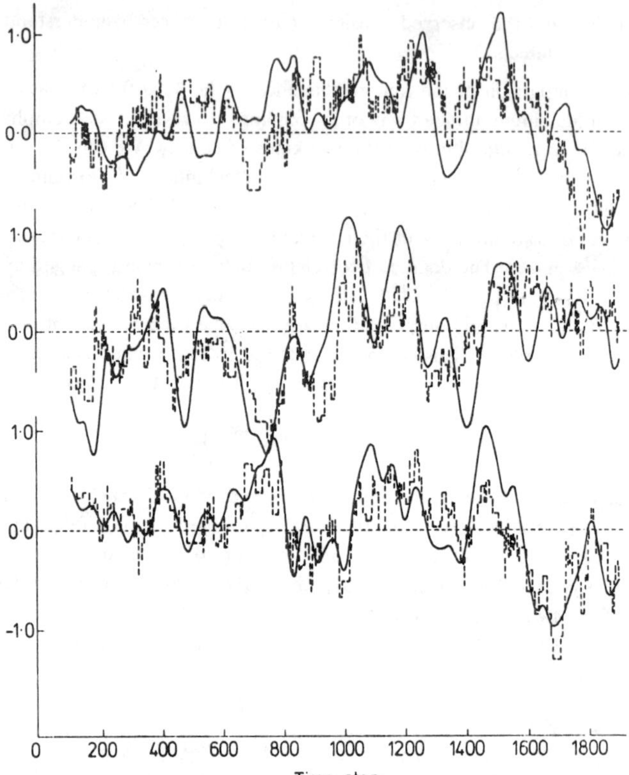

Figure 29. Direct comparison for simulated CaF_2 at 1611 K of the frequency-filtered anion density and a linear combination of the vacancy and interstitial densities, this linear combination beign chosen to optimize the correlation with the anion density. The three graphs show the comparison for three equivalent wavevectors $\mathbf{q} = (2\pi/a_0)$ (7/3,0,0), (0,7/3,0) and (0,0,7/3). From Gillan 1986b.

We show in Figure 30 the spectra and cross-spectra involving the anion density $\hat{\rho}_-(\mathbf{q})$ and the vacancy and interstitial densities for simulated CaF_2 at $T = 1611$ K with $\mathbf{q} = (2\pi/a_0)$ (7/3,0,0). These results show that in the quasielastic region, the anion density is strongly correlated with the interstitial density, but rather weakly correlated with the vacancy density: the quasielastic peak is mainly due to the motion of interstitials at this $\mathbf{q}$. From the cross-spectra, we can determine the values of the form factors $f_\alpha^a(\mathbf{q})$ that yield the maximum correlation between the linear combination of the defect densities defined in equation (83), and the actual $\hat{\rho}_\alpha(\mathbf{q})$ in the quasielastic region; this is explained in detail in Gillan (1986b). It turns out that over a wide range of $\mathbf{q}$, the $f_\alpha^a(\mathbf{q})$ can be chosen so that the correlation is nearly perfect in the quasielastic region. This means that the fluctuating variable responsible for the quasielastic peak is conclusively identified as a linear combination of defect densities. In this sense, we can assert that the peak is a manifestation of defect motions.

From this analysis, we obtain the values of the defect form factors $f^a_\alpha(\mathbf{q})$ as a function of $\mathbf{q}$. The results for two different temperatures are summarized in Table 5. We note that these depend strongly on $\mathbf{q}$. For example, $f^i_-(\mathbf{q})$ varies by an order of magnitude as we go from $\mathbf{q} = (2\pi/a_0)$ (1/3,0,0) to $(2\pi/a_0)$ (7/3,0,0).

At this point, we want to return to the question about the $\mathbf{q}$–dependence of the quasielastic intensity. We showed in equation (86) that this could come either from the $\mathbf{q}$–dependence of the form factors or from correlations between the defects, contained in $\hat{\sigma}_{ab}(\mathbf{q})$. We want to show now that it comes almost entirely from the strong variation of the $f^a_\alpha(\mathbf{q})$ displayed in Table 5. This can be seen most clearly from a calculation of the $\hat{\sigma}_{ab}(\mathbf{q})$. Note from the definition (equation (87)) of these quantities that if there were no correlations between the defects then all the cross terms in this formula involving pairs of distinct defects would vanish and we would

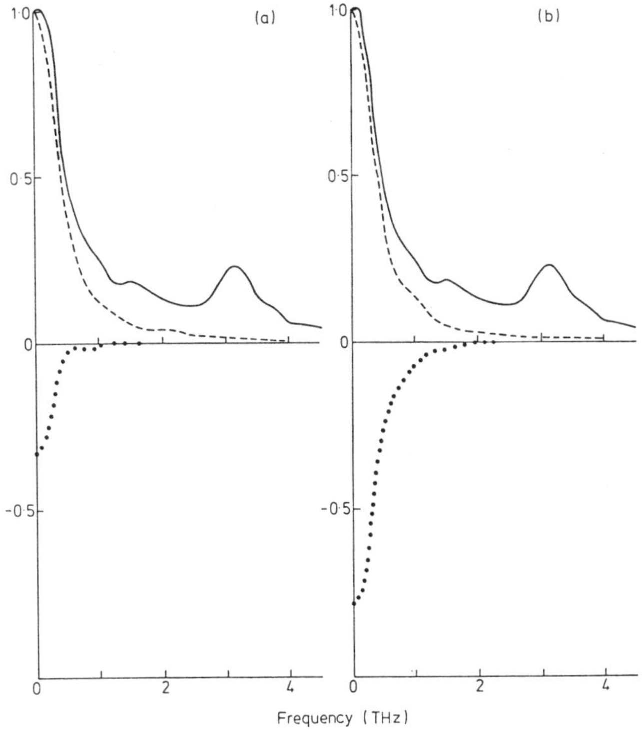

Figure 30. Autocorrelation and cross-correlation spectra $\langle AB^* \rangle_\omega$ for anions and defects calculated from a simulation of CaF_2 at 1611 K for the wavevector $\mathbf{q} = (2\pi/a_0)$ (7/3,0,0); (a)____ anion-anion, — vacancy-vacancy, ... anion-vacancy; (b)____ anion-anion, — interstitial-interstitial, ... anion-interstitial. The graphs show the normalized quantities $\langle AB^* \rangle_\omega / [\langle |A|^2 \rangle_\omega \langle |B|^2 \rangle_\omega]^{\frac{1}{2}}$. From Gillan 1986b.

κ	f^v_+	f^i_+	f^v_-	f^i_-
1/3	−0.23	0.14	−0.51	0.33
	−0.26	0.13	−0.57	0.32
2/3	−0.27	0.03	−0.55	0.11
	−0.29	0.04	−0.60	0.14
1	−0.38	−0.01	−0.63	0.00
	−0.41	−0.05	−0.72	0.00
4/3	−0.50	0.04	−0.84	0.10
	−0.56	0.04	−0.93	0.07
5/3	−0.95	0.61	−1.60	0.42
	−1.10	0.61	−1.84	0.32
7/3	1.21	−0.71	1.03	−3.80
	1.54	−0.61	0.09	−4.49
8/3	0.69	−0.07	0.26	−1.90
	0.98	−0.20	0.12	−2.08
3	0.66	−0.07	0.17	−1.27
	0.92	0.11	−0.22	−0.73

Table 5. Defect form factors $f^a_\alpha(\mathbf{q})$ for wavevectors $\mathbf{q} = (2\pi/a_0)\times(\kappa,0,0)$ in the $\langle 100\rangle$ direction. The upper and lower values for each wavevector are the results obtained from the simulations at 1611 and 1390 K respectively.

$(\kappa_1,\kappa_2,\kappa_3)$	σ_{vv}	σ_{ii}	σ_{vi}
$(\frac{1}{3}00)$	0.77	0.97	0.45
	0.91	1.06	0.43
$(\frac{2}{3}00)$	0.88	1.04	0.24
	0.93	1.00	0.23
(100)	1.04	1.14	0.26
	0.96	1.02	0.17
$(\frac{1}{3}\frac{1}{3}0)$	0.77	1.06	0.23
	0.89	1.02	0.18
$(\frac{2}{3}\frac{2}{3}0)$	0.96	0.98	−0.05
	0.98	0.97	−0.02
(110)	0.86	1.04	−0.05
	0.91	0.97	−0.09
$(\frac{1}{3}\frac{1}{3}\frac{1}{3})$	0.85	1.00	0.03
	0.92	0.99	0.03

Table 6. Static structure factors $\sigma_{ab}(\mathbf{q})$ for defects at different wavevectors $\mathbf{q} = (2\pi/a_0)\times(\kappa_1,\kappa_2,\kappa_3)$. Upper and lower values for each wavevector are results from the simulations at 1611 and 1390 K respectively.

get $\hat{\sigma}_{vv}(\mathbf{q}) = \hat{\sigma}_{ii}(\mathbf{q}) = 1$, and $\hat{\sigma}_{vi}(\mathbf{q}) = 0$ for all $\mathbf{q}$. The $\hat{\sigma}_{ab}(\mathbf{q})$ are straightforward to calculate, since we know the defect positions at each time step. We show results for simulated CaF_2 at low temperatures in Table 6. These imply that the correlations between defects are rather weak, especially at higher wavevectors. It is clear that the $\mathbf{q}$–dependence of the $\hat{\sigma}_{ab}(\mathbf{q})$ gives only a minor contribution to the variation of the quasielastic intensity, which therefore comes almost entirely from the defect form factors.

We saw in §10 that the quasielastic peak appears in all three of the dynamical structure factors $S_{\alpha\beta}(\mathbf{q},\omega)$, in spite of the fact that diffusion is occurring only on the anion sublattice. The reason for this is now clear. The distortion surrounding an anion defect affects both the cation and anion sublattices, as is clear from the form factors given in Table 5. Since the peak arises mainly from this distortion, one expects to see it in all three $S_{\alpha\beta}(\mathbf{q},\omega)$.

The calculations summarized in this section yield rather clear conclusions. The coherent quasielastic peak observed experimentally and in the simulations is due to the motion of the point defects identified in the hopping analysis. The lattice distortion surrounding the defects plays a key role. The strong $\mathbf{q}$–dependence of the quasielastic intensity is mainly associated with the form factors that characterize the distortion pattern.

13. Long-Wavelength Fluctuations

We have seen in the previous section that the presence of the quasielastic peak is intimately connected with the conduction process itself. It arises from the motion of the defects, which are responsible for the diffusion of the ions. This means that it should be possible to relate the frequency width of the peak to such things as the rates of hopping and of creation and annihilation of the defects. These relations have not yet been fully explored. However, it turns out to be possible to draw important conclusions by studying the peak at small q (long wavelengths). In this regime, the frequency width can be expressed rather simply in terms of the d.c. conductivity and the static dielectric constant.

It will be convenient to begin by considering the spectrum of charge fluctuations in the small-q limit. There is a general theory of this for ionic systems, which is valid regardless of whether the material is solid or liquid (Giaquinta *et al.* 1976, Giaquinta and Parrinello 1977, Zeyher 1978, Parrinello and Tosi 1979, Zehnlé and Baus 1981). Let $\hat{\rho}_Q(\mathbf{q})$ denote the charge density in the system at wavevector $\mathbf{q}$, with charges expressed in units of e. We shall define this with the following normalization:

$$\hat{\rho}_Q(\mathbf{q}) = (N_+ z_+^2 + N_- z_-^2)^{-\frac{1}{2}} \sum_\alpha z_\alpha \sum_{i=1}^{N_\alpha} e^{i\mathbf{q}\cdot\mathbf{r}_{\alpha i}} \; . \tag{89}$$

We define the charge-charge dynamical structure factor $S_{QQ}(\mathbf{q},\omega)$ as

$$S_{QQ}(\mathbf{q},\omega) = \frac{1}{2\pi} \int_{-\infty}^{\infty} dt \, e^{i\omega t} \langle \hat{\rho}_Q(\mathbf{q},t) \hat{\rho}_Q(\mathbf{q},0)^* \rangle \; . \tag{90}$$

This represents the spectrum of charge fluctuations, in the same way that the $S_{\alpha\beta}(\mathbf{q},\omega)$ represent the spectra of ion density fluctuations. In fact, the two are simply related:

$$S_{QQ}(\mathbf{q},\omega) = (N_+z_+^2 + N_-z_-^2)^{-1} \sum_{\alpha\beta} (N_\alpha N_\beta)^{\frac{1}{2}} z_\alpha z_\beta S_{\alpha\beta}(\mathbf{q},\omega) \ . \tag{91}$$

Now there is a general statistical-mechanical argument based on the fluctuation-dissipation theorem (Kubo 1957, Kubo 1966), which shows that $S_{QQ}(\mathbf{q},\omega)$ can be exactly expressed in terms of the wavevector- and frequency-dependent dielectric function $\varepsilon(\mathbf{q},\omega)$. The relation is (Parrinello and Tosi 1979):

$$S_{QQ}(\mathbf{q},\omega) = -(q^2/\pi\omega\kappa_D^2)\ \mathscr{I}m\ \varepsilon(\mathbf{q},\omega)^{-1} \ , \tag{92}$$

where κ_D is the Debye wavevector, given by

$$\kappa_D^2 = \frac{4\pi e^2}{k_B T} \sum_\alpha \bar{\rho}_\alpha z_\alpha^2 \ . \tag{93}$$

In the small-q limit, $\varepsilon(\mathbf{q},\omega)$ goes to the frequency-dependent dielectric function $\varepsilon(\omega)$ discussed in §9. Using the relation (56), we have:

$$S_{QQ}(\mathbf{q},\omega) \to -(q^2/\pi\kappa_D^2)\ \mathscr{I}m\ \frac{1}{\omega+4\pi i\sigma(\omega)} \ . \tag{94}$$

This means that at small q, we can predict $S_{QQ}(\mathbf{q},\omega)$, knowing only $\sigma(\omega)$. We can verify this directly, using the results for $\sigma(\omega)$ described in §9 and the results for the

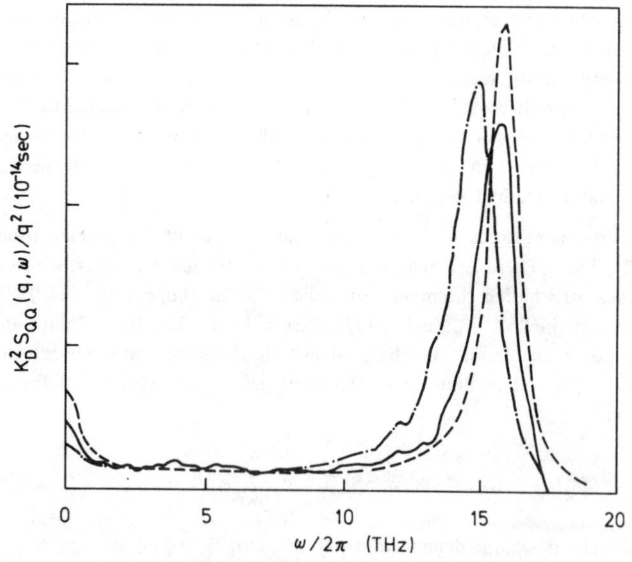

Figure 31. The charge-charge dynamical structure factor $S_{QQ}(\mathbf{q},\omega)$ multiplied by κ_D^2/q^2 for the two smallest available wavevectors $\mathbf{q} = (2\pi/a_0)$ (1/3,0,0) and (1/3,1/3,0) in simulated CaF_2 at 1611 K (solid and dotted curves respectively), compared with the $q \to 0$ limiting form calculated from the frequency-dependent conductivity (broken line).

$S_{\alpha\beta}(\mathbf{q},\omega)$ of §10 substituted into equation (91). Figure 31 compares the quantities on either side of equation (94) for the simulation of CaF_2 at 1611 K at the smallest available wavevectors $\mathbf{q} = (2\pi/a_0)\ (1/3,0,0)$ and $(2\pi/a_0)\ (1/3,1/3,0)$. The rather close agreement shows that this wavevector is small enough for equation (94) to be valid. The relevant condition is, in fact, that q should be smaller than the screening wavevector, but we shall not discuss this here.

The quasielastic peak in $S_{QQ}(\mathbf{q},\omega)$ has a rather small intensity, for a reason to be mentioned shortly. However, the width of the peak in all three $S_{\alpha\beta}(\mathbf{q},\omega)$ has the same value for this $\mathbf{q}$, which is therefore the value of the width in $S_{QQ}(\mathbf{q},\omega)$. What determines this width? For small ω, we have

$$\varepsilon(\omega) \simeq 4\pi i\sigma_0/\omega + \varepsilon_0 \ . \tag{95}$$

This will be accurate enough, if we can assume that $\mathcal{R}e\ \sigma(\omega)$ is constant and $\mathcal{I}m\ \sigma(\omega)$ is linear in ω over the range of the peak. Putting this into equation (94), we have:

$$\begin{aligned} S_{QQ}(\mathbf{q},\omega) &\simeq -(q^2/\pi\kappa_D^2)\ \mathcal{I}m\ \frac{1}{\varepsilon_0\,\omega+4\pi i\sigma_0} \\ &= \frac{4\sigma_0 q^2}{\kappa_D^2\varepsilon_0^2}\ \frac{1}{\omega^2+(4\pi\sigma_0/\varepsilon_0)^2} \ . \end{aligned} \tag{96}$$

The width of the peak is thus $\Delta\omega = 4\pi\sigma_0/\varepsilon_0$. Let us check this. From the results for σ_0 and ε_0 from the simulation of CaF_2 at $T = 1390$ K, we obtain the estimate $\Delta\omega/2\pi = 0.19$ THz. Going now to Table 4, we see that this value agrees well with the width observed in the simulation at $\mathbf{q} = (2\pi/a_0)\ (1/3,0,0)$. As we go to higher temperatures, the agreement becomes less good, mainly because the assumption that $\mathcal{R}e\ \sigma(\omega)$ is constant becomes less adequate.

The reason why the quasielastic peak in $S_{QQ}(\mathbf{q},\omega)$ is small can be understood from equation (92). This shows that its intensity is $q^2/\kappa_D^2\varepsilon_0$. However, the *total* intensity of $S_{QQ}(\mathbf{q},\omega)$ is q^2/κ_D^2 at small q – this is known as the Stillinger-Lovett sum rule (Stillinger and Lovett 1968a,b). The quasielastic peak thus accounts for only $1/\varepsilon_0$ of the total.

It is instructive to consider how the width arises from the motion of the point defects. Since the charges carried by vacancies and interstitials are equal and opposite, charge fluctuations are associated with the difference of concentration of the two kinds of defect. This difference, $P_d(\mathbf{r})$, varies because of the diffusion of the defects. Let the flux of vacancies minus the flux of interstitials be denoted by $\mathbf{J}_d(\mathbf{r})$. Then we have

$$\partial P_d/\partial t + \nabla\cdot\mathbf{J}_d = 0 \ . \tag{97}$$

This 'conservation equation' arises because the difference of the *total* numbers of vacancies and interstitials is a conserved quantity (equal to zero). Now the fluctuations of defect densities gives rise to fluctuations of the electric field $\mathbf{E}(\mathbf{r})$. At long wavelengths, $\mathbf{E}(\mathbf{r})$ will be related to $P_d(\mathbf{r})$ by Poisson's equation

$$\nabla\cdot\mathbf{E} = 4\pi e P_d/\varepsilon_0 \ , \tag{98}$$

where we have noted that the charges on the defects are $\pm e$; we include a factor $1/\varepsilon_0$ to account for the dielectric response of the lattice. This electric field will cause a drift of defects. Since we are at long wavelengths, the relation between the two will be given by the macroscopic d.c. conductivity:

$$e\mathbf{J}_d = \sigma_0 \mathbf{E} \tag{99}$$

If we solve equations (97-99) in Fourier space, we find that the time dependence of $\hat{P}_d(\mathbf{q})$, the difference of defect densities at wavevector $\mathbf{q}$ is

$$\partial \hat{P}_d(\mathbf{q})/\partial t = -(4\pi\sigma_0/\varepsilon_0)\hat{P}_d(\mathbf{q}) \tag{100}$$

The relaxation rate here is exactly the frequency width of $S_{QQ}(\mathbf{q},\omega)$. This is as it should be: the slow fluctuations of charge density are associated with variations of the diffence of vacancy and interstitial densities.

Now let us consider instead the *sum* of the vacancy and interstitial concentrations. The total number of defects in the system is not constant. The creation and annihilation of vacancy-interstitial pairs causes their number to fluctuate. Let $\bar{n}_s$ be the mean number of defects in the system and $\bar{n}_s + \delta n_s$ be the number present at any instant. The fluctuations of this number can be studied by examining the correlation function

$$s(t) = \langle \delta n_s(t) \delta n_s(0) \rangle / \langle \delta n_s^2 \rangle \ . \tag{101}$$

The decay time of $s(t)$ provides us with a measure of the mean lifetime τ_l of the defects.

The correlation function $s(t)$ can be calculated by simulation, since our defect analysis (§11) yields the value of n_s at each time step. Figure 32 shows results for $s(t)$ for CaF_2 at 1390 K. We see that the defect lifetime at this temperature is very short – of the order of 0.9 psec. This is what we should expect from the discussion of §11. There we discussed the contributions to the anion diffusion coefficient due to the hopping of defects and due to their creation and annihilation. We found that these contributions are comparable at high temperatures: the defects are being created and destroyed nearly as fast as they are hopping.

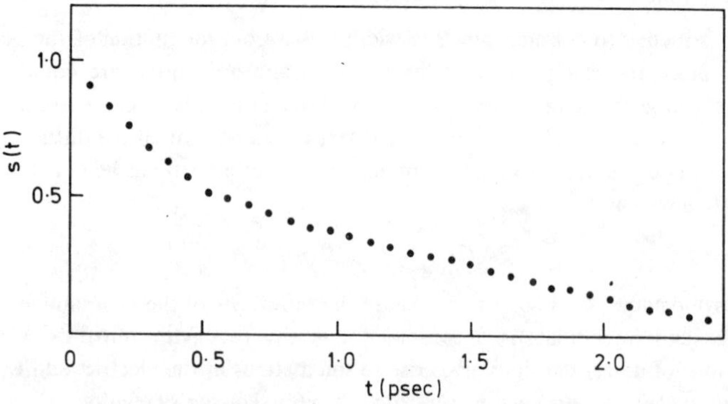

Figure 32. Normalized defect-number autocorrelation function $s(t)$ (equation (101)) in simulated CaF_2 at 1390 K.

It will be noted that $s(t)$ is the zero-q limit of the correlation function $\langle \hat{P}_s(\mathbf{q},t)\hat{P}_s(\mathbf{q},0)^* \rangle$ describing fluctuations of the sum $\hat{P}_s(\mathbf{q})$ of the vacancy and interstitial concentrations at wavevector $\mathbf{q}$. The general picture, then, is that the fluctuations of defect concentration at long wavelengths are governed by two relaxation times: (i) the conductivity relaxation time $\tau_c = (4\pi\sigma_0/\varepsilon_0)^{-1}$ for the *difference* of vacancy and interstitial concentrations, and (ii) the defect lifetime τ_l for the *sum* of the concentrations. Although these two times are physically independent, the results we have already presented imply that they are numerically almost identical in the superionic state. This follows from the fact that $\hat{P}_d(\mathbf{q})$ and $\hat{P}_s(\mathbf{q})$ can be written as

$$\begin{aligned} \hat{P}_d(\mathbf{q}) &= \hat{P}_v(\mathbf{q}) - \hat{P}_i(\mathbf{q}) \\ \hat{P}_s(\mathbf{q}) &= \hat{P}_v(\mathbf{q}) + \hat{P}_i(\mathbf{q}) \cdot \end{aligned} \tag{102}$$

The inverse times τ_c^{-1} and τ_l^{-1} are thus simply the frequency widths of the peaks in the two functions $\hat{\sigma}_{vv}(\mathbf{q},\omega) + \hat{\sigma}_{ii}(\mathbf{q},\omega) \mp 2\hat{\sigma}_{vi}(\mathbf{q},\omega)$. But we know that the widths in the three functions $\hat{\sigma}_{ab}(\mathbf{q},\omega)$ for given $\mathbf{q}$ are all the same (see Figure 30 and Table 4). The reason for the approximate equality $\tau_c \simeq \tau_l$ remains to be explained, but the conclusion is that there is essentially a single quasielastic width at long wavelengths, which is given by $4\pi\sigma_0/\varepsilon_0$.

It is unfortunate that although the quasielastic peak is readily observable in the simulations at small q, it cannot be seen experimentally in this region. The reason is simply that its intensity becomes very small there, as is clear from Figure 24. However, according to the defect interpretation, the width observed in the ionic dynamical structure factors $S_{\alpha\beta}(\mathbf{q},\omega)$ should be the same as that of the peak in the defect dynamical structure factors $\hat{\sigma}_{ab}(\mathbf{q},\omega)$. We have already seen that this is true. Now since the defects reside on the regular anion sites, the $\hat{\sigma}_{ab}(\mathbf{q},\omega)$ are by definition identical for wavevectors differing by any reciprocal lattice vector $(4\pi/a_0)\ (h,k,l)$ of the anion simple cubic lattice. The same periodicity in $\mathbf{q}$–space should therefore be found for the observed widths. Examination of Table 4 shows that this is true. For example, the quasielastic widths at the three wavevectors $(\mathbf{q}a_0/2\pi) = (\frac{1}{3},0,0)$, $(\frac{5}{3},0,0)$ and $(\frac{7}{3},0,0)$ are all the same. This means that the $q \to 0$ width should be the same as the width observed for $\mathbf{q} \to (2\pi/a_0)\ (2,0,0)$, which should therefore be non-zero and given by $4\pi\sigma_0/\varepsilon_0$. According to the neutron-scattering measurements on PbF_2 (Hutchings *et al.* 1984), the width does indeed appear to go to a constant at this reciprocal lattice vector. In CaF_2, the situation is not quite so clear, and there seems to be an indication that the width is tending to zero as $\mathbf{q} \to (2\pi/a_0)\ (2,0,0)$. This would be very hard to reconcile with the theory we have given above; it would also be strange if the two materials behaved in qualitatively different ways. Let us leave aside this problem with CaF_2, and ask whether the experimental width in PbF_2 is indeed equal to the experimental value of $4\pi\sigma_0/\varepsilon_0$. To test this, we use the measured widths for different temperatures at $\mathbf{q} = (2\pi/a_0)\ (2.15,0,0)$, which is the experimental $\mathbf{q}$ nearest to $(2\pi/a_0)\ (2,0,0)$. The conductivity is taken from the measurements of Carr *et al.* (1978). No measurements of ε_0 appear to be available above 300 K, but extrapolation of Samara's data (Samara 1976) indicates that, in the region of T_c, $\varepsilon_0 \simeq 25$. The comparison is presented in Figure 33. Above T_c, the agreement is quite good, but there seems to be a noticeable discrepancy below this temperature. However, the widths become very small and subject to larger relative errors at the lower temperatures. In addition, it is not quite clear that the wavevector is close enough to the reciprocal lattice vector. It is therefore hard to draw a definitive conclusion at this stage. Our explanation of the width nevertheless seems to us consistent with the data.

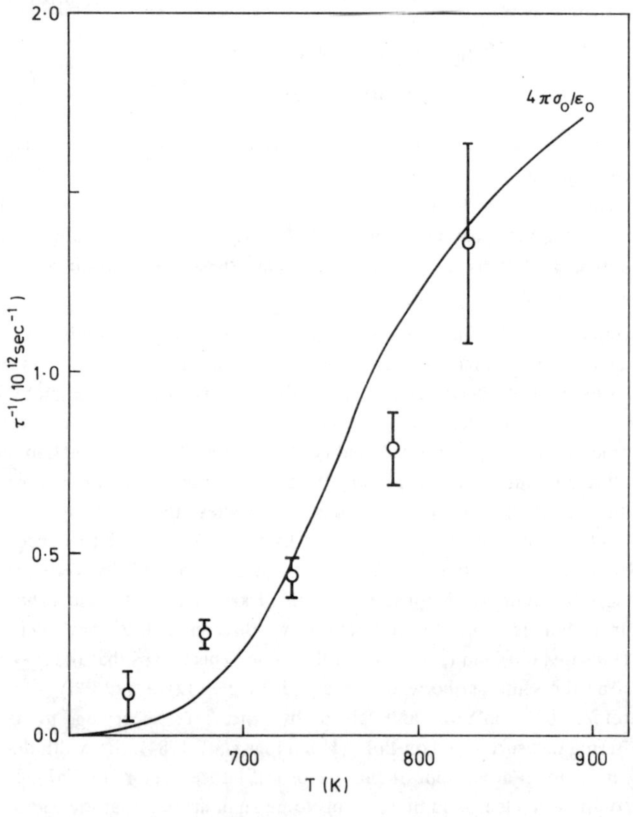

Figure 33. Comparison of the experimental quasielastic width $\Delta\omega$ (Hutchings *et al.* 1984) in PbF_2 at $\mathbf{q} = (2\pi/a_0)\ (2.15,0,0)$ (circles with error bars) with the quantity $4\pi\sigma_0/\varepsilon_0$. Experimental values of the d.c. conductivity σ_0 are taken from Carr *et al.* (1978), and the value of the static dielectric constant ε_0 is extrapolated from data of Samara (1976).

14. Discussion

Two related themes run through the work we have described: (i) the validation of simulation by comparison with experimental results; and (ii) the use of simulation to construct an interpretation of the observations.

14.1 *Comparison with experiment*

One reason why MD simulation is important is that it allows one to calculate many different physical quantities that can be measured experimentally. The quantities that we have discussed are: the tracer diffusion coefficient, the van Hove self-correlation function, the density distribution of the ions, the radial distribution functions, the frequency-dependent electrical conductivity and the dynamical structure factors. The comparisons with experiment that we have discussed indicate that in all essential respects the simulation models faithfully reproduce the behaviour of the real materials on the atomic scale. Most importantly, the anion diffusion coefficient rises to liquid-like values in the correct region of temperature and its magnitude in the superionic state agrees semi-quantitatively with experimental measurements. The experimental and simulation results for the van Hove self function both agree that diffusion occurs by jumping of the anions mainly between first- and second-neighbour regular sites, and they yield almost the same values for the proportions of the two kinds of jumps. The anion density distribution $\rho_-(\mathbf{r})$ calculated by simulation of PbF_2 agrees satisfactorily with the results of diffraction experiments. We attach particular importance to the comparison of the simulated dynamical structure factors with the results of quasielastic coherent neutron scattering. The observed quasielastic peak appears in the simulations at the correct temperature, and the temperature- and wavevector-dependence of its intensity agree very satisfactorily with experiment; the simulated and experimental frequency widths are also in reasonable accord at high temperature, though there seems to be a difference below T_c. There is also a significant difference with experiment for the Haven ratio in this region of temperature. We have suggested that these two discrepancies both arise from a size effect in the simulation; this appears when the mean number of vacancy-interstitial pairs in the simulated system is less than unity.

It is important to appreciate that the information that goes into the simulations consists solely of the model interionic potentials. Once these have been chosen, the simulated system is, in a sense, autonomous – in the same sense, indeed, as a real system is autonomous. The trajectories of the ions are dictated entirely by the prescribed equation of motion. The quantities that are compared with experiment are calculated by simple averaging over these trajectories. The calculation involves no adjustment of parameters, nor any selection of 'significant' events.

14.2 *Interpretation*

Although the fluorites are among the simplest superionic conductors, an atomistic interpretation of their behaviour on the basis of experimental data alone is not straightforward. The great advantage of MD simulation is that it provides full information about the atomic motions. This means that proposed interpretations can be exhaustively tested.

The results that we have described lead to a simple and unified description of the diffusion mechanism. Diffusion occurs by discrete sudden hops of the anions between their regular sites – sudden in the sense that the flight time τ_f is much less than the residence time τ_r of the ions: typically, $\tau_f = 0.6$ psec, $\tau_r = 7$ psec in the superionic state. Most of the hops (~ 80 %) are between first-neighbour regular sites, the remainder being almost all between next-neighbour sites. The Chudley-Elliott model gives an excellent description of single-ion diffusion. The diffusing ions cause an increase of ion density in the region between neighbouring regular anion sites. The diffusion can be interpreted in terms of vacancy and interstitial defects, both of which jump between the regular sites. The concentration of defects is small (up to 3 % of vacancies per regular site), and their mobility is extremely high. The defect lifetimes are short: typically less than 1 psec above T_c. The motion of the defects, as they are created, jump and are annihilated, is responsible for the quasielastic peak observed in coherent neutron scattering. The defects cause a distortion of the surrounding lattice; the strong variation of the quasielastic intensity with wavevector comes mainly from the **q**–dependence of the form factor describing this distortion. The frequency width of this peak at small q is simply related to the d.c. conductivity.

An important conclusion from the simulations is that in the superionic state, vacancies and interstitials contribute roughly equally to the diffusion of the ions. It has often been assumed in the past (see e.g. Mościński and Jacobs 1985, Allnatt *et al.* 1987) that since diffusion occurs entirely by jumps between the anion regular sites, a vacancy mechanism must be dominant. We want to stress here the reason why this assumption is unjustified. As we pointed out in § 2.1, conductivity measurements indicate that as the temperature rises towards T_c, the jump rates of vacancies and interstitials become comparable, and that the residence time for both defects becomes exceedingly short – hardly greater, in fact, than the typical vibrational period. This must imply that interstitials cease to occupy the cube-centre interstitial site in a well-defined manner, and we have seen that the observed density distribution in the superionic state supports this conclusion. The fact that the interstitials no longer occupy a distinct site does not, of course, mean that they stop being interstitials. It does mean, though, that the ion jumps associated with their motion are now effectively jumps between the regular sites. The observed jump mechanism of the anions is therefore entirely compatible with the roughly equal mobility of vacancies and interstitials.

We now consider briefly the relation between the picture we have established and the cluster picture proposed by Catlow and Hayes (1982) and by Hutchings *et al.* (1984). At first sight, it appears that the two pictures are radically different. One of the essential points demonstrated by our results is that there is an intimate connection between the coherent quasielastic peak and the superionic conduction process itself: they both arise from the same thing, namely the motion of point defects. The two types of defect have essentially the same dynamic behaviour and play a rather symmetrical role, although the interstitial happens to contribute more to the intensity in the region of **q**–space where the observed intensity is large. The cluster picture, by contrast, is often presented as though '..the interstitials are tied up in clusters, while the vacancies carry most of the current..' (Mościński and

Jacobs 1985), so that the fast conduction and the coherent quasielastic effects have little connection with each other. Our results show that that particular view of the cluster model is definitely incorrect.

We believe nevertheless that the version of the cluster model advocated by Hutchings *et al.* (1984) can be largely reconciled with the mechanism we have described. We have in mind particularly the so-called 3:1:2 cluster and its close relative the 9:1:8 cluster, which Hutchings *et al.* find give a reasonable interpretation of their results. The first of these consists of an interstitial near the mid-point of a cube edge, the two neighbouring anions being relaxed away from it, together with a vacancy at a more distant site; the 9:1:8 cluster, which the experimentalists find to be the most satisfactory of those they have considered, differs from this only in including further relaxed ions. We believe that the additional distant vacancy has little effect on the scattering from these clusters, and can be ignored. Then both clusters consist essentially of a single interstitial with some surrounding lattice distortion, and correspond almost exactly to what we refer to simply as an interstitial. Whether one chooses to assign this interstitial to a regular site, as we do, or to a mid-site position, as the experimentalists do, is almost irrelevant, given its exceedingly high mobility. From this point of view, though, we regard the term 'cluster' as unfortunate: one should be clear that it does not refer to a cluster of defects, but to a single defect surrounded by lattice distortion.

It will be remarked that the picture outlined in this article differs strongly from the description that is often suggested in the literature. It has often been assumed in the past that the defect structure of fluorites in the superionic state is the same as at lower temperatures, and in particular that a substantial fraction of the anions occupy the cube-centre 'interstitial' position. We have seen that this idea is incompatible with the simulation results. Early diffraction data appeared to support cube-centre occupation above T_c, but it is now recognized that the interpretation of the data was incorrect. The good agreement between the simulation results for the anion distribution $\rho_-(\mathbf{r})$ for PbF_2 and recent diffraction measurements confirm the correctness of simulation on this point. In fact the absence of significant cube-centre occupation should be no surprise. As we pointed out in §2, the conductivity measurements on PbF_2 indicate that even substantially below T_c the residence time of cube-centre interstitials is only a few vibration periods. It therefore seems quite reasonable that the cube-centre site will cease to be occupied in a well-defined manner above T_c. It has been pointed out that in any case interactions between vacancies and interstitials will render this site unstable when the defect concentration rises to the percent level (Gillan and Richardson 1979). It is unfortunate that much interpretative work on superionic fluorites has been based on the importance of cube-centre interstitials. In our view, the value of such work now seems questionable, to say the least.

Acknowledgment

The preparation of this paper was supported in part by the Underlying Research Programme of the United Kingdom Atomic Energy Authority.

References

Allen, M. P. and Tildesley, D. J., 1987, *Computer Simulation of Liquids,* (Oxford: Clarendon)

Allnatt, A. R., Chadwick, A. V. and Jacobs, P. W. M., 1987, *Proc. R. Soc. A,* **410**, 385

Ashcroft, N. W. and Mermin, N. D., 1976, *Solid State Physics,* (New York: Holt, Rinehart and Winston), ch.27

Azimi, A., Carr, V. M., Chadwick, A. V., Kirkwood, F. G. and Saghafian, R., 1984, *J. Phys. Chem. Solids,* **45**, 23

Bachmann, R. and Schulz, H., 1983, *Proc. Int. Conf. on Solid State Ionics, Grenoble 1983,* (Amsterdam: North-Holland), p.521

Baker, J. M., 1974, in *Crystals with the Fluorite Structure,* ed. W. Hayes, (Oxford: Clarendon), p.341

Baker, J. M., Davies, E. R. and Hurrell, J. P., 1968, *Proc. R. Soc. A,* **308**, 403

Bates, J. B. and Farrington, G. C. (ed.), 1981, *Proc. Int. Conf. on Fast Ion Transport in Solids, Gatlinburg,* (Amsterdam: North-Holland)

Berendsen, H. J. C. and van Gunsteren, W. F., 1986, in *Simulazione di sistemi statistico-meccanici con la dinamica molecolare,* (Amsterdam: North-Holland), p.43

Brass, A., 1987, Ph. D. thesis, University of Edinburgh (unpublished)

Carr, V. M., Chadwick, A. V. and Saghafian, R., 1978, *J. Phys. C,* **11**, L637

Catlow, C. R. A., 1980, *Comments Solid State Phys.,* **9**, 157

Catlow, C. R. A. and Norgett, M. J., 1973, *J. Phys. C,* **6**, 1325

Catlow, C. R. A., Norgett, M. J. and Ross, T. A., 1977, *J. Phys. C,* **10**, 1627

Catlow, C. R. A. and Hayes, W., 1982, *J. Phys. C,* **15**, L9

Catlow, C. R. A., Dixon, M. and Mackrodt, W. C., 1982, in *Computer Simulation of Solids,* ed. C. R. A. Catlow and W. C. Mackrodt, Lecture Notes in Physics, Vol. 166 (Berlin: Springer), p.130

Chadwick, A. V., 1983a, *Solid State Ionics,* **8**, 209

Chadwick, A. V., 1983b, *Radiat. Effects,* **74**, 17

Chudley, C. T. and Elliott, R. J., 1961, *Proc. Phys. Soc. (London),* **77**, 353

Clausen, K., Hayes, W., Hutchings, M. T., Kjems, J. K., Schnabel, P. and Smith, C., 1981, *Solid State Ionics,* **5**, 589

Compaan, K. and Haven, Y., 1956, *Trans. Faraday Soc.,* **52**, 786

Compaan, K. and Haven, Y., 1956, *Trans. Faraday Soc.,* **54**, 1498

Cooper, M. J., 1970, *Acta Cryst. A,* **26**, 208

Corish, J. and Jacobs, P. W. M., 1973, in *Surface and Defect Properties of Solids, Vol. 2,* (London: Chemical Society)

Dawson, B., Hurley, A. C. and Maslen, V. W., 1967, *Proc. Roy. Soc.,* A**298**, 289

de Leeuw, S. W., Perram, J. W. and Smith, E. R., 1980, *Proc. Roy. Soc.,* A**373**, 27

Derrington, C. E., Lindner, A. and O'Keeffe, M., 1975, *J. Solid State Chem.,* **15**, 171

Dick, B. G. and Overhauser, A. W., 1958, *Phys. Rev.,* **112**, 90

Dickens, M. H., Hutchings, M. T., Kjems, J. K. and Lechner, R. E., 1978, *J. Phys. C,* **11**, L583

Dickens, M. H., Hayes, W., Smith, C., Hutchings, M. T. and Kjems, J. K., 1979, in *Fast Ion Transport in Solids,* ed. P. Vashishta, J. N. Mundy and G. K. Shenoy, (Amsterdam: Elsevier), p.229

Dickens, M. H., Hayes, W., Hutchings, M. T. and Smith, C., 1982, *J. Phys. C,* **15**, 4043

Dickens, M. H., Hayes, W., Schnabel, P., Hutchings, M. T., Lechner, R. E. and Renker, B., 1983, *J. Phys. C,* **16**, L1

Dixon, M. and Gillan, M. J., 1978, *J. Phys. C,* **11**, L165

Dixon, M. and Gillan, M. J., 1980a, *J. Phys. C,* **13**, 1919

Dixon, M. and Gillan, M. J., 1980b, *J. Phys. (Paris),* **41**, C6-24

Enderby, J. E. and Neilson, G. W., 1981, *Rep. Prog. Phys.,* **44**, 593

Evans, D. J. and Morriss, G. P., 1984, *Comput. Phys. Rep.,* **1**, 297

Ewald, P., 1921, *Ann. Phys.,* **64**, 253

Gear, C. W., 1971, *Numerical Initial Value Problems in Ordinary Differential Equations,* (Englewood Cliffs N. J.: Prentice-Hall)

Giaquinta, P. V., Parrinello, M. and Tosi, M. P., 1976, *Phys. Chem. Liquids,* **5**, 305

Giaquinta, P. V. and Parrinello, M., 1977, *Lett. Nuovo Cim.,* **19**, 215

Gillan, M. J., 1985, *Physica B,* **131**, 157

Gillan, M. J., 1986a, *J. Phys. C,* **19**, 3391

Gillan, M. J., 1986b, *J. Phys. C,* **19**, 3517

Gillan, M. J. and Richardson, D. D., 1979, *J. Phys. C,* **12**, L61

Gillan, M. J. and Dixon, M., 1980a, *J. Phys. C,* **13**, L835

Gillan, M. J. and Dixon, M., 1980b, *J. Phys. C,* **13**, 1901

Gillan, M. J. and Dixon, M., 1981, *Solid State Ionics,* **5**, 593

Gillan, M. J. and Holloway, R. W., 1985, *J. Phys. C,* **18**, 4903

Gillan, M. J. and Wolf, D., 1985, *Phys. Rev. Letts.,* **55**, 1299

Hansen, J.-P. and McDonald, I. R., 1986, *Theory of Simple Liquids,* (New York: Academic)

Hayes, W., 1978, *Contemp. Phys.,* **19**, 469

Hayes, W., 1980, *J. Phys. (Paris),* **41**, C6-7

Hodby, J. W., 1974, in *Crystals with the Fluorite Structure,* ed. W. Hayes, (Oxford: Clarendon), p.1

Howard, R. E. and Lidiard, A. B., 1964, *Rep. Prog. Phys.,* **27**, 161

Hutchings, M. T., Clausen, K. Dickens, M. H., Hayes, W., Kjems, J. K., Schnabel, P. G. and Smith, C., 1984, *J. Phys. C,* **17**, 3903

Jacobs, P. W. M. and Ong, S. H., 1976, *J. Physique (Paris),* **37**, *C7,* 331

Jacucci, G. and Rahman, A., 1978, *J. Chem. Phys.,* **69**, 4117

Kleitz, M., Sapoval, B. and Chabre, Y. (ed.), 1983, *Proc. Int. Conf. on Solid State Ionics, Grenoble, 1983,* (Amsterdam: North-Holland)

Koto, K., Schulz, H. and Huggins, R. A., 1980, *Solid State Ionics,* **1**, 355

Kubo, R., 1957, *J. Phys. Soc. Japan,* **12**, 570

Kubo, R., 1966, *Rep. Prog. Phys.,* **29**, 255

Landau, L. D. and Lifshitz, E. M., 1980, *Statistical Mechanics,* (Oxford: Pergamon)

Lidiard, A. B., 1957, in *Handb. d. Phys.,* **20**, 246, (Berlin: Springer)

Lidiard, A. B., 1974, in *Crystals with the Fluorite Structure,* ed. W. Hayes, (Oxford: Clarendon), p.101

Lidiard, A. B., 1980, Harwell Report AERE-TP.841

Lovesey, S. W., 1984, *Theory of Neutron Scattering from Condensed Matter,* (Oxford: Clarendon)

Matar, S. F., Reau, J. M., Hagenmuller, P. and Catlow, C. R. A., 1984, *J. Phys. Chem. Solids,* **45**, 453

Mościński, J. and Jacobs, P. W. M., 1985, *Proc. R. Soc. A,* **398**, 173

Parrinello, M. and Tosi, M. P., 1979, *Riv. Nuovo Cim.,* **2**, 1

Pauling, L., 1960, *Nature of the Chemical Bond, 3rd edition,* (Ithaca, N. Y.: Cornell Univ. Press)

Perram, J. W. (ed.), 1983, *The Physics of Superionic Conductors and Electrode Materials, NATO ASI, Odense,* (New York: Plenum)

Rahman, A., 1976, *J. Chem. Phys.,* **65**, 4845

Rowe, J. M., Sköld, K., Flotow, H. E. and Rush, J. J., 1971, *J. Phys. Chem. Solids,* **32**, 41

Salamon, M. B. (ed.), 1979, *Physics of Superionic Conductors,* (Berlin: Springer)

Samara, G. A., 1976, *Phys. Rev. B,* **13**, 4529

Sangster, M. J. and Dixon, M., 1976, *Adv. Phys.,* **25**, 247

Schoonman, J., 1980, *Solid State Ionics,* **1**, 123

Schröter, W. and Nölting, J., 1980, *J. Phys. (Paris)*, **41**, C6-20

Shapiro, S. M. and Reidinger, F., 1979, in *Physics of Superionic Conductors*, ed. M. B. Salamon, (Berlin: Springer), p.45

Stillinger, F. H. and Lovett, R., 1968a, *J. Chem. Phys.*, **48**, 3858

Stillinger, F. H. and Lovett, R., 1968b, *J. Chem. Phys.*, **49**, 1991

Verlet, L., 1967, *Phys. Rev.*, **159**, 98

Walker, A. B., Dixon, M. and Gillan, M. J., 1982, *J. Phys. C*, **15**, 4061

Willis, B. T. M., 1963a, *Proc. R. Soc. A*, **274**, 122

Willis, B. T. M., 1963b, *Proc. R. Soc. A*, **274**, 134

Willis, B. T. M., 1965, *Acta Cryst.*, **18**, 75

Yarnell, J. L., Katz, M. J., Wenzel, R. G. and Koenig, S. H., 1973, *Phys. Rev. A*, **7**, 2130

Zehnlé, V. and Baus, M., 1981, *Z. Phys. B*, **44**, 265

Zeyher, R., 1978, *Z. Phys. B*, **31**, 127

Appendix 1: A Note on the Defect Scheme

Since the defect analysis described in §11 is relatively unfamiliar, we feel it may be worth adding here some supplementary comments and illustrating how it works for a simple one-dimensional model.

The notion of vacancies needs no comment, but the idea of an interstitial jumping between regular sites may require clarification. We have found it helpful to think about this in terms of the so-called Frenkel-Kontorova model, which is a linear chain of particles coupled by springs, the particles being acted on by a periodic potential. We assume here that there is one particle per period. The correspondence with superionic conductors is that the particles correspond to the mobile ions, the springs to the interactions between these ions, and the periodic potential to the interaction with the immobile counter-ions. This model can obviously not be literally applied to fluorites; we use it here merely to discuss the defect analysis. We apply this as we did for the fluorites, but instead of site spheres we now have 'site segments' of half-length λd centred on the potential minima (Figure 34).

At V_1 in Figure 34, we show the stable configuration of a vacancy in the model. In the remainder of the sequence V_2 ... we show the vacancy jumping to a neighbouring site. At V_2 the particle leaves segment 1 and at V_4 it arrives at segment 2. Configuration V_3 is supposed to be at a time half-way between V_2 and V_4; this is the instant of the jump according to our definition: at this instant, by definition, the occupancy of site 1 decreases by unity and that of site 2 increases by unity. Remark that because of statistical fluctuations in thermal equilibrium, the instant of a jump will not generally be the moment at which the particle crosses the maximum of the potential. All this correponds to the usual idea of vacancy diffusion. Now consider the jump motion of the interstitial, which is shown in the sequence I_1 to I_5. At I_2, the jumping particle leaves segment 1 and at I_4 it arrives at segment 2. The instant of the jump is at V_3. Before this instant the interstitial is on site 1 and after it on site 2.

We stress once more that the presence of a vacancy on a site is not signalled by the absence of a particle from the corresponding segment (or site sphere). Similarly, the presence of an interstitial on a site does not necessarily have anything to do with two particles being in the segment. The presence of the defects is defined in terms of the intersite jumps – it is in the definition of the latter that we need the segments (or spheres).

Appendix 2: A Note on the Haven Ratio

The results for the d.c. conductivity σ_0 presented in §9 indicate that the Haven ratio H_R is underestimated by the simulations for $T \leqslant T_c$. The quasielastic widths appear to be overestimated in this region (§10). We want to suggest that the two effects are related, and are an artefact due to the limited size of the simulated system.

Let us recall the meaning of the Haven ratio. In real systems, if the defect concentration is very low, the defects move independently of one another and

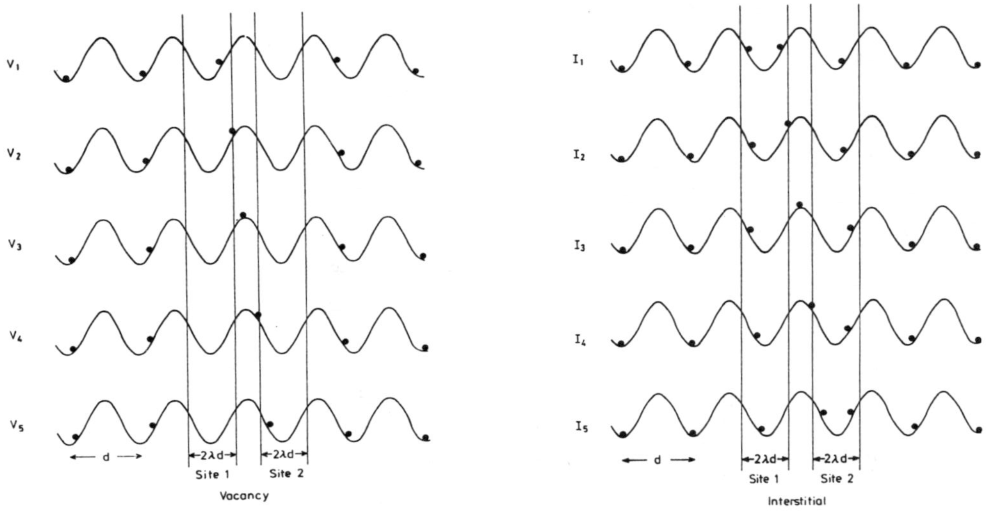

Figure 34. Schematic depiction of vacancy and interstitial motion in the Frenkel-Kontorova model, to illustrate the operation of the defect analysis used for studying the motion of defects in simulations of fluorites. Sequence $V_1,\ldots V_5$ shows the jump of a vacancy from site 2 to site 1, detected by the passage of a particle from segment 1 to segment 2; the particle leaves segment 1 at V_2 and enters segment 2 at V_4. Sequence $I_1,\ldots I_5$ shows the analogous process where an interstitial jumps from site 1 to site 2.

consecutive jumps of each defect are uncorrelated. Consecutive jumps of each ion, however, are not uncorrelated. For example, when a vacancy jumps between sites, there is a well-defined probability that the next jump of the ion involved will be caused by the same vacancy; this causes a correlation between consecutive jumps of the ions. The calculation of the Haven ratio, which gives a measure of this effect, is a well-understood problem.

If the jumps of the defects themselves are not uncorrelated, the usual calculation of the Haven ratio will not be correct. This is what will happen when the concentration of defects is high enough for interactions between them to become significant. The electric field produced by each defect will bias the jumping of other defects. Because of this bias, consecutive jumps of a defect will tend to be in the same direction. For a given concentration of defects, and a given defect jump rate, the electrical conductivity will be higher than it would be if the defect motion were random. This mechanism will therefore cause the Haven ratio to decrease as we approach T_c in fluorites.

A simulation will overestimate this bias effect at low defect concentrations. Suppose the concentration is so low that the mean number of vacancies (and interstitials) in the simulated system is much less than unity. Most of the time, there will be no defects at all. Occasionally, there will be one vacancy and one interstitial. Then, though the mean concentration of defects may be correctly given by the simulation, the mean distance between the defects will be quite incorrect. The exaggerated interaction between the defects will produce a correlation in their motion which would not happen in the real system. Consequently, the simulation will underestimate the Haven ratio if the defect concentration is low. The underestimation will become significant when the mean number of vacancies in the system is less than unity. For the system of 324 ions used for most of the present work, this means $T \lesssim T_c$.

We have pointed out in §13 that the coherent quasielastic width at long wavelengths is proportional to the d.c. conductivity σ_0. Therefore, if σ_0 is too large, the width will also be too large. We believe that this is at least partly responsible for the disagreement between the simulated and experimental widths below T_c.

Chapter 4

IONIC DISORDER IN CRYSTALS AT HIGH TEMPERATURES WITH EMPHASIS ON FLUORITES

W. Hayes* and M.T. Hutchings**

*Clarendon Laboratory, Parks Road, Oxford, U.K.
**Materials Physics and Metallurgy Division,
Harwell Laboratory, Didcot, Oxon OX11 0RA, U.K.

In this chapter we shall be concerned with the thermally-induced cooperative build up of anomalously large concentrations of ionic defects in crystals, primarily those with the fluorite structure shown in figure 4.1. The build up occurs at a temperature T_c substantially below the melting temperature T_m; it gives rise to a specific heat anomaly and is sometimes referred to as premelting (Ubbelohde 1978; Hayes 1986). The unusual ionic disorder results in unusually large ionic conductivity and materials in this disordered state are referred to as superionics or fast-ion conductors (Hayes 1986).

The Ehrenfest criterion defines the order of a phase transition at a temperature T_c by the temperature dependence of the Gibbs free energy G = H - TS at T_c, the transition is said to be first order. For a second- or higher-order transition the structure varies continuously through T_c, and such transitions are referred to as continuous. For continuous-melting transitions, premelting obviously occurs. However, the occurrence of premelting does not necessarily imply that melting is a

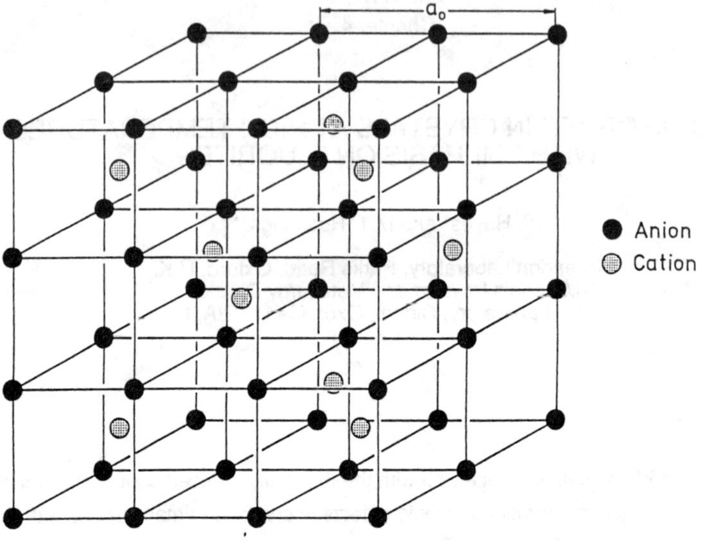

Figure 4.1 The fluorite crystal structure, Fm3m, with the non-primitive lattice constant a_o (after Hutchings et al 1984).

second-order or higher-order transition. In some solids one finds a continuous change in structure at $T \rightarrow T_c$ and also a final discontinuous change in structure at $T = T_c$. Indeed, one expects that many melting transitions if examined sufficiently close to T_m would show premelting, being nearly first order rather than precisely first-order.

It is well established that a single vibrational mode in a crystal may fall dramatically in frequency because of thermally-induced instability in a particular set of force constants. Such unstable modes are referred to as *soft modes* and play an important role in many types of structural phase changes in solids (Cummins and Levenyuk 1983). Although all vibrational modes in solids are anharmonic to some degree, it is possible that extreme vibrational anharmonicity may develop for specific vibrational modes with increasing temperature, leading to enhanced amplitudes of displacement of atoms from their normal lattice sites (see e.g. Boyer 1980). Such large amplitudes have analogies with those of the soft modes mentioned above but give rise to thermally induced lattice defects rather than symmetry rearrangement. The distinction between point defect formation and large anharmonic displacement is not always clearcut, and it is not always possible to establish the difference using a single type of measurement alone, e.g. specific heat.

Frenkel and Schottky defects are the two principal types of point defect in ionic solids (Hayes and Stoneham 1985; Harding 1989, Chapter 2). One generally finds that, for a given crystal, the energies of formation of Frenkel and Schottky defects are sufficiently different that one type is dominant. Although there are no experimental values for formation energies of Schottky defects in fluorites, calculation suggests (see e.g. Hayes and Stoneham 1985) that the energy required to produce a Schottky pair in CaF_2 is $\gtrsim$ 5 eV and to produce a cation Frenkel pair is $\sim$ 6 eV. These large values indicate that such defects are not normally of consequence, even at high temperatures. On the other hand, the anion Frenkel pair formation energy is $\sim$ 2 eV in fluorites such as CaF_2, and it is the anion sublattice which becomes dynamically disordered at high temperatures. It is worth keeping in mind that the concentration of point defects in thermal equilibrium at 1000K for a formation energy of 2 eV is $\sim$ 10 ppm.

In defining the work necessary to create a single defect we recognise that most experiments are done at constant temperature and constant pressure, and that the Gibbs free energy, g, for a single defect is relevant (see above and also Harding 1989, Chapter 2). This, again, may be split into two components, $g = h - Ts$, where h is the enthalpy and s the nonconfigurational entropy of a single defect, largely vibrational in origin. For the fluorites, orders of magnitude are $h \sim 1$ eV and $s \sim 10^{-4}$ eV/K, so that the entropy term is only important at high temperatures, and is sometimes ignored. It should be noted that higher temperatures may cause reduction in h because of lattice expansion. We may write the enthalpy as $h = u + Pv$ where u is the defect formation energy and v is the defect volume. The term Pv only becomes important in situations where the pressure reaches a few kilobars (10^8 N/m^2) and normally the enthalpy and energy of defect formation are not significantly different.

Although we shall concentrate on fluorites, it is useful to describe briefly the behaviour of silver halides at high temperatures, since these make an interesting comparision due to the difference in the character of the disorder. Details of the behaviour of AgI are given by Salamon (1979). At room temperature the silver halides are moderately good ionic conductors ($\sigma \approx 10^{-4} (\Omega\ \mathrm{cm})^{-1}$) due to the presence of cationic Frenkel defects. However, AgI differs from AgCl and AgBr in that its co-ordination is tetrahedral in its room-temperature wurtzite (β) form whereas the coordination of both AgCl and AgBr is octahedral (rocksalt structure). In the case of AgI there is a first-order structural phase transition to a body-centred cubic structure (α-AgI) at T_c = 420K. The DC electrical conductivity σ changes dramatically at T_c. The value of σ just below T_c is ∿ 3 x $10^{-4}(\Omega\ cm)^{-1}$ whereas σ increases to 1.3 $(\Omega\ cm)^{-1}$ just above T_c, an increase of about four orders of magnitude. Thereafter σ does not change appreciably with increasing temperature until the crystal melts at T_m = 825K, when σ actually falls by 12%.

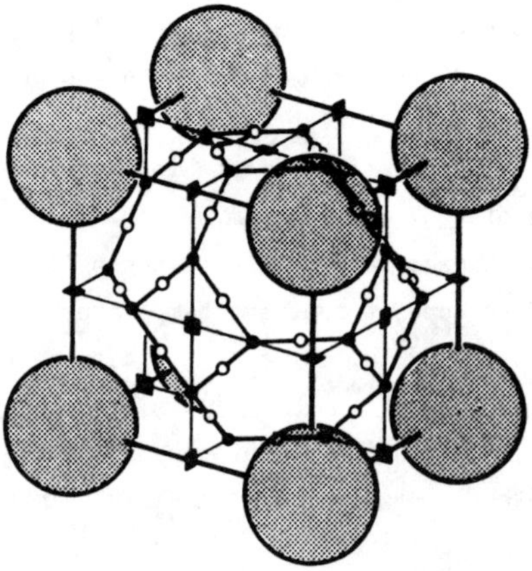

Figure 4.2 The unit cell of the bcc crystal structure of α-AgI, α-Ag_3SI, and β-Ag_2S. The large shaded circles are the iodine or sulphur atoms. The small symbols refer to the 42 possible silver ion sites first suggested by Strock (1934, 1936): the octahedrally coordinated 6 b sites (■); the tetrahedrally coordinated 12 d sites (•) and the triangularly coordinated 24 h sites (O) (after Shapiro and Reidinger 1979).

Structural studies show that, in the α phase of AgI, the iodine ions constitute a fairly rigid bcc lattice and that the Ag^+ ions occupy some of the large number of tetrahedrally-coordinated interstitial sites shown in figure 4.2. A variety of experiments suggest that motion of Ag^+ ions between neighbouring tetrahedral sites is the basic step in migration. The mechanism which drives the transition at 420K is by no means fully understood but it may be related to the mixed ionic-covalent bonding of AgI, and in particular to weak bond-bending forces. It is customary to use the expression 'sublattice melting' in connection with the silver sublattice in α-AgI and, indeed, the Ag^+ ions in α-AgI have many of the properties of a melt. Results of neutron scattering experiments are discussed in section 4.3.6.

Neither AgCl nor AgBr show the extreme conductivity behaviour of AgI. However, both materials show an anomalous increase in σ beginning at about 100 degrees below T_m (T_m = 728K for AgCl and 701K for AgBr); in the case of AgBr $\sigma = 10^{-1}(\Omega\ cm)^{-1}$ at 600K, rising to $1(\Omega\ cm)^{-1}$ at T_m (Aboagye and Friauf 1975). Experimental information currently available suggests that there are ∿ 2% of silver Frenkel interstitials in AgBr just below T_m and ∿ 0.2% in AgCl. These defects lead to an easily-observable anomaly in the thermal expansion coefficient as $T \rightarrow T_m$ (Lawson 1950). There is also an anomalous softening of elastic constants (Tannhauser et al 1956) and an anomalous increase in specific heat (Christy and Lawson 1951) as $T \rightarrow T_m$. There is however no peak in the specific heat for $T < T_m$ corresponding to that observed for halides with the fluorite structure (section 4.2.1).

The charged point defects in ionic solids are influenced by long-range Coulomb interactions in the same manner as ions in a liquid electrolyte, with positive defects surrounded by a cloud of negative defects and vice versá. The resultant (Debye-Hückel) screening reduces the average energy of point defect formation for low defect concentrations. However, the calculation of defect formation energies for defect concentrations above 1 mole % is difficult because of convergence problems (see Harding 1989, Chapter 2). It seems, nevertheless, that screening effects are not enough to account in full for the large concentration of Ag^+ interstitials in AgBr near T_m. It has been suggested by Aboagye and Friauf (1975) that there is a general softening of the lattice near T_m, a view consistent with greatly enhanced thermal diffuse scattering of X-rays. However, we shall see later (section 4.3) that there is no corresponding anomalous increase of thermal diffuse scattering of neutrons by fluorites as $T \rightarrow T_c$, although resistance to shear does decrease dramatically as T_c is approached (section 4.2.5 and figure 4.8).

We shall discuss in section 4.2 a variety of measurements on fluorites showing anomalous behaviour as $T \rightarrow T_c$ in specific heat, electrical conduction and elasticity because of the build-up of lattice disorder. In section 4.3 we give a detailed description of neutron scattering studies on

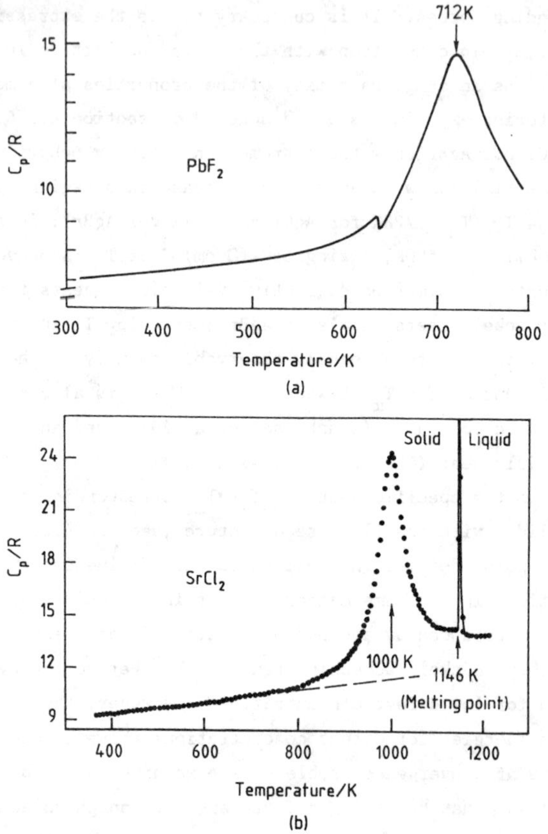

Figure 4.3 Temperature dependence of the Molar heat capacity of (a) PbF_2 and (b) $SrCl_2$ at constant pressure (after (a) Nölting, unpublished and (b) Schröter and Nölting 1980).

fluorites at high temperatures, with particular emphasis on quasielastic scattering (QES) measurements. We also discuss briefly neutron scattering measurements on other compounds. Finally, in section 4.4 we draw conclusions and attempt to give a picture of the structure and dynamics of the superionic state of fluorites, based principally on neutron studies, but with input too from the molecular dynamics simulations discussed in Chapter 3.

4.2 Some Physical Properties of Fluorities at High Temperatures

4.2.1 Specific Heat

It is now well known that many materials with the fluorite structure show an anomalous increase in the specific heat at a temperature T_c substantially below the melting temperature T_m, such that $T_c \sim 0.8\ T_m$ (table 4.1). Early measurements were carried out by Naylor (1945) and Dworkin and Bredig (1963, 1968). It was also established that these materials have anomalously low values of both the heat and entropy of fusion. The most recent detailed measurements of specific heat have been made for $SrCl_2$ and PbF_2 (see Schröter and Nölting 1980 and figure 4.3). The specific heat anomaly at T_c is associated with development of extensive disorder in the anion sublattice. The broad features of the behaviour in the highly disordered phase of fluorites can be described phenomenonologically by a Gibbs free energy of the form

$$G = xh^F - h_1(x) - xS_vT - S_c(x)T \quad . \qquad (4.2.1)$$

In this approximation it is assumed that the disorder arises from a rapid, cooperative, build-up of anion Frenkel pairs at T_c, largely due to attractive interactions between defects of opposite charge (vacancies and interstitials), and that the build-up ceases when repulsive interactions between defects of the same sign become significant. In equation (4.2.1) x is the interstitial concentration, h^F the Frenkel pair formation enthalpy at $x = 0$ (table 4.2), $h_1(x)$ the defect interaction enthalpy which reduces the effective Frenkel enthalpy to a minimum as x increases, S_v is the defect vibrational entropy, assumed independent of x, and $S_c(x)$ is the defect configurational entropy (Catlow et al 1978).

It is very difficult to make good theoretical estimates of $h_1(x)$ in highly disordered structures (see section 4.1, Catlow et al 1978, and Chapter 2). It is also difficult to make reliable estimates of defect concentrations from specific heat measurements (Schröter and Nölting 1980). However, a qualitative application of equation (4.2.1) is consistent with an interstitial concentration of defects rising rapidly near T_c owing to attractive defect interactions; repulsive defect interactions developing at

Table 4.1

Approximate melting temperature T_m of fluorite crystals and temperatures T_c of specific heat maxima

Crystal	T_m(K)	T_c(K)	T_c/T_m
CaF_2	1696	1430	0.84
SrF_2	1723	1400	0.81
BaF_2	1550	1230	0.79
$SrCl_2$	1146	1000	0.87
PbF_2	1128	712	0.63
UO_2	3120	-	-
ThO_2	3640	-	-
Li_2O	1705	-	-

$x \cong 0.1$ suppress further interstitial generation until T reaches T_m. Approaches of this sort to describe the specific heat anomaly have been used by Belosludov et al (1974), Oberschmidt (1981), and Andersen et al (1983).

It has been known for many years that the specific heat of the fluorite UO_2 behaves anomalously at high temperatures (Kerrisk and Clifton 1972), and this behaviour is of particular interest because of use of UO_2 as a nuclear fuel. However, because of experimental problems associated with steady-state calorimetry at high temperatures direct measurement of the specific heat is restricted to $T \lesssim 1200K$. At higher temperatures, the technique of drop calorimetry is commonly used, giving the enthalpy as a function of temperature. Differentiation with respect to temperature is then required to extract $C_p(T)$, a notoriously inaccurate procedure. We show in figure 4.4 the

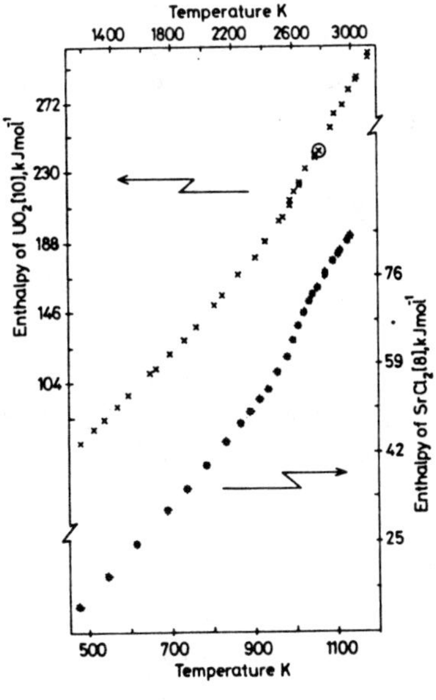

Figure 4.4 Comparison of enthalpy data of $SrCl_2$ and UO_2 (after Ralph and Hyland 1985).

measured enthalpy H(T) of UO_2 in comparison with that of $SrCl_2$. Below 1500K the specific heat of UO_2 has a value expected from vibrational excitation and from population of the crystal field levels of the ground electronic state of $U^{4+}(5f^2, {}^3H_4)$. The 9-fold degeneracy of ${}^3H_4$ is raised by the cubic crystal field of UO_2 giving a singlet Γ_1, a doublet Γ_3 and two triplets Γ_4 and Γ_5. Four transitions from the ground Γ_5 state have recently been observed within this manifold by inelastic neutron scattering in the region of 160 meV by Osborn et al (1987, 1988). In addition, as we shall see later (section 4.3.4), dynamic disorder develops in the anion sublattice at T > 2000K, and this may make a substantial contribution to the specific heat. It has been suggested that between 1500 and 2300K electronic excitations also make a substantial contribution to the specific heat (MacInnes and Catlow 1980; Harding et al 1980; Ralph and Gillan 1984; Browning et al 1983), an important process being the reversible reaction

$$2\,U^{4+} \rightleftarrows U^{3+} + U^{5+}$$

which gives rise to the formation of electron (U^{3+}) and hole (U^{5+}) small polarons.

Ralph and Hyland (1985) conclude from a careful analysis of the enthalpy data of figure 4.4 that there is a specific heat peak for UO_2 at 2610K and suggest that disorder in the anion sublattice may be involved. There is some evidence for onset of plasticity or high creep rates near this temperature (Ackerman et al 1956; Clausen et al 1984a; Slagle 1983).

4.2.2 Ionic Conductivity

Measurement of the temperature dependence of the ionic conductivity of undoped crystals, and of crystals deliberately doped with aliovalent ions, gives reliable values for the free energy of formation and motion of intrinsic ionic defects (Hayes and Stoneham 1985). The ionic conductivity of a cubic crystal may be written

$$\sigma = \Sigma_i\, q_i\, c_i\, \mu_i \quad , \tag{4.2.2}$$

where q_i is the effective ionic charge, μ_i is the mobility and $c_i \sim \exp(-h_i^F/2)$ is the temperature-dependent concentration of defects of species i. If the species i, once formed, has only one important class of diffusion jump, the mobility takes the form

$$\mu_i = \frac{q_i R_i^2}{k_B T}\, d_i\, \nu_{eff}\, \exp(-h_i^M/k_B T) \quad , \tag{4.2.3}$$

where R_i is the displacement distance in a jump, h_i^M is the enthalpy of motion and d_i is a degeneracy factor determined by the number of possible equivalent jump directions. The effective frequency ν_{eff} is determined by a number of independent factors but the major component is an attempt frequency ν_o (see Chapter 2), i.e. the number of times per second that the thermal motion of the defect carries it toward the barrier to be surmounted. The entropy of motion contributes a factor, S^M/k_B to ν_{eff}. A further factor is mobility drag, arising from Coulomb interactions between the various defect species i. In the absence of detailed information ν_{eff} is often approximated by the Debye frequency, but recently atomistic methods have given accurate predictions of both ν_{eff} and h_i^M (see Chapter 2).

In analysing conductivity data it is normal to plot log σT against 1/T. Figure 4.5 shows such plots for PbF_2 doped with Na^+ and La^{3+} and for pure PbF_2 (Azemi et al 1984). The method of data analysis commonly used for these

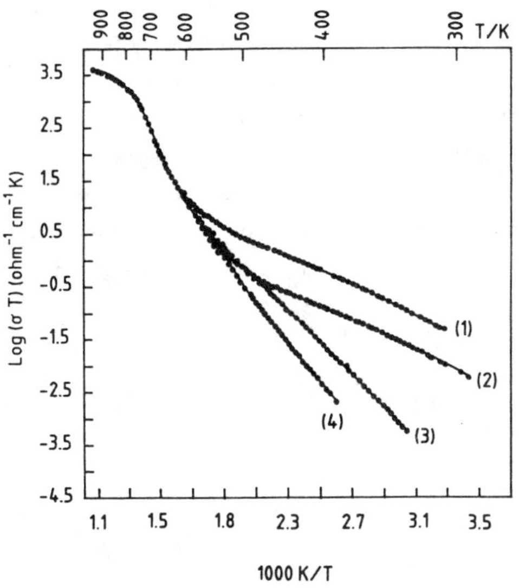

Figure 4.5 Comparison of the ionic conductivity of four PbF_2 crystals with the nominal dopant levels: (1) 1000 ppm Na^+, (2) 100 ppm Na^+, (3) 100 ppm La^{3+}, (4) pure (after Azemi et al 1984).

rather complex plots has been described by Jacobs and Ong (1980). In the analysis of the conductivity plots of Azemi et al (1984) it was assumed that the predominant thermally-induced point defects below T_c were anion Frenkel pairs, and that both Na^+ and La^{3+} dissolve in cation sites with the generation of charge-compensating anion vacancies and anion interstitials respectively. When the impurity and its charge-compensating defect are on nearest-neighbour sites, they are assumed to form neutral complexes which do not contribute to the conductivity. On this basis, expressions for the concentrations of isolated anion vacancies and interstitials are derived in terms of the anion Frenkel formation enthalpy h^F and formation entropy s^F, the total concentration of dopant, the energy of association of the impurity-defect complex and the corresponding entropy. The total conductivity is then evaluated from equation (4.2.2) with c_i dependent on formation enthalpies and entropies. The individual defect mobilities (equation 4.2.3) are assumed to be governed by migrational enthalpies h_v^M for vacancies and h_i^M for interstitials and the corresponding entropies S_v^M and S_i^M. Mobility drag is allowed for, and up to nine defect parameters are used to fit the measured log σT against $1/T$ curve. The temperature dependence of the lattice parameter and the electrical permittivity must also be allowed for. Results indicate that $\mu_v \gtrless \mu_i$ at lower temperatures in fluorites, but that at the upper temperature limit of the model used for analysis ((T_c - 100)K) the interstitials are more mobile ($\mu_i > \mu_v$) because $S_i^M > S_v^M$ compensates $h_i^M > h_v^M$. Some measured parameters are given in table 4.2.

Table 4.2

Activation enthalpies (eV) for formation of anion Frenkel pairs (h^F) and for motion of vacancies (h_v^M) and interstitials (h_i^M) in fluorite crystals well below T_c (Chadwick 1983).

Crystal	h^F	h_v^M	h_i^M
CaF_2	2.7	0.4	0.8
SrF_2	2.4	0.5	0.7
BaF_2	1.8	0.5	0.7
$SrCl_2$	1.9	0.3	0.8
PbF_2	1.0	0.2	0.5

In general, plots of log σT against $1/T$ for fluorites begin to rise rapidly as $T \rightarrow T_c$ with an apparent Arrhenius energy of a few eV, followed by a slower increase at $T \sim T_c$ as seen in figure 4.5. The curve settles down to a linear fast-ion region above T_c with a small Arrhenius energy of a few tenths of an eV and with σ close to that of the molten salt. There seems little doubt that the rapid increase in σ as $T \rightarrow T_c$ is due to a rapid increase in defect concentration. However, analysis of conductivity data in the region of interacting defects ($T \lesssim T_c$) is highly problematic (see e.g. Salamon 1979; Schoonman 1980).

4.2.3 Nuclear Magnetic Resonance (nmr)

Nmr techniques are useful for the study of the dynamics of disordered systems. In a rigid lattice the magnetic dipole interaction between neighbouring nuclei gives a Gaussian linewidth to a good approximation, and the width is of the order of the local field produced at the site of the spin by its neighbours ie $10^{-3} - 10^{-4}$ T. However, in liquids and gases where fast relative motion of the ions occurs, linewidths are an order of magnitude smaller than in a rigid lattice, as only the average value of the fluctuating field is felt, and the lineshape is more nearly Lorentzian than Gaussian. These time-varying fields cause nuclear spin-flip transitions, and produce a nuclear spin-lattice relaxation. However, nmr does not distinguish between local and long-range motion, and cannot measure ionic jump distances.

In order to give a motional narrowing of nmr lines, the fluctuations caused by ionic jumps should be fast compared with the Larmor precession in the instantaneous local field. The rate of fluctuation in the local field can be associated with a jump time τ_c identical to the average jump time in diffusing systems (see section 4.2.4). Hence, to get motional narrowing we require

$$\frac{1}{\tau_c} > (\overline{\Delta\omega_o^2})^{\frac{1}{2}} , \qquad (4.2.4)$$

where $(\overline{\Delta\omega_o^2})$ is the second moment of the line shape in the absence of diffusion. The hopping rate, τ_c^{-1}, is usually given by an Arrhenius relation,

$$\frac{1}{\tau_c} = \frac{1}{\tau_o} e^{-U/k_B T} . \qquad (4.2.5)$$

where the prefactor τ_o^{-1} is an effective attempt frequency, as discussed in equation (4.2.3), expected to be of the order of an optical phonon frequency (10^{12} - 10^{13} s^{-1}), and U is the activation energy for motion. Thus a measurement of nmr linewidth as a function of temperature gives the ionic hopping-rate, so the effective attempt frequency, τ_o^{-1}, and the barriewr, height, U, can be found. The ultimate resolution of line-width measurements is determined by magnetic field homogeneity, and conventional nmr techniques limit hopping-rate measurements to $\lesssim 10^7$ s^{-1}. However, in superionic fluorites hopping rates can be as fast as $\sim 10^{12}$ s^{-1}, and a study of nmr *relaxation rates* provides a more appropriate means of investigating T_2 and the ion dynamics.

The nmr relaxation data that have been used to determine ion hopping times in superionic conductors are the spin-lattice relaxation time T_1, the spin-spin relaxation time T_2, and the spin-lattice relaxation time in the rotating frame $T_{1\rho}$. The functional form of each of these times depends on details of the relaxation process. The relaxation process is determined (a) by magnetic interaction with the nuclei of other lattice ions or with the electronic moments of paramagnetic impurities (manganese is a common impurity in fluorites) and (b) by quadrupole interactions with the nuclear electric quadrupole moment when the nuclear spin I > ½. Detailed expressions for relaxation times are available in specific cases (see e.g. Boyce and Huberman 1979). For a diffusing system we may write

$$\frac{1}{T_1} \sim \frac{H_p^2 \tau_c}{1 + \omega^2 \tau_c^2} , \qquad (4.2.6)$$

where H_p is the local magnetic field. This gives a maximum in T_1^{-1} when $\omega\tau_c \sim 1$. It also predicts that, for $\omega\tau_c > 1$ (low temperature), T_1 varies as $\omega^2\tau_c$, and that for $\omega\tau_c < 1$ (high temperature) T_1 varies as τ_c^{-1}.

Detailed studies of ^{19}F (I = ½) nmr relaxation have been made in PbF_2 up to temperatures greater than T_c = 712K (Boyce et al 1977; Gordon and Strange 1978). The latter authors determine the self-diffusion coefficient (see section 4.2.4) of ^{19}F over six orders of magnitude (figure 4.6). They also estimate the anion Frenkel pair formation energy to be 0.88 eV, assuming that a single vacancy diffusion mechanism operates up to 500K. Interpretation of

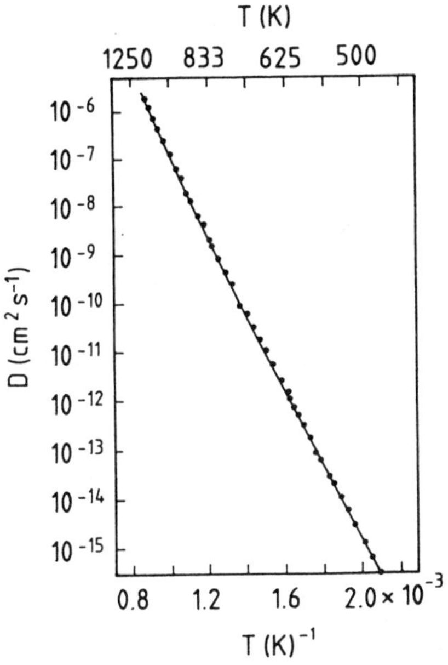

Figure 4.6 Fluorine self-diffusion coefficients in undoped BaF_2 single crystals obtained by a variety of techniques (after Figueroa et al 1978; see text).

results near T_c is again problematic because of effects of correlated ionic motion, and also because of influence of paramagnetic impurities.

4.2.4 Diffusion

For a single diffusing species, the mean square displacement in a time t is given by

$$r^2(t) = 6\ Dt \quad , \tag{4.2.7}$$

where D is the diffusion coefficient (cm^2/s). If a is the average jump distance of a particle, it follows that

$$D = \frac{1}{6}\Gamma a^2 \quad , \tag{4.2.8}$$

where Γ is the jump rate for diffusion (i.e. total jump probability of an ion per unit time from an initial site to all other sites). Equation (4.2.8) ignores correlation effects betwen successive jumps (see eg Howard and Lidiard 1964, and Chapter 2). If ionic conductivity and diffusion occur by the same mechanism, then σ (equation 4.2.2) and μ (equation 4.2.3) are related to D (equation 4.2.7) by the Nernst-Einstein relation:

$$\sigma = \left(\frac{Nq^2}{k_B T}\right) D \ , \tag{4.2.9a}$$

$$\mu = \left(\frac{q}{k_B T}\right) D \ . \tag{4.2.9b}$$

Correlation effects give rise to factors of order unity which are again omitted from equation (4.2.9a,b). For a simple vacancy transport mechanism, σT and D obey relations of the form

$$\sigma T = A \exp\left(- Q'/k_B T\right) \ , \tag{4.2.10a}$$

$$D = D_o \exp\left(- Q'/k_B T\right) \ , \tag{4.2.10b}$$

where A and D_o are constants assuming the entropy is constant , and $Q' = \frac{h^F}{2} + h^M$.

For anion vacancy motion in fluorites the jump time τ_c measured by nmr is $\tau_c = \frac{1}{\Gamma}$, where Γ is the rate constant of equation (4.2.8). Hence the diffusion coefficient measured by nmr (section 4.2.3) is

$$D_v^{nmr} = \frac{a_o^2}{6\tau_c} \ , \tag{4.2.11}$$

where a is set equal to a_o, the nearest neighbour anion-anion separation. The analogous expression for anion interstitial migration in fluorites is given by $a = a_o\sqrt{(3/2)}$,

$$D_i^{nmr} = \frac{a_o^2}{4\tau_c} \ . \tag{4.2.12}$$

A comparison of diffusion coefficients obtained from conductivity and nmr measurements in BaF_2 at temperatures of up to 1200K is given by Figueroa et al (1978). In this case surprisingly good agreement between the two diffusion coefficients is obtained by assuming a fluorine vacancy diffusion model.

Radioactive tracer techniques may also be used to determine anion diffusion coefficients in fluorites (Benière et al 1979). These measurements have confirmed that the cation is effectively immobile at high temperatures and that the predominant disorder is on the anion sublattice. It is helpful to combine tracer diffusion coefficients, D*, and diffusion coefficients, D_σ, obtained from σ (see equation 4.2.2). The ratio $H_R = D^*/D_\sigma$ depends on lattice structure and the mechanism of diffusion and can be calculated from a model (Le Claire 1970). For vacancy and noncollinear interstitialcy methods of diffusion in fluorites (Lidiard 1974), H_R = 0.65 and 0.74 respectively.

A brief review of diffusion, nmr and conductivity data for fluorites at high temperatures has been given by Chadwick (1983), (see also Allnatt et al 1986).

4.2.5 Light scattering

Effects of ionic defects on lattice dynamics and ionic transport can be studied using Raman scattering, Brillouin scattering and quasielastic light scattering (Hayes 1982). Each of these techniques has been applied to the study of fluorites at high temperatures. A detailed study of the effect of anharmonicity ($T < T_c$) and lattice disorder ($T > T_c$) on the Raman spectrum of CaF_2, SrF_2, BaF_2, $SrCl_2$ and PbF_2 was carried out by Elliott et al (1978). The first-order Raman spectrum of crystals with the fluorite structure consists only of a zone-centre phonon with T_{2g} symmetry, with the two anion sublattices vibrating in antiphase and with stationary cations. However, defects in the crystals can induce scattering of phonons with other symmetry configurations owing to the breakdown of wavevector conservation.

We shall describe here only results obtained for PbF_2 between room temperature and 980K, although similar data were obtained for the other materials mentioned above. At room temperature the T_{2g} phonon, at 257 cm^{-1}, is already broad (full-width at half maximum, FWHM = 50 cm^{-1}) and has a symmetric shape. As the temperature is raised the T_{2g} phonon continues to broaden, and at T_c and above an additional scattering is observed on the low energy side. The position and shape of the bands below T_c were calculated using third- and fourth-order anharmonicity. The theory requires a third and fourth order anharmonic parameter, (see equation 4.3.36), and the values required to give the calculated curves (figures 4.7a,b) are comparable with

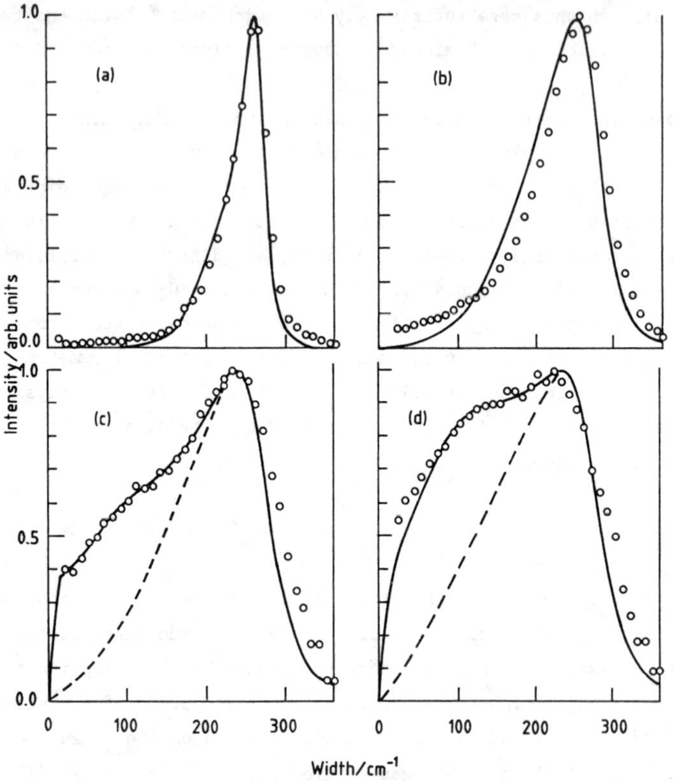

Figure 4.7 Measured (dots) and calculated (full lines) Raman susceptibilities of PbF_2 at (a) 290K (b) 580K (c) 775K and (d) 880K. The dashed lines represent the theoretical contribution of anharmonicity alone (after Elliott et al 1978).

values of third- and fourth-order anharmonic parameters obtained by other experimental techniques (see Elliott et al 1978). The additional scattering on the low energy side of the T_{2g} phonon at T_c and above (figure 4.7c,d) is accounted for by a theory of defect-induced scattering which includes effects of anion vacancies and anion interstitials. It should be emphasised that such calculations are not sensitive to the precise configuration of vacancies and interstitials.

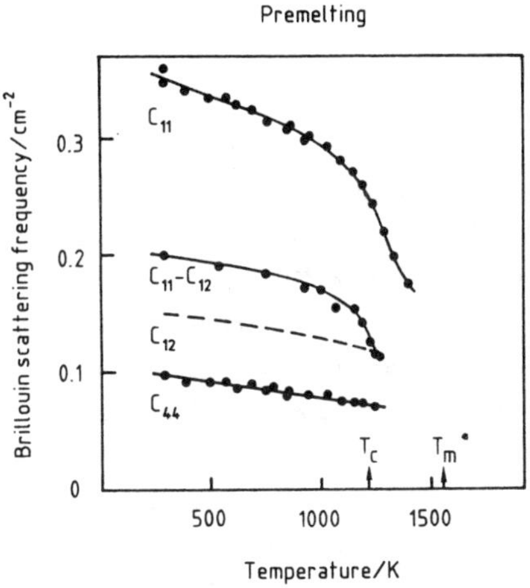

Figure 4.8 Temperature dependence of the square of Brillouin scattering frequencies (proportional to elastic constants) in BaF_2 (after Catlow et al 1978).

At the highest temperatures studied the Raman scattering of fluorites is similar to the single-phonon density of states. In the case of PbF_2 the defect-induced scattering above T_c (figure 4.7d) almost develops into a subsidiary peak below the Raman peak because of a large difference between the Raman peak and the peak in the smoothed-out, high-temperature, single-phonon density of states (see figure 4.20b, Dickens et al 1980).

Effects of disorder in the superionic state of fluorites on elastic constants were studied by Catlow et al (1978) using Brillouin scattering techniques. Results for PbF_2 (figure 4.8) show a linear decrease of C_{11} and ($C_{11}-C_{12}$) (shear) with increasing temperature up to T_c and a dramatic fall at T_c. This behaviour is similar to that found for AgBr (see section 4.1). The linear region is accounted for by anharmonicity and the fall at T_c by anion disorder (Catlow et al 1978). The Brillouin scattering studies show that the elastic constants C_{12} and C_{44} are not appreciably affected by disorder. The different behaviour of these elastic constants may be explained to some extent by the fact that the contributions to C_{11} from short-range and Coulomb forces have the same sign whereas they have opposite signs for C_{12} and C_{44}. It appears that the effect of defects on C_{12} and C_{44} is small because the defect-induced changes in Coulomb and short-range forces largely cancel each other. Again, as in the case of Raman scattering, the theory used to account for the Brillouin scattering is not sensitive to the precise configuration of vacancies and interstitials.

The effect of doping with trivalent cations on T_c of superionic fluorites has been studied by Catlow et al (1981) using Brillouin scattering methods. It was found, for example, that doping CaF_2 with 9 mole % of YF_3 reduces T_c from 1430K (table 4.1) to $\sim$ 1200K. The Y^{3+} ions dissolve in Ca^{2+} sites and are charge-compensated by F_i^- interstitials close to the centre of empty fluorine cubes in close association with Y^{3+} ions. Calculation (Catlow et al 1981) shows that Y^{3+} - F_i^- complexes act as traps for thermally-generated anion interstitials, thus reducing the energy of formation of anion Frenkel pairs and also reducing T_c.

The mobile ion in a fast-ion conductor may stay at a lattice site sufficiently long to execute a number of cycles of vibration. A parameter of some consequence is the ratio of the time spent at a lattice site, τ_d, to the time spent in flight, τ_f. From the simple diffusion theory of section 4.2.4,based on the motion of uncorrelated particles, we get $\tau_d = a^2/6D$ (see equation 4.2.8). If the time of flight is the jump distance divided by the velocity then

$$\tau_f \sim (ma^2/k_BT)^{1/2} \quad , \qquad (4.2.13)$$

where m is the mass of the diffusing ion. If we take a = 2Å and D = 10^{-10} m^2 s^{-1} as typical values for superionics we find $\tau_d \simeq 3$ ps and $\tau_f \simeq 0.5$ ps. To the extent that $\tau_d > \tau_f$ it is appropriate to describe the dynamics of the moving ion by a hopping, or jump-diffusion, model (Beyeler et al 1979).

Light scattering is caused by changes in ionic polarizability (Hayes and Loudon 1978). In superionics, polarizability changes associated with ion movement from site to site cause dynamic quasielastic scattering. This scattering can be studied with very high resolution laser-scattering techniques. These techniques have been applied to $(ZrO_2)_{1-y}(Y_2O_3)_y$ by Perry and Feinberg (1980) and by Suemoto and Ishigama (1983). Stabilised zirconia (ZrO_2) containing $\gtrsim$ 10 mole % of Y_2O_3, has the fluorite structure, whereas pure ZrO_2 has a lower symmetry. Anion vacancies are present in the mixed crystal in large numbers because replacement of Zr^{4+} by Y^{3+} requires charge compensation. These vacancies become mobile above $\sim$ 700°C giving high ionic conductivity. The ionic conductivity peaks at intermediate dopant levels, y, and this is reflected in the measured intensity of quasielastic scattering

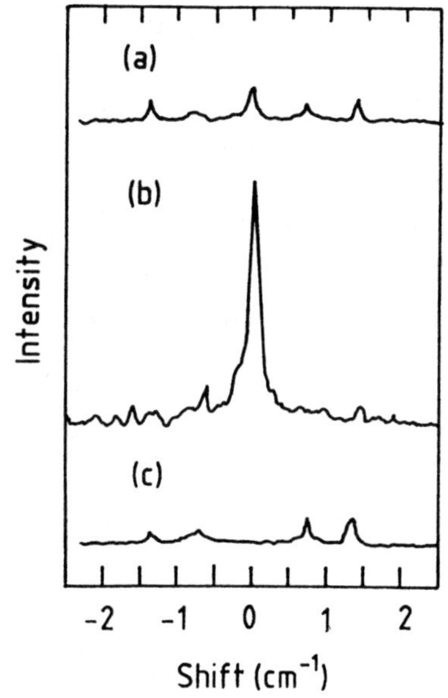

Figure 4.9 Quasielastic (Brillouin) scattering spectra of $(ZrO_2)_{1-y}(YO_{1.5})_y$ at 1127K for (a) y = 0.33 (b) y = 0.18 and (c) y = 0. The spectrum (c) is multiplied by a factor of 5 (after Suemoto and Ishigama 1983).

(figure 4.9). From the temperature dependence of ionic conductivity of sintered samples containing 18 mole % of Y_2O_3 (Dixon et al 1963) the hopping rate at 1127K is estimated to be 1.3 x 10^{10} s^{-1}, with an activation energy of 0.96 eV (see section 4.2.4). This hopping rate would suggest a FWHM = 0.07 cm^{-1} for the quasielastic light scattering, smaller than observed (figure 4.9). Suemoto and Ishigama (1983) suggest that better agreement with the measured quasielastic light scattering can be obtained by assuming a distribution of barrier heights and attempt frequencies, both of which are temperature dependent. Neutron scattering measurements on a range of concentrations of Y_2O_3 in ZrO_2 give further details, as described in section 4.3.5.

4.3 Neutron and x-ray Scattering of Ionic Solids at High Temperatures

4.3.1 Introduction

In this section the use of neutron and x-ray scattering to study the development of thermally induced ionic disorder in solids will be discussed. The emphasis will be on neutron scattering techniques and their use to investigate crystals with the fluorite structure. Although x-ray sources of varying intensity are more universally available, neutrons provide the more appropriate probe for the investigation of ionic disorder at high temperature. This is because of their greater relative penetration of the metal sample containers, heater elements and heat shields of the furnace necessary for high temperature studies, and because of their ability to probe the dynamical behaviour of the ions. The latter property stems from the fact that neutrons from reactor or accelerator sources are characterised by both wavelengths and energies which are of the same order of those of the motion of the ions in a crystal. In the case of x-rays, the energy of photons with the appropriate wavelength is very much higher than that of the ions, necessitating extremely good resolution in order to detect an inelastic scattering event. Nevertheless with the advent of high intensity synchrotron sources inelastic scattering of x-rays is becoming a feasible experiment. A further advantage of neutron scattering is its ability to probe either the behaviour of individual nuclei or ions, or that of their correlated positions or motion, by the observation of incoherent or coherent scattering respectively, whereas x-rays only scatter coherently from ions. The terms incoherent and coherent in this context are explained in the next section, but stem from the dependence of the interaction of the neutron with a nucleus of an ion in the crystal on its isotopic constitution and on its spin. This interaction is described by a neutron scattering length, b, which varies randomly over a limited range of values with the atomic number of the scattering nucleus, in contrast to the interaction of x-rays which increase monotonically with atomic number. Neutron scattering can therefore often probe the behaviour of light atoms in the presence of heavy atoms much more easily than x-rays.

Although most aspects of neutron scattering theory and techniques are touched upon, three types of scattering experiment have been used to provide most of the information on ionic solids at high temperatures: diffraction, diffuse scattering, and inelastic scattering. Diffraction techniques give direct information on the symmetry and size of the unit cell, the time-averaged distribution of nuclear or electronic density within the average cell, and the mean thermal motion of the ions. At ambient temperature and below, most ionic crystals can be described by an ordered lattice of ions, with all the ions located on the ideal lattice sites executing thermal motion which can be described by isotropic or anisotropic harmonic thermal parameters. As the temperature is raised 'normal' solids, such as NaCl expand linearly. The mean thermal motion is now described by harmonic thermal parameters which usually increase linearly with temperature, and by anharmonic thermal parameters. Crystals exhibiting low activation energies of thermally induced defects and ionic motion, that is fast-ion conductors such as the fluorites, are characterised by an increase in nuclear or electron densities well away from the regular lattice sites. These can be described either by increasingly large anharmonic thermal parameters, or by the occupation of additional lattice sites, or both. An alternative description of these solids is in terms of the probability scattering density function, $\rho_c(\underline{r})$, which gives the time-averaged distribution of coherent scattering amplitudes throughout the unit cell.

Thermally-induced disorder, which is usually non-periodic, can also be examined by the coherent diffuse scattering to which it gives rise. This increases in intensity with the degree of disorder, and the distribution of the intensity in reciprocal space, integrated over energy transfer, gives direct information on the average instantaneous correlations between the positions of the defective ions, a 'snapshot' picture. The integration over energy transfer is automatic in the case of x-ray scattering, but inelastic scattering from phonons is also included and is difficult to separate out. Indeed conventionally diffuse scattering is the term given to all non-Bragg scattering. In our discussion of neutron scattering we shall separate out the phonon scattering and term only the disorder scattering as diffuse scattering. If the correlations have a finite lifetime, τ, the coherent diffuse scattering is quasielastic. The energy width may be readily resolved by a neutron scattering experiment if τ is less than $\sim 10^{-10}$s, and by using specialised spectrometers if τ is less than 10^{-6}s. The wavevector dependence of the energy width can then give information on the translational motion of the correlated ions. If the disordered ions have nuclei which exhibit incoherent scattering, this may be used to give very direct information on the diffusive motion of individual ions provided it is sufficiently rapid, and gives a diffusion constant analogous to that from tracer diffusion methods. Unfortunately the experimental separation of coherent and incoherent quasielastic diffuse neutron scattering is not always straightforward. In some cases, particularly if crystals are doped or alloyed, coherent diffuse scattering arises which also peaks near zero wavevector transfer. It is then usually called 'small angle' scattering and can be related to the shape, volume fraction and size distribution of the inhomogeneities in the scattering density caused by the disordered ions. Small angle scattering has yet to be used extensively to study thermally-induced disorder, mainly because it examines inhomogeneities of larger size than usually occur.

Finally, inelastic coherent neutron scattering can give detailed information on the lattice vibrational mode energies, their wavevector dependence, and their lifetimes. The temperature variation of low lying optic modes may be related directly to the vibration of mobile ions. A knowledge of long wavelength acoustic mode energies enables the elastic constants to be determined, and it is important for the correction for their contribution (thermal diffuse scattering or TDS), to diffraction intensity. The measured energy dispersion relation may be used to help to determine the interatomic potentials used in molecular dynamics simulation of the thermally induced disorder.

Although in principle all the information about the system contained in the time-dependent total pair correlation function, $G(\underline{r},t)$, may be extracted from an inelastic scattering experiment which determines the scattering function $S(\underline{Q},\omega)$, in practice all three main types of scattering must be studied in order to obtain a complete picture of the nature of thermally induced disorder in ionic crystals. The data from each may be then used to develop a model of the disorder which can explain the main observations and which can provide other predictions. However, for such a complex state as the fast-ion phase of a material, any model is likely to be an idealised picture of the true situation. An alternative method of analysis is to compare the data directly with a computer simulation using molecular dynamics techniques. The simulation may also be thought of as an experiment, but one in which one can control the vital parameters on a microscopic scale and can examine, albeit with some difficulty, the type of disorder and motion which give rise to the scattering in terms of the actual time dependent positions of the ions.

In the following subsection some main points in the theory of neutron scattering and its application to the study of ionic compounds at high temperature will be given. This application will then be illustrated mainly by reference to materials with the fluorite or antifluorite structure, as these provide an ideal system in which to study fast-ion behaviour. Their structure is well defined and relatively simple, and makes them amenable to study by both experimental and theoretical techniques. Large single crystals of most fluorite compounds, necessary for scattering experiments, can readily be grown, and they can be doped with relative ease to produce extrinsically defective materials with either an excess or deficiency of anions. There is no structural phase change on raising the temperature into the fast-ion conduction phase. Furthermore, the fluorites have many technological uses and the most important nuclear fuel materials, UO_2 and PuO_2, have the fluorite structure as mentioned in section 4.2.1 . A considerable amount of work using scattering techniques has also been carried out on other ionic materials exhibiting fast ion conduction at high temperature, such as AgI, Ag_3SI, Ag_2S, $RbAg_4I_5$, β-alumina, NASICON and Li_3N, and this will be referred to where appropriate. In all this work, single crystal samples are favoured over powders. Although powders are more readily available, the crystallites tend to orient preferentially and grow in size at temperatures approaching their melting point. Even without these difficulties, scattering intensities from powders cannot give detailed information on the anisotropy of the scattering, and the analysis of powder diffraction data faces difficulties of overlapping reflections and lack of intensity, making it mandatory to assume a trial model of the structure in the interpretation.

4.3.2 Theory of Neutron Scattering

In this section we shall outline the basic theoretical concepts and equations used in the analysis of neutron scattering data from ionic solids at high temperatures. Corresponding relations for elastic x-ray scattering follow from the replacement in the expressions of the neutron scattering length by the x-ray electron form factor for the ion. However, as the effect of the form factor is to reduce the scattering intensity at large scattering vectors, the amount of data available from x-ray work may be correspondingly reduced.

4.3.2.1 Coherent and Incoherent Scattering

We shall denote the incident, and final, neutron wavevector and energy by $\underline{k}_i$, and $\underline{k}_f$, and E_i, and E_f, respectively. The scattering vector is described by $Q = \underline{k}_i - \underline{k}_f$, and the energy transfer by $\hbar\omega = h\nu = E_i - E_f$. The commonly used units are 1meV = 0.242THz = 8.07cm^{-1} = 11.6k_BK, where k_B = Boltzmann's constant. The general notation adopted will be essentially that of the books to which the reader is referred for the full theory of neutron scattering (Squires 1978; Lovesey 1984; Price and Sköld 1986).

It is important at the outset to make clear the distinction between *coherent* and *incoherent* neutron scattering cross sections and their relationship to the pair correlation functions of the nuclei in the system (Van Hove 1954, Squires 1978). In the terminology of neutron scattering these cross section refer only to the effects of the distribution of scattering length, b_α, of a system of nuclei of one element, due either to the isotopic or spin state, α. If ⟨b⟩ denotes the mean scattering length averaged over the different nuclear states, and $\langle b^2\rangle$ the mean square scattering length, we define a coherent cross section $\sigma_{coh} = 4\pi\langle b\rangle^2$ proportional to the square of the mean scattering length, and an incoherent cross section $\sigma_{incoh} = 4\pi(\langle b^2\rangle-\langle b\rangle^2)$, proportional to the mean square deviation in scattering length. Values of ⟨b⟩, σ_{coh}, σ_{incoh} and σ_{abs}, the absorption cross section, are given in table 4.3 for the isotopes discussed in this Chapter. We shall first illustrate the information which can be gained from the two types of scattering, involving each of these expressions as coefficients in the differential scattering cross section, by reference to a system of nuclei of a single element (see Hutchings et al 1983, 1984). Most of the diffuse scattering experiments on fast ion systems at high temperatures have been interpreted in terms of correlations between defective ions of a single mobile species. However, all the ions need to be considered in the analysis of diffraction and inelastic scattering data, and in an *a priori* analysis of diffuse scattering data. Where appropriate, reference will be made to the expressions for more than one species of nuclei following Andersen et al (1987). Although the theoretical distinction between coherent and incoherent scattering is clear, as we have mentioned it is often difficult to make the separation experimentally.

The general expression for the inelastic scattering differential cross section from a system of N nuclei of a single element is given by

$$\frac{d^2\sigma}{d\Omega dE_f} = \frac{k_f}{k_i}\,\frac{1}{2\pi\hbar}\sum_{j,j'}\langle b_j b_{j'}\rangle \int\langle \exp[-i\underline{Q}.\underline{R}_j(0)]\exp[i\underline{Q}.\underline{R}_{j'}(t)]\rangle e^{-i\omega t}dt, \tag{4.3.1}$$

Table 4.3

Values of mean bound scattering lengths <b> (in units of 10^{-15}m), and coherent, incoherent and absorption cross sections (in barns, 1b = $10^{-28}m^2$) of relevant elements and isotopes (after Sears 1986).

Z	A	Element	<b>	$\sigma_{coh}=4\pi\langle b\rangle^2$	σ_{incoh}	σ_{abs}
Mobile Ions						
1		H	-3.739	1.757	79.9	0.333
1	2	D	6.671	5.592	2.04	0.0005
3		Li (nat)	-1.90	0.454	0.83	70.5
3	7	Li	-2.22	0.619	0.68	0.0454
8		O	5.803	4.232	0.00	0.0002
9		F	5.654	4.017	0.001	0.0096
11		Na	3.63	1.66	1.62	0.530
17		Cl	9.577	11.53	5.2	33.5
17	35	Cl	11.66	17.07	4.5	44.1
29		Cu	7.719	7.486	0.52	3.78
47		Ag	5.922	4.407	0.58	63.3
Lattice Ions						
6		C	6.646	5.551	0.0	0.0035
12		Mg	5.375	3.631	0.077	0.063
13		Al	3.449	1.495	0.0092	0.231
14		Si	4.149	2.163	0.015	0.171
16		S	2.847	1.019	0.007	0.53
20		Ca	4.90	3.02	0.03	0.43
34		Se	7.97	7.98	0.33	11.7
35		Br	6.795	5.80	0.10	6.9
38		Sr	7.02	6.19	0.04	1.28
39		Y	7.75	7.55	0.15	1.28
40		Zr	7.16	6.44	0.16	0.185
53		I	5.28	3.50	0.00	6.2
56		Ba	5.25	3.46	0.01	1.2
58		Ce	4.84	2.94	0.0	0.63
82		Pb	9.405	11.12	0.003	0.171
90		Th	10.63	14.20	0	7.37
92		U	8.417	8.903	0.006	7.57
Encapsulation Materials						
42		Mo	6.95	6.07	0.28	2.55
74		W	4.77	2.86	2.00	18.4
78		Pt	9.63	11.65	0.13	10.3

where $\underline{R}_j(t)$ denotes the position of nucleus j at time t, and $\langle b_j b_{j'} \rangle$ is the ensemble average over the scattering lengths of nuclei at j and j'. Equation (4.3.1) may be written

$$\frac{d^2\sigma}{d\Omega dE_f} = \sum_{j,j'} \langle b_j b_{j'} \rangle \gamma_{jj'} = \langle b^2 \rangle \sum_j \gamma^s_{jj} + \langle b \rangle^2 \sum_{j \neq j'} \gamma^d_{jj'}, \qquad (4.3.2)$$

where $\gamma_{jj'}$ is written for the prefactor and integral in equation (4.3.1) and the suffixes s and d emphasise the summation is over self and distinct terms. Equation (4.3.2) can be rewritten as the sum of a coherent and an incoherent term:

$$\frac{d^2\sigma}{d\Omega dE_f} = \langle b \rangle^2 \sum_{j,j'} \gamma_{jj'} + [\langle b^2 \rangle - \langle b \rangle^2] \sum_j \gamma_{jj}$$

$$= \frac{\sigma_{coh}}{4\pi} \sum_{j,j'} \gamma_{jj'} + \frac{\sigma_{incoh}}{4\pi} \sum_j \gamma_{jj} \quad . \qquad (4.3.3)$$

The two terms are usually written in the form:

$$\frac{d^2\sigma}{d\Omega dE_f} = \frac{\sigma_{coh}}{4\pi} \frac{k_f}{k_i} N\, S(\underline{Q},\omega) + \frac{\sigma_{incoh}}{4\pi} \frac{k_f}{k_i} N\, S_i(\underline{Q},\omega) , \qquad (4.3.4)$$

where $S(\underline{Q},\omega)$ is the (total) Van Hove scattering function, $S_i(\underline{Q},\omega)$ is the incoherent scattering function and N is the number of particles in the system. These scattering functions are related by a double Fourier transform to the time-dependent total pair-correlation function $G(\underline{r},t)$, and to the time dependent self-correlation function $G_s(\underline{r},t)$, respectively. Corresponding intermediate functions $I(\underline{Q},t)$ may also be defined, related by only a space-coordinate Fourier transform. Their relationships are:

$$S(\underline{Q},\omega) = \frac{1}{2\pi\hbar} \int G(\underline{r},t) \exp[i(\underline{Q}.\underline{r} - \omega t)] d\underline{r} dt \qquad (4.3.5a)$$

$$G(\underline{r},t) = \frac{1}{(2\pi)^3} \int I(\underline{Q},t) \exp[-i\underline{Q}.\underline{r}] d\underline{Q} \qquad (4.3.5b)$$

$$I(\underset{\sim}{Q},t) = \frac{1}{N} \sum_{j,j'} \langle \exp[-i\underset{\sim}{Q}.\underline{R}_j(0)].\exp[i\underset{\sim}{Q}.\underline{R}_{j'}(t)] \rangle \ , \qquad (4.3.5c)$$

for the total functions. Self and distinct functions, S_i, G_s, I_i and S_d, G_d and I_d may be defined via partial summations $\sum_{j=j'}$, and $\sum_{j\neq j'}$, respectively in equation (4.3.5c). Thus $S(\underset{\sim}{Q},\omega) = S_i(\underset{\sim}{Q},\omega)+S_d(\underset{\sim}{Q},\omega)$, $G(\underline{r},t) = G_s(\underline{r},t)+G_d(\underline{r},t)$ and $I(\underset{\sim}{Q},t) = I_i(\underset{\sim}{Q},t)+I_d(\underset{\sim}{Q},t)$. $S_d(\underset{\sim}{Q},\omega)$ is seldom discussed alone.

The *coherent* scattering thus gives direct information on $G(\underline{r},t)$. This time-dependent total pair-correlation function may be visualised physically in the classical limit, when $G(\underline{r},t)d\underline{r}$ gives the probability that, given a nucleus at the origin at t=0, *any* nucleus (the same one or a different one) will be found within volume $d\underline{r}$ at $\underline{r}$ and at time t. *Incoherent* scattering, on the other hand, gives information on the motion of *individual* nuclei through $G_s(\underline{r},t)$, which in the classical limit gives the probability that given a nucleus is at the origin at t=0 the same nucleus is to be found within $d\underline{r}$ at $\underline{r}$ at time t. Clearly in the same approximation $G_d(\underline{r},t)$ is the corresponding function for different nuclei only.

If one concentrates now on the coherent scattering, and integrates $S(\underset{\sim}{Q},\omega)$ over all energy transfers ω at constant $\underset{\sim}{Q}$, one obtains the quasistatic cross section:

$$\sigma^{qs}_{coh}(\underset{\sim}{Q}) = \frac{\sigma_{coh}}{4\pi} N \hbar \int S(\underset{\sim}{Q},\omega)d\omega = \frac{\sigma_{coh}}{4\pi} N\, S(\underset{\sim}{Q}). \qquad (4.3.6)$$

The structure factor $S(\underset{\sim}{Q})$ is given by

$$S(\underset{\sim}{Q}) = \hbar \int S(\underset{\sim}{Q},\omega)d\omega = I(\underset{\sim}{Q},0) = \int G(\underline{r},0)e^{i\underset{\sim}{Q}.\underline{r}}d\underline{r} = 1+\int g(\underline{r})e^{i\underset{\sim}{Q}.\underline{r}}d\underline{r} \ , \qquad (4.3.7)$$

where $G(\underline{r},0) = \delta(\underline{r}) + g(\underline{r})$ and $g(\underline{r})$ is the static pair distribution function, giving the average nuclear density at $\underline{r}$ with respect to any nucleus at the origin. From equation (4.3.5c) it is seen that $I(\underset{\sim}{Q},0)$ describes the instantaneous correlations between the nuclei, and so $S(\underset{\sim}{Q})$ gives a 'snapshot' picture of their arrangement. In contrast, the purely elastic coherent cross section, giving the scattering at $\omega = 0$, is related to $I(\underset{\sim}{Q},\infty)$, the Fourier Transform of $G(\underline{r},\infty)$:

$$\sigma_{coh}^{e\ell}(\underline{Q}) = \frac{\sigma_{coh}}{4\pi} \; N \; I(\underline{Q},\infty) \quad . \tag{4.3.8}$$

If the correlation between $\underline{R}_j(0)$ and $\underline{R}_{j'}(t)$ becomes independent of t as t → ∞, $I(\underline{Q},\infty)$ may be written as:

$$I(\underline{Q},\infty) = \frac{1}{N} \sum_{jj'} \langle \exp[-i\underline{Q}.\underline{R}_j] \rangle \langle \exp[i\underline{Q}.\underline{R}_{j'}] \rangle \quad . \tag{4.3.9}$$

The elastic cross section therefore gives a time-averaged picture of the arrangement of nuclei and, in the case of a regular lattice of nuclei, this is obtained from the Bragg peak intensities. Note that $I(\underline{Q}\ \infty)$ is a part of $S(\underline{Q})$.

4.3.2.2 Coherent Scattering from a System of More than one Species of Ion

For a system of nuclei consisting of a number of species of elements the coherent term in equation (4.3.4) becomes

$$\frac{d^2\sigma}{d\Omega dE_f} = \frac{k_f}{k_i} \sum_{s,s'} \sqrt{N_s N_{s'}} \; \langle b_s \rangle^* \; \langle b_{s'} \rangle \; S^{ss'}(\underline{Q},\omega) \quad , \tag{4.3.10}$$

where there are N_s ions of each element (or species) s, and the summation over j,j' in the expression $S^{ss'}(\underline{Q},\omega)$, equation (4.3.3), is over all ions of all species. The corresponding incoherent term is simply the sum over the expression, equation (4.3.4), for each species. It is convenient to discuss the coherent scattering in terms of a new coherent scattering function $S_c(\underline{Q},\omega)$ and intermediate function $I_c(\underline{Q},t)$, per unit cell, which include the scattering lengths or, in the case of x-ray scattering, the form factors (Andersen et al 1987). However, this new function loses some of its generality as it is now probe dependent.

$$S_c(\underline{Q},\omega) = \frac{1}{2\pi\hbar} \frac{1}{N_c} \sum_{s,s'} \langle b_s \rangle^* \langle b_{s'} \rangle \int_{-\infty}^{\infty} \sum_{\substack{j\,\epsilon s \\ j'\epsilon s'}} \langle e^{-i(\underline{Q}.\underline{R}_j(0)} e^{i\underline{Q}.\underline{R}_{j'}(t)} \rangle e^{-i\omega t} dt$$

$$= \frac{1}{2\pi\hbar} \int I_c(\underline{Q},t) e^{-i\omega t} dt \quad , \tag{4.3.11}$$

so that

$$\frac{d^2\sigma}{d\Omega dE_f} = \frac{k_f}{k_i} N_c S_c(\underline{Q},\omega) \quad , \tag{4.3.12}$$

where N_c is the number of unit cells in the system.

In order to calculate analytically the correlations expressed by $I_c(\underline{Q},t)$ for a highly disordered system, such as an ionic solid at high temperature, one must adopt a model for the ionic positions $\underline{R}_j(t)$. These are defined in terms of *positions* in the unit cell of the low temperature structure, and to evaluate the summation over *ions* it is necessary to define a probability $p_{\underline{\ell d}j}(t)$ which is +1 or 0 depending on whether or not at time t ion j occupies the site at position $\underline{d}$ in the cell whose origin is at $\underline{\ell}$. The summation over ions may be then changed to one over sites. Defining $\underline{u}_{\underline{\ell d}j}(t)$ as the displacement of ion j from the equilibrium site at time t, we may write

$$\underline{R}_j(t) = \sum_{\underline{\ell},\underline{d}} p_{\underline{\ell d}j}(t)\ (\underline{\ell} + \underline{d} + \underline{u}_{\underline{\ell d}j}(t)). \tag{4.3.13}$$

The coherent cross section depends only on the elemental species s of the ion at a given position, so that a corresponding probability $p_{\underline{\ell ds}}(t)$ is defined, which is +1 or 0 depending on whether the site $(\underline{\ell} + \underline{d} + \underline{u}_{\underline{lds}}(t))$ is occupied by an ion of species s or not:

$$\sum_{j\epsilon s} \underline{R}_j(t) = \sum_{\underline{\ell},\underline{d}} p_{\underline{\ell ds}}(t)\ (\underline{\ell} + \underline{d} + \underline{u}_{\underline{\ell ds}}(t)) \quad . \tag{4.3.14}$$

The modified scattering function now becomes,

$$S_c(\underline{Q},\omega) = \frac{1}{2\pi\hbar} \frac{1}{N_c} \sum_s \sum_{s'} \langle b_s\rangle^* \langle b_{s'}\rangle \int_{-\infty}^{\infty} \sum_{\substack{\underline{\ell},\underline{d} \\ \underline{\ell}',\underline{d}}} \langle p_{\underline{\ell ds}}(0) p_{\underline{\ell}'\underline{d}'s'}(t\ ) \times$$

$$e^{-i\underline{Q}.(\underline{\ell} + \underline{d} + \underline{u}_{\underline{\ell ds}}(0))}\ e^{i\underline{Q}(\underline{\ell}'+\underline{d}'+ \underline{u}_{\underline{\ell}'\underline{d}'s'}(t))}\rangle e^{-i\omega t} dt \ . \tag{4.3.15}$$

The time-averaged occupation of species s on sites $\underline{d}$ in the lattice is defined as

$$c_{\underline{ds}} = \langle \frac{1}{N_c} \sum_{\ell} p_{\underline{\ell ds}}(t) \rangle_t \quad , \qquad (4.3.16)$$

and deviations from this average occupancy as;

$$\Delta p_{\underline{\ell ds}}(t) = p_{\underline{\ell ds}}(t) - c_{\underline{ds}} \quad . \qquad (4.3.17)$$

Substitution of equation (4.3.17) into equation (4.3.15), leads to an expression involving terms in (a) $c_{\underline{ds}} c_{\underline{d}'s'}$, (b) $\Delta p_{\underline{\ell ds}}(o)\, \Delta p_{\underline{\ell}'\underline{d}'s'}(t)$, and (c) cross terms. The displacements $\underline{u}_{\underline{\ell ds}}(t)$ are treated using the standard relation for operators U,V (Squires, 1978).

$$\begin{aligned} \langle \exp U . \exp V \rangle &= \exp\,[\tfrac{1}{2}(U^2+V^2)].\exp \langle UV \rangle \\ &= \exp\,[\tfrac{1}{2}(U^2+V^2)].(1 + \langle UV \rangle + ..). \end{aligned} \qquad (4.3.18)$$

The exponential term on the right hand side of equation (4.3.18) gives rise to the harmonic temperature factor or Debye-Waller factor $\exp\,[-W_{\underline{ds}}(\underline{Q})]$ which multiplies the terms in $\exp[i\underline{Q}.\underline{d}]$ in both (a), and also in (b) since the time scale difference of $\underline{u}_{\underline{\ell ds}}(t)$ and $\Delta p_{\underline{\ell ds}}(t)$ allows the correlations between these terms to be decoupled. The second, and higher, order terms in $\langle UV \rangle$ occurring in (a) give the scattering from phonons. The cross terms (c) are negligible in the approximation decoupling the lattice vibrations. The coherent scattering function is then given by

$$S_c(\underline{Q},\omega) = S_c^B(\underline{Q},0) + S_c^D(\underline{Q},\omega) + S_c^P(\underline{Q},\omega) \quad . \qquad (4.3.19)$$

Bragg scattering. The first term in equation (4.3.19) gives the Bragg scattering contribution:

$$\begin{aligned} S_c^B(\underline{Q},0) &= \frac{(2\pi)^3}{v_o} \sum_{\underline{\tau}} \left| \sum_{\underline{d},s} \langle b_s \rangle c_{\underline{ds}} e^{-W_{\underline{ds}}(\underline{Q})} e^{i\underline{Q}.\underline{d}} \right|^2 \delta(\underline{Q} - \underline{\tau}) \\ &= \frac{(2\pi)^3}{v_o} \sum_{\underline{\tau}} \left| F(\underline{Q}) \right|^2 \delta(\underline{Q} - \underline{\tau}). \end{aligned} \qquad (4.3.20)$$

Here $\underline{\tau}$ is a reciprocal lattice vector, v_o the volume of the unit cell, and $F(\underline{Q})$ the cell structure factor. A more general temperature factor than the harmonic Debye-Waller term is given by $T_{\underline{ds}}(\underline{Q}) = \langle \exp[-i\underline{Q}.\underline{u}_{\underline{ds}}] \rangle$ expanded to higher terms in $\underline{u}_{\underline{d}s}$, and

$$F(\underline{Q}) = \sum_{\underline{d},s} (\langle b_s\rangle\, c_{\underline{d}s} T_{\underline{d}s}(\underline{Q}) e^{i\underline{Q}\cdot\underline{d}}) \quad . \tag{4.3.21}$$

Clearly an experimental determination of $F(\underline{Q})$ will give the average occupancy of sites in the cell as well as their mean thermal vibration from the temperature factor $T_{\underline{d}s}(\underline{Q})$. It will be seen that, in general, $T_{\underline{d}s}(\underline{Q})$ gives the Fourier transform of the infinite-time self-correlation function $G_s(\underline{r},\infty)$ of the vibrating ion.

Diffuse scattering. The diffuse scattering is given by the second term in equation (4.3.19):

$$S_c^D(\underline{Q}.\omega) = \frac{1}{2\pi\hbar} \frac{1}{N_c} \sum_s \sum_{s'} \langle b_s\rangle^* \langle b_{s'}\rangle \sum_{\substack{\underline{\ell},\underline{d} \\ \underline{\ell}',\underline{d}'}} e^{i\underline{Q}.(\underline{\ell}'-\underline{\ell})} e^{i\underline{Q}(\underline{d}'-\underline{d})} \; x$$

$$e^{-W_{\underline{d}s}(\underline{Q})} e^{-W_{\underline{d}'s'}(\underline{Q})} \; x \int_{-\infty}^{\infty} \langle \Delta p_{\underline{\ell d}s}(0).\Delta p_{\underline{\ell}'\underline{d}'s'}(t)\rangle \, e^{-i\omega t} dt \;. \tag{4.3.22}$$

This therefore gives as much information as it is possible to obtain about the time dependence of the correlations between defects, or about deviations from the average occupancy of sites. More limited, but more accessible, data are given by $S_c^D(\underline{Q})=I_c^D(\underline{Q},0)$, which is related to the instantaneous (t=0) disordered configuration:

$$S_c^D(\underline{Q}) = \frac{1}{N_c} \left| \sum_{\underline{\ell d}s} (p_{\underline{\ell d}s}\langle b_s\rangle - c_{\underline{d}s}\langle b_s\rangle)\, e^{-W_{\underline{\ell d}s}(\underline{Q})}\, e^{i\underline{Q}.(\underline{\ell}+\underline{d})} \right|^2 \tag{4.3.23}$$

It is clearly more straightforward to model $S_c^D(\underline{Q})$ than $S_c^D(\underline{Q},\omega)$, in terms of an instantaneous picture of the disorder at high temperatures.

Inelastic Phonon Scattering. The one-phonon scattering is given by the third term in equation (4.3.19):

$$S_c^{P\pm 1}(\underline{Q},\omega) = \frac{2\pi}{v_o} \frac{\hbar^2}{N_c} \sum_m \frac{(\langle n_m\rangle + \tfrac{1}{2} \pm \tfrac{1}{2})}{2\hbar\omega_p^m(\underline{q})} \left| \sum_{\underline{d},s} \frac{\langle b_s\rangle c_{\underline{d}s}}{\sqrt{M_s}}\, e^{-W_{\underline{d}s}(\underline{Q})}\, e^{i\underline{Q}.\underline{d}.}(\underline{Q}_m.\underline{\hat{e}}_m^{\underline{d}}) \right|^2 \; x$$

$$\sum_{\underline{\tau}} \delta(\underline{Q} \mp \underline{q} - \underline{\tau})\ \delta(\hbar\omega \mp \hbar\omega_p^m(\underline{q})), \qquad (4.3.24)$$

where M_s is the ionic mass of species s, and for a mode m with energy $\hbar\omega_p^m(\underline{q})$, $\langle n_m \rangle$ is the Bose population factor, and $\hat{e}_{\underline{m}}^{\underline{d}}$ is a unit polarisation vector normalised so that $\sum_{\underline{d}} \hat{e}_{\underline{m}}^{\underline{d}*} . \hat{e}_{\underline{m}}^{\underline{d}} = 1$.

We shall now consider each of these types of scattering, given by equations (4.3.20), (4.3.22) and (4.3.24), in more detail.

4.3.2.3 Diffraction - Fourier Synthesis

In studying the diffraction of x-rays or neutrons from ionic solids at high temperatures we are primarily concerned with the changes in their structure from that at ambient temperature. The difference may involve a change in the size and symmetry of the unit cell, such as in the case of the silver conductors, or simply a change in the distribution of ionic density within the unit cell, as in the case of the fluorites. Even if the symmetry does not change, a relatively simple measurement of the lattice constant variation with temperature can yield useful information. The variation can be compared with that calculated using assumed interatomic potentials, and gives a useful test of the potential. A comparison between the variation of lattice constant determined from diffraction with expansion measured by a dilatometer (Roberts and White, 1986) can give information on the formation of Schottky defects, which affect only the latter. The density is required in many thermodynamic relations and may be calculated from the lattice constant. Theoretical predictions, such as those of Tallon and Roberts (1985), for the effect of thermally-induced Frenkel defect formation on the lattice constant may be tested.

If we set aside for the moment a number of practical problems related to the correction of the measured Bragg reflection intensities, we may discuss the information which can be gained. From equation (4.3.20) the intensity of the reflection at $\underline{Q} = \underline{\tau}$ is given by

$$S_c^B(\underline{\tau},0) = I_c(\underline{\tau},\infty) = \int G_c(\underline{r},\infty)\, e^{i\underline{\tau}\cdot\underline{r}} d\underline{r} = \frac{(2\pi)^3}{v_o} \mid F(\underline{\tau}) \mid^2 \quad , \qquad (4.3.25)$$

where we now include the scattering lengths in the pair correlation function, $G_c(\underline{r},\infty)$, which is also known as the 'Patterson function' by crystallographers. It can be shown (e.g. Squires 1978, p77) that $G_c(\underline{r},\infty)$ is related to the 'probability scattering density function', $\rho_c(\underline{r})$, which gives the average distribution of coherent scattering amplitudes within the crystal, by the self-convolution:

$$G_c(\underline{r},\infty) = \int \rho_c^*(\underline{r}')\, \rho_c(\underline{r} + \underline{r}')d\underline{r}'. \qquad (4.3.26)$$

By substitution, it can be seen that

$$\rho_c(\underline{r}) = \frac{1}{v_o} \sum_{\underline{\tau}} F(\underline{\tau})e^{-\underline{\tau}.\underline{r}} \qquad (4.3.27)$$

and

$$F(\underline{Q}) = \int_{cell} \rho_c(\underline{r})\, e^{i\underline{Q}.\underline{r}}d\underline{r} \quad . \qquad (4.3.28)$$

From these expressions it is in principle possible to obtain $\rho_c(\underline{r})$ from the measured Bragg reflection structure factors $F_{obs}(\underline{\tau})$. However it is the intensity $S_c^B(\underline{\tau},0)$ which is measured from the cross section and, whereas the Patterson function may be deduced unambiguously by Fourier inversion of equation (4.3.25), the probability scattering density function necessitates knowledge of the phase factor, $e^{i\phi(\tau)}$, of $F(\underline{\tau})$. For centrosymmetric structures this is ±1, and may be deduced from either a knowledge of the unperturbed low temperature structure, from calculations based on the immobile ions positions or from calculations based on a model of the high temperature structure, as discussed below. However in practice it is not possible to measure all the Bragg reflections in order to carry out a complete Fourier transform. If $\underline{\tau}_{max}$ is the maximum reciprocal lattice vector at which a reflection is measured, this results in an accuracy in r of $\Delta r = 0.715 \times 2\pi/\tau_{max}$ (James 1948). Furthermore corrections must be made to the measured intensities to obtain $F_{obs}(\underline{\tau})$, for absorption, extinction and thermal diffuse scattering (TDS). The latter is the component of phonon scattering, and in the case of fast-ion disorder possible diffuse scattering, falling within the instrumental resolution function. Whereas absorption may be corrected for on the basis of densities of ions, the extinction and TDS corrections necessitate some sort of model of the lattice to be made.

4.3.2.4 Diffraction - Modelling the Unit Cell

In view of the practical difficulties mentioned above, it is usual first to analyse diffraction data in terms of a model of the unit cell. At low temperatures this presents no conceptual problem, since a certain structure is tried with assumptions made as to the sites occupied and the ions occupying them. Again the phase problem must first be solved. The parameters describing the positions of the ions are varied to obtain the best fit of the calculated structure factor to that observed. Even a model which does not fit perfectly may be used to calculate the correction factors due to extinction, TDS, etc, and these corrections may be improved iteratively. Knowing these corrections, $\rho_c(\underline{r})$ may then be determined by Fourier synthesis, equation (4.3.27).

At elevated temperature, although the structure may be sufficiently close to a known low temperature structure to determine the phase factor, an analysis in terms of a model must inevitably involve approximations in order to give a time averaged picture of what is often a complex dynamical situation. The model must be developed by drawing on other data, such as that provided by the 'snapshot picture' given by diffuse scattering, to supplement the diffraction data. The model can only be judged by how well it explains all these data, and how it predicts new phenomena. It provides a visual picture of the complex high temperature state. Even if the model does not represent the exact 'true' situation, it can usually be used to calculate a $\rho_c(\underline{r})$, by summation of the contributions from the individual ionic species, which is a good approximation to the 'true' distribution. In this way an approach using a model may be a better method of determining $\rho_c(\underline{r})$, in cases of limited data, than a Fourier synthesis.

Site Occupation: The Ionic Probability Density Function. The model approach involves choosing sites $\underline{d}$ in the unit cell, and determining their occupation, $c_{\underline{d}s}$, and temperature factor $T_{\underline{d}s}(\underline{Q})$, from the structure factor

$$F(\underline{Q}) = \sum_{\underline{d},s}^{\text{cell}} \langle b_s \rangle \, c_{\underline{d}s} \, T_{\underline{d}s}(\underline{Q}) \, e^{i\underline{Q}\cdot\underline{d}} = \sum_{\underline{d}s}^{\text{cell}} F_{\underline{d}s}(\underline{Q}) \quad , \tag{4.3.29}$$

where $F_{\underline{d}s}(\underline{Q})$ is the component of the cell structure factor from ionic species s at site $\underline{d}$. Substituting in equation (4.3.27) it is seen that

$$\rho_c(\underline{r}) = \sum_{\underline{ds}} c_{\underline{ds}} \langle b_s \rangle \, f_{\underline{ds}}(\underline{r}) = \sum_{\underline{ds}} \rho_{\underline{ds}}(\underline{r}) \quad , \qquad (4.3.30)$$

where $\rho_{\underline{ds}}(\underline{r}) = c_{\underline{ds}}\langle b_s\rangle f_{\underline{ds}}(\underline{r})$ is the probability scattering length density of ionic species $\underline{s}$ at site $\underline{d}$, and the 'ionic probability density function', $f_{\underline{ds}}(\underline{r})$, is given by

$$f_{\underline{ds}}(\underline{r}) = \frac{1}{v_o} \sum_{\underline{\tau}} T_{\underline{ds}}(\underline{\tau}) e^{-\underline{\tau}\cdot\underline{r}} \quad . \qquad (4.3.31)$$

Clearly

$$T_{\underline{ds}}(\underline{Q}) = \int f_{\underline{ds}}(\underline{r}) e^{i\underline{Q}\underline{r}} d\underline{r}, \qquad (4.3.32)$$

so that the temperature factor, $T_{\underline{ds}}(\underline{Q})$, is related by Fourier transform to the ionic probability density function.

It should be noted that the above expressions relating the structure factor to the product of functions of $\underline{Q}$ arise from the general relation which expresses the Fourier transform of a convolution of two functions as the product of the Fourier transform of each function, and its extension to more than two functions. The scattering density function may be split into a series of convolutions of densities from the electron distribution in the case of x-rays, from the smearing effect of thermal vibration, from the distribution of sites in the cell, and from the arrays of lattice points in the crystal.

<u>The Independent Average Potential</u>. The 'total, or 'joint', probability density function for species s is given by

$$f_s(\underline{r}) = \sum_{\underline{d}} c_{\underline{ds}} \, f_{\underline{ds}}(\underline{r}). \qquad (4.3.33)$$

This joint probability density function for an ion may be related to the single particle potential in which it moves, if one makes the approximation that the total potential energy of the crystal can be decoupled into a system of independent, anharmonic, average potentials $V_s(\underline{r})$ (Willis 1970) defined by:

$$f_s(\underline{r}) = e^{-V_s(\underline{r})/k_BT)} / \int e^{-V_s(\underline{r})/k_BT} d\underline{r} \quad . \qquad (4.3.34)$$

Hence $V_s(\underline{r})$ may be determined in principle from the measured $f_s(\underline{r})$ using

$$V_s(\underline{r}) = V_{so} - k_BT \ln f_s(\underline{r}) \quad . \qquad (4.3.35)$$

The behaviour of the anions in compounds with the fluorite structure provides a good illustration of the applications of these relations, particularly as they have now been studied extensively at high temperatures. If the potential in which the anion vibrates is expanded beyond the isotropic harmonic term, which gives rise to the harmonic Debye-Waller factor, to include the anharmonic anisotropic cubic term, it may be written

$$V(u_x,u_y,u_z) = V_o + \frac{1}{2}\alpha_{\underline{d}s}(u_x^2 + u_y^2 + u_z^2) + \beta_{\underline{d}s}(u_xu_yu_z) + O(4\text{th degree}). \qquad (4.3.36)$$

The anion probability density function is found (Dawson et al 1967; Cooper et al 1968; Willis 1969) by substitution into the equation (4.3.34) as

$$f_{\underline{d}s}(\underline{u}) = \frac{\alpha_{\underline{d}s}^{3/2}}{(2\pi)^{3/2}(k_BT)^{3/2}} \exp\left[-\tfrac{1}{2}\frac{\alpha_{\underline{d}s}}{k_BT}(u_x^2 + u_y^2 + u_z^2)\right]\left(1 - \frac{\beta_{\underline{d}s}}{k_BT}u_xu_yu_z\right) \quad . \qquad (4.3.37)$$

The temperature factor $T_{\underline{d}s}(Q)$ is then given by the Fourier transform, equation (4.3.32), or directly from $V_{\underline{d}s}(\underline{u})$, as

$$T_{\underline{d}s}(Q) = \int e^{i\underline{Q}\cdot\underline{u}}\left(e^{-V_{\underline{d}s}(\underline{u})/k_BT}\right) d\underline{u} \Big/ \int e^{-V_{\underline{d}s}(\underline{u})/k_BT} d\underline{u} \quad . \qquad (4.3.38)$$

For terms up to cubic, the value at the Bragg reflection (hkl) is given by:

$$T_{\underline{d}s}(Q) = \left[\exp\left(-\frac{k_BT}{2\alpha_{\underline{d}s}^2}Q^2\right)\right]\left[1 - \frac{i\,\beta_{\underline{d}s}}{\alpha_{\underline{d}s}^3}(k_BT)^2 Q_hQ_kQ_\ell\right] \quad , \qquad (4.3.39)$$

or

$$T_{\underline{ds}}(\underline{\tau}) = [\exp -(B_{\underline{ds}} \frac{\sin^2\Theta}{\lambda^2})] \; [1 + i \; c^o_{123} \;\; hk\ell] \; , \qquad (4.3.40)$$

where $Q_h = \frac{2\pi h}{a_o}$ etc; here we have the usual parameters

$$B_{\underline{ds}} = \frac{8\pi^2 k_B T}{\alpha^2_{\underline{ds}}} = \frac{8\pi^2}{3} \langle u^2 \rangle \; , \qquad (4.3.41)$$

and

$$C^o_{123} = - \frac{8\pi^3 (k_B T)^2}{a^3_o \;\; \alpha^3_{\underline{ds}}} \;\; \beta_{\underline{ds}} \; , \qquad (4.3.42)$$

where a_o is the lattice parameter.

A better approximation to anharmonic vibrations is to expand the anharmonic potential to further terms than equation (4.3.37). Dawson (1970) and Willis (1970) have considered terms up to fourth degree for the anion site in fluorites, and have given expressions for the corresponding ionic probability density function $f_{\underline{ds}}(\underline{r})$ and temperature factor $T_{\underline{ds}}(\underline{Q})$.

An alternative approach, adopted by Johnson (1970), and by Schulz and Zucker (1981) and Bachmann and Schulz (1983) is to expand $f_{\underline{ds}}(\underline{u})$ directly in the form of a series expansion of partial differential operators acting on a Gaussian (harmonic) function $\hat{g}_{\underline{ds}}(\underline{u})$. They show that this expansion may best be written in the form of a Gram-Charlier expansion involving tensor coefficients, $c_{\ell'}$, and multidimensional Hermite polynominals, $H_{\ell'}$, written symbolically as,

$$f_{\underline{ds}}(\underline{u}) = \hat{g}_{\underline{ds}}(\underline{u}) \;\; [1 + \sum_{\ell'} \; c_{\ell'} H_{\ell'}] \qquad . \qquad (4.3.43)$$

This is then Fourier transformed to give an expansion of the form, written symbolically,

$$T_{\underline{ds}}(\underline{Q}) = T^h_{\underline{ds}}(\underline{Q}) \;\; [1 + \sum_{\ell'} \; a_{\ell'} c_{\ell'} Q_h Q_k Q_\ell \;] \; , \qquad (4.3.44)$$

involving the same tensor coefficients $c_{\ell'}$ and products of components of the scattering vector of degree equal to the rank of the tensor. $T^h_{\underline{ds}}(\underline{Q})$ is the harmonic temperature factor. The tensor coefficients c_ℓ can then be fitted to the Bragg diffraction data, but care must be taken to ensure that they are physically meaningful. One test of this is to ensure that the corresponding $f_{\underline{ds}}(\underline{u})$ is positive.

Different Model Approaches. The time-averaged density of ions in a fast-ion conductor, in which the density of some ions is smeared out over large regions of the unit cell, may be interpreted in terms of a model through either of the above two approaches. One may either assume a model which involves a number of non- regular sites for the mobile ions with a simple approximation for the thermal parameters, or one may try to express the distribution in terms only of regular sites, i.e. the low temperature structure, but with highly-anharmonic temperature factors. Which model is physically more realistic may depend on the system in question and the nature of its diffusion process, which may or may not involve the occupation of interstitial sites. A time-averaged picture may not by itself be able to distinguish between these, since it does not give information on the time of occupancy of regions of the cell. Either approach involves the determination of a large number of parameters in the model, and it must be ensured that these are physically reasonable. It may be shown that the two approaches are equivalent in simple cases, such as the equivalence of a model of two nearby sites with harmonic temperature factors to that of one site with an anharmonic temperature factor (Andersen et al 1987). This equivalence illustrates the tendency to get strong correlations between site occupancy and temperature factor parameters when fitting a model to diffraction. However, if the ultimate aim is to determine ionic joint probability distribution functions $f_s(\underline{r})$, it is frequently found that this is relatively independent of the model approach chosen.

Schulz and Zucker (1981) have used the joint probability density function $f_s(\underline{r})$ to determine the effective one particle potential $V_s(\underline{r})$ using equation (4.3.35). In some cases they find that the potential for the diffusing ion has minima only on regular sites and is nearly independent of temperature, and the barrier between sites is close to the activation energy for conduction. They suggest that for these cases their approach is a more realistic picture of the situation in which ions diffuse than one involving interstitial sites. Although extra sites may have been necessary initially to determine $f_s(\underline{r})$ from experiment, this ionic density can give a $V_s(\underline{r})$ which has minima only at regular sites. In using this approach one must, of course, be sure that the density distribution does arise from a time-average of dynamic disorder, rather than from an average of more static positional disorder, as may well be the case for α-Ag_3SI. Didisheim et al (1986) interpret their neutron data differently, in terms of positional disorder, from the analysis of x-ray data by Perenthaler et al (1981) who derived a single potential. There are however a number of 'philosophical' questions which might be posed challenging this approach, particularly regarding the time scales on which the densities are averaged and the approximation used to decouple the total crystal potential. In the case of fast-ion conductors the diffusing ions will certainly affect the potential of their neighbours on certain short timescales which are just those involved in the conduction process. Andersen et al (1987) have pointed out that before $V(\underline{r})$ is deduced from diffraction data, the TDS correction needs to be treated very accurately, since it has a similar effect on Bragg intensities as the temperature factor.

A comparison of the analysis of diffraction data, corrected for TDS, absorption and extinction, in terms of the Fourier synthesis method or of a model of the structure may be made by comparing the probability scattering length density $\rho_c(\underline{r})$ which each method produces. A useful test of a particular model may be made by making a 'Difference Fourier' synthesis which gives a map of $\Delta\rho_c(\underline{r})$, the error in scattering density arising from an imperfect fit of calculated structure factors $F_c(\underline{\tau})$ to the observed structure factors $F_o(\underline{\tau})$. $\Delta\rho_c(\underline{r})$ is given by

$$\Delta\rho_c(\underline{r}) = \frac{1}{v_o} \sum_{\underline{\tau}} [F_o(\underline{\tau}) - F_c(\underline{\tau})]\, e^{i\Phi(\underline{\tau})} e^{-i\underline{\tau}.\underline{r}}, \qquad (4.3.45)$$

where the phase terms in the structure factor, $e^{i\Phi(\underline{\tau})}$, is taken taken out as a factor. A perfect fit will give $\Delta\rho_c(\underline{r})$ uniformly zero, whereas any regular features in $\Delta\rho_c(\underline{r})$ may well indicate omissions from the model.

The deliberate omission of some of the ions in the model, giving a $F_c(\underline{\tau})$ from only a part of the structure gives a 'Partial Fourier' synthesis map of $\Delta\rho_c(\underline{r})$ from equation (4.3.45). The partial Fourier synthesis may be used to isolate the scattering observed from the less well defined components of a system. If the immobile ions of a two component fast-ion conductor, whose positions are well defined, are used to calculate $F_c(\underline{\tau})$ for example, the scattering density from the mobile ions will be given by $\Delta\rho_c(\underline{r})$. Examples will be given in figures 4.27a,b,c. An alternative approach, useful if there is no crystal structure change, is to use the known low-temperature structure to calculate the $F_c(\underline{\tau})$, when the $F_o(\underline{\tau})$ determined from high temperature data will then give the change in probability scattering length density arising from the change in temperature.

4.3.2.5 Diffuse Scattering

It is seen from equation (4.3.5a) that a Fourier transform of the inelastic coherent diffuse scattering $S^D(\underline{Q},\omega)$ will yield the time dependent total pair correlation function $G(\underline{r},t)$ of the disordered ions. However, in practice, even though the diffuse neutron scattering can be isolated relatively easily from the phonon-, Bragg-, and sample environmental background- scattering, it is not practical to measure data over an adequate range of $\underline{Q}$ and ω. Indeed even if an integration over ω can be made experimentally to give $S^D(\underline{Q})$, the $\underline{Q}$ range is unlikely to be sufficient to yield $G(\underline{r},0)$ with good resolution. Such an integration is automatic for x-ray measurement, but in this case the separation from phonon scattering is difficult and the total $S(\underline{Q})$ is usually measured. It is therefore necessary to interpret $S^D(\underline{Q},\omega)$ or $S^D(\underline{Q})$ in terms of a model of the disordered structure, developed using additional data from as many sources as possible.

The most basic approach is to model the ionic potentials and then calculate the spatial and temporal behaviour of each ion on a computer using molecular dynamics techniques. This approach is described fully by Gillan (1989), Chapter 3. $S_c(\underline{Q},\omega)$ and $S_c(\underline{Q})$ may be obtained by Fourier transform of the correlation functions and compared with experiment. However the method is time-consuming, and it is difficult to present the results in the form of a quantitative picture of the disorder described by a few parameters. Such a picture can be visualised if the disorder is modelled more directly in terms of the time-dependent occupation of certain sites in the unit cell. The average 'instantaneous' or 'snapshot' picture of the disorder, given by $S_c^D(\underline{Q})$, is more amenable to modelling than $S_c^D(\underline{Q},\omega)$. Even so, diffraction and other data are usually also required to help build up the basis of a suitable model. Molecular dynamics may prove helpful in this respect, for example, by indicating which ionic correlations contribute most to $S_c^D(\underline{Q})$ and $S_c(\underline{Q},\omega)$.

The diffuse scattering from dynamic disorder in ionic solids at high temperature is quasielastic in nature. Whereas quasielastic, incoherent, diffuse scattering has received a great deal of theoretical attention, with increasing sophistication (Chudley and Elliott 1961; Springer 1972; Wolf 1977; Tahir-Kheli and Elliott 1983), the more complex case of quasielastic, coherent, diffuse scattering has not been considered in as much detail (Vineyard 1958; Sköld 1967; Hutchings et al 1983; Gillan and Wolf 1985; Ross et al 1986; Faux and Ross 1987). The exceptionally large incoherent scattering cross section of hydrogen has led to a great deal of experimental work on the motion of hydrogen in solids, and the individual motion of Li^+, Ag^+, and $C\ell^-$ ions with their smaller incoherent cross section has been studied to a lesser extent. The coherent QES from ionic solids at high temperatures in which these ions rapidly diffuse has also been studied, but it is most easily isolated in systems such as some fluorites where the fast ions are O^{2-} or F^- which have no incoherent cross section.

The quasielastic nature of the incoherent and coherent scattering cross sections for randomly diffusing ions or vacancies may be illustrated by reference to a simple Bravais lattice on which there are N_ℓ sites (Hutchings et al 1983). Let there be N_p ions (particles) and N_v vacancies, so that $N_v + N_p = N_\ell$, and let the scattering length of the ions be $\langle b\rangle$. We assume that the transit time of the random jumps of the particles (vacancies) is negligible compared with their residence time, and that there are no correlations between the jumps. This assumption of random motion is clearly only valid for the motion of a vacancy in an otherwise fully occupied lattice, and will be a poor approximation to the particle diffusion in a fully occupied lattice. A better approach is given by Wolf's *encounter model* discussed briefly in section 4.3.4.3.

Coherent Diffuse Scattering. As we only consider one species, we relate to equations (4.3.5) to (4.3.7). Let us consider the case $N_p \gg N_v$ and follow the procedure outlined in section 4.3.2.2. We can define site occupation parameters $P_\ell(t) = \sum_j p_{\underline{\ell}j}(t)$, such that $\underline{P}_\ell(t)$ equals unity if lattice site $\underline{\ell}$ is occupied by *any* particle at time t and is zero otherwise, and we use this to calculate $S^{D'}(\underline{Q})$, and $I^{D'}(\underline{Q},t)$ $(\underline{Q} \neq \underline{\tau})$, where the prime indicates that we omit the scattering at $\underline{\tau}$. We then write

$$I(\underline{Q},t) = \frac{\langle b\rangle^2}{N_p} \sum_{j,j'} \langle e^{-i\underline{Q}.\underline{R}_j(0)} . e^{i\underline{Q}.\underline{R}_{j'}(t)} \rangle$$

$$= \frac{\langle b\rangle^2}{N_p} \sum_{\underline{\ell},\underline{\ell}'} \sum_{j,j'} \langle P_{\underline{\ell}j}(0)\, e^{-i\underline{Q}.\underline{\ell}}\; P_{\underline{\ell}'j'}(t)\, e^{i\underline{Q}.\underline{\ell}'}\rangle . e^{-2W(\underline{Q})} + I^P(\underline{Q},t)$$

$$= \frac{\langle b\rangle^2}{N_p} \sum_{\underline{\ell},\underline{\ell}'} \langle P_{\underline{\ell}}(0)\, P_{\underline{\ell}'}(t)\, e^{i\underline{Q}.(\underline{\ell}'-\underline{\ell})}\rangle . e^{-2W(\underline{Q})} + I^P(\underline{Q},t). \qquad (4.3.46)$$

By writing

$$P_{\underline{\ell}}(0)P_{\underline{\ell}'}(t) = [P_{\underline{\ell}}(0) -1][P_{\underline{\ell}'}(t) -1] + P_{\underline{\ell}}(0) + P_{\underline{\ell}'}(t) - 1 \qquad (4.3.47)$$

in equation (4.3.46), we see that the first term is zero unless there are vacancies at both $(\underline{\ell},0)$ and $(\underline{\ell}',t)$ and will then account for diffuse scattering. The remaining terms will give rise to contributions at the Bragg points $Q = \underline{\tau}$, or to phonons.

We can therefore write, for $\underline{Q} \neq \underline{\tau}$:

$$I^{D'}(\underline{Q},t) = \frac{\langle b\rangle^2}{N_p} \sum_{\underline{R}_v,\underline{R}_{v'}} \langle e^{i\underline{Q}.(\underline{R}_{v'}(t) - \underline{R}_v(0))}\rangle , \qquad (4.3.48)$$

where $\underline{R}_v$ is the position of vacancy v on the lattice. Equation (4.3.48) shows that the scattering may be expressed as a sum over the vacancy positions only.

In order to consider the vacancy motion we define $T_v(n,t)$ as the probability that a vacancy has made n jumps by time t, and $F_v(\underline{Q})$ as the Fourier transform of the spatial distribution of the vacancy after one jump (see for example Gissler and Rother 1970):

$$T_v(n,t) = \frac{1}{n!} \left[\frac{|t|}{\tau_v}\right]^n e^{-|t|/\tau_v} ,$$

$$F_v(\underline{Q}) = \sum_{k=1}^{N_{s'}} M_k\, e^{i\underline{Q}.\underline{L}_{ok}} , \qquad (4.3.49)$$

where τ_v is the average vacancy residence time and M_k is the mean probability that a vacancy originally at $\underline{R}_o$ can be found at $(\underline{R}_o + \underline{L}_{ok})$ after one jump. In the simplest approximation $M_k = 1/N_{s'}$, where $N_{s'}$ is the total number of possible sites. We can then write

$$e^{i\underline{Q}\cdot\underline{R}_v(t)} = e^{i\underline{Q}\cdot\underline{R}_v(0)}.e^{-W(\underline{Q})}.\sum_{n=o} T_v(n,t)\ [F_v(\underline{Q})]^n$$

$$= e^{i\underline{Q}\cdot\underline{R}_v(0)}.e^{-[1-F_v(\underline{Q})]|t|/\tau_v}.e^{-W(\underline{Q})} \quad . \tag{4.3.50}$$

Thus, using the subscript v to denote the case of dilute vacancies, we get

$$I_v^{D'}(\underline{Q},t) = I_v^{D'}(\underline{Q},0).e^{-[1-F_v(\underline{Q})]|t|/\tau_v} \quad , \tag{4.3.51}$$

and writing $S_v^{D'}(\underline{Q}) = I_v^{D'}(\underline{Q},0)$,

$$S_v^{D'}(\underline{Q},\omega) = \frac{1}{2\pi\hbar}\int I^{D'}(\underline{Q},t)e^{-i\omega t}dt = \frac{S_v^{D'}(\underline{Q})}{2\pi\hbar}\ \frac{2\tau_v[1-F_v(\underline{Q})]}{[1-F_v(\underline{Q})]^2+(\omega\tau_v)^2} \quad . \tag{4.3.52}$$

The coherent QES in the case of a few mobile vacancies will therefore have a Lorentzian lineshape in energy transfer with width given by the mean vacancy residence time τ_v and the distribution of jump sites $F_v(\underline{Q})$. Since $F_v(\underline{Q}) \to 1$ as $\underline{Q} \to 0$ or a reciprocal lattice vector, the energy width will tend to zero at these points. $S_v^{D'}(\underline{Q})$ gives an instantaneous picture of the vacancy correlations.

A second simple case is that of a dilute concentration of particles, $N_p \ll N_v$ on a Bravais lattice. In this case, by starting with the basic expression for $I(\underline{Q},t)$ given in equation (4.3.46), and now considering the random particle motion in the same manner as equation (4.3.49) to (4.3.52), but using the subscript p, we find

$$S_p^{D'}(\underline{Q},\omega) = \frac{S_p^{D'}(\underline{Q})}{2\pi\hbar}\ \frac{2\tau_p[1-F_p(\underline{Q})]}{[1-F_p(\underline{Q})]^2+(\omega\tau_p)^2} \quad , \tag{4.3.53}$$

where τ_p is the mean particle residence time and $F_p(\underline{Q})$ gives the distribution of particle jump sites.

Incoherent Diffuse Scattering. The incoherent diffuse scattering function in this case is given in a similar manner by summation over the individual particles:

$$I_p^S(Q,t) = \frac{1}{N_p} \sum_j \langle e^{iQ.R_j(0)}.e^{-iQ.R_j(t)}\rangle$$

$$= e^{-2W(Q)} e^{-[1-F_p(Q)]|t|/\tau_p} , \qquad (4.3.54)$$

so that we obtain the expression of Chudley and Elliott (1961),

$$S_i^p(Q,\omega) = \frac{e^{-2W(Q)}}{2\pi\hbar} \quad \frac{2\tau_p[1-F_p(Q)]}{[1-F_p(Q)]^2 + (\omega\tau_p)^2} \quad . \qquad (4.3.55)$$

Relationship. We therefore see that for this dilute particle case

$$S_p^{D'}(Q,\omega) = S_p^{D'}(Q).S_i^p(Q.\omega).\ e^{2W(Q)} \ . \qquad (4.3.56)$$

This relation, without the temperature factor, was first given by Vineyard (1958) for scattering from a liquid. Sköld (1967) has pointed out that equation (4.3.56) does not satisfy the second moment relationship, and has suggested that the expression

$$S_p^{D'}(Q,\omega) = S_p^{D'}(Q).S_i^p\left(\frac{Q}{\sqrt{S_p^{D'}(Q)}},\omega\right) e^{2W(Q)} \qquad (4.3.57)$$

should be used. He found that this gave a good account of data on liquid argon. However, as we shall use in section 4.3.4.2 this is not found to be the case for diffuse scattering from fluorites. It should be noted that we are here considering only the QES scattering and we exclude phonon contributions.

Returning now to the case of a few vacancies $N_p \gg N_v$ in a concentrated Bravais lattice, we see that if we make the (poor) assumption of random uncorrelated particle diffusion and obtain expression (4.3.55) for $S_i^p(Q,\omega)$, we may relate the coherent and incoherent scattering functions in this case. We note that in this approximation $F_v(Q) = F_p(Q)$, and that since each jump involves the interchange of a vacancy and particle,

$$N_p\tau_p^{-1} = N_v\ \tau_v^{-1}, \qquad (4.3.58)$$

so that we have

$$S_v^{D'}(\underline{Q},\omega) = S_v^{D'}(\underline{Q}) \frac{N_v}{N_p} S_i^P(\underline{Q}, \frac{N_v}{N_p}\omega) e^{2W(\underline{Q})} . \quad (4.3.59)$$

This expression suggests that the coherent scattering from a concentrated system will have the same spectral shape as the incoherent scattering, but that the energy scale is increased by N_p/N_v, giving coherent scattering energy widths broader than those of incoherent scattering by this factor.

In more complex cases, such as mobile clusters comprising Frenkel interstitials, vacancies and relaxed ions, (see below), the scattering will be less straightforward in general. However, in practice the observed scattering can be accounted for by the general form of equation (4.3.56), which is a single Lorentzian in energy transfer. The simplest Lorentzian form is given if the clusters form and decay exponentially without any translational correlation, in which case the energy width will determine the lifetime of the correlated ions and be independent of $\underline{Q}$. Both the simple concentrated and dilute examples discussed above give a coherent scattering function whose energy width tends to zero at $\underline{Q} = \underline{\tau}$. In the more complex cases with interstitials present and where the particles or vacancies jump over distances which are not lattice vectors this may not be so.

It should be noted that at low $|\underline{Q}|$, corresponding to correlation over long distances, equation (4.3.55) for the incoherent QES scattering function yields a Lorentzian spectral function with half width half maximum (HWHM) $\Gamma_{incoh} = D_{incoh} Q^2$, where $D_{incoh} = D^*$, the single particle or tracer diffusion constant. A similar low $|\underline{Q}|$ limit results from equations (4.3.52 and 53) for the coherent scattering function, with HWHM $\Gamma_{coh} = D_{coh} Q^2$. Springer (1972) relates $D_{coh} = D_{chem}$, the collective or chemical diffusion constant giving the particle current in terms of the particle density gradient. Hempelmann et al (1988) show that for an interacting lattice gas system at low $|\underline{Q}|$, $\Gamma_{coh} = \Gamma_{incoh} (f_m/S(\underline{Q})f_t)$, where f_t and f_m are 'tracer' and 'mobility' correlation factors; a result in agreement with Sköld's (1967) *ad hoc* relation (4.3.57) if the correlation factors are set to unity. This relation has been recently tested empirically by Hemplemann et al (1989) and Ross et al (1989) for diffusion of deuterium in α'-NbD_x. It has yet to be extended to include the anisotropy of the lattice.

The Diffuse Structure Factor. The calculation of $S_c^D(\underline{Q})$ requires fewer assumptions than that of $S_c^D(\underline{Q},\omega)$ and, as seen above, it is the first step in calculating the latter for diffusion of ions by hopping between sites on a Bravais lattice. Starting from equation (4.3.23), we make the distinction between sites $\underline{d}_B$ on a regular lattice basis, and sites $\underline{d}_I$ at an interstitial position to which ions may be excited, for example by the creation of a Frenkel pair defect. If the occupation of site $\underline{d}_B$ is written in terms of its occupation by a vacancy $(1-p_{\ell \underline{d}_B}(t))$, the terms in equation (4.3.23) may be split into those in $p_{\ell \underline{d}_I}(t)$ and $(1 - p_{\ell \underline{d}_B}(t))$, and those in c_{d_I} and $c_{\underline{d}_B}$. The latter terms in c can be seen to contribute to the diffuse scattering only at the Bragg points $\underline{Q} = \underline{\tau}$, whereas the terms in p give diffuse scattering at all $\underline{Q} \neq \tau$ with a scattering function which we characterise below with a prime. This scattering function may be written as a summation over the interstitial, $\underline{R}_I$, and vacancy $\underline{R}_V$, sites:

$$S_c^{D'}(\underline{Q}) = \frac{\langle b\rangle^2}{N_c} \left| \sum_{\underline{R}_I}^{\substack{\text{Intl.}\\ \text{occ.}}} e^{i\underline{Q}.\underline{R}_I(0)} - \sum_{\underline{R}_V}^{\substack{\text{Vac.}\\ \text{occ.}}} e^{i\underline{Q}.\underline{R}_V(0)} \right|^2 e^{-2W(\underline{Q})} \quad . \qquad (4.3.60)$$

A single Frenkel defect will contribute one vacancy and one interstitial to this summation, whereas a Schottky defect will contribute only vacancies. It should be noted that a relaxed ion is represented by a vacancy at its regular site associated with an interstitial at the relaxed ion position.

As a simple example we may consider an isolated vacancy on a simple cubic lattice, when the six nearest neighbours ions are relaxed towards the vacancy giving seven vacancy positions and six interstitial positions in equation (4.3.60). We use the term 'cluster' to denote such an arrangement of correlated defective ions. If the concentration of such clusters is low and they are randomly dispersed in the lattice, then the correlations between them may be neglected and the scattering from each cluster added incoherently so that $S_c^{D'}(\underline{Q})$ will be given by the number of clusters, $N_{c\ell}$, times the coherent intensity from one cluster as given by summation over the cluster in equation (4.3.60). In the case of an isolated cluster the vectors $\underline{R}_I$ and $\underline{R}_V$ may be referred to any origin such as the centre of the cluster $\underline{R}_c$, since the term in $\underline{R}_c$ factorises as a phase factor from the summation. In this approximation, when substituted into equation (4.3.58), the temporal behaviour as given by the energy width of the QES, will reflect the geometry of sites to which the cluster effectively jumps as a whole, and its average residence time. Such a model might be expected to give a first approximation to the scattering from vacancy doped fluorites, such as ZrO_2/Y_2O_3. However, in practice, the scattering in this case is more complex, as we shall see in section 4.3.5, due to the interaction between the clusters.

In general, when more complex defects are thermally-induced, the instantaneous picture given by $S_c^D(\underline{Q})$ is an average over all types of clusters, i, orientations of symmetry axis relative to the crystal, j, and configurations of displaced or relaxed ions, k. In the dilute limit, we then have

$$S_c^{D'}(\underline{Q}) = \sum_{ijk} \frac{N_{c\ell}^{ijk}\langle b\rangle^2}{N_c} \left| \mathfrak{F}_{ijk}(\underline{Q}) \right|^2 e^{-2W(\underline{Q})} , \qquad (4.3.61)$$

where

$$\mathfrak{F}_{ijk}(\underline{Q}) = \left[\sum_{\underline{R}_I}^{\text{Intl}} e^{i\underline{Q}.(\underline{R}_I - \underline{R}_c)} - \sum_{\underline{R}_v}^{\text{Vac}} e^{i\underline{Q}.(\underline{R}_v - \underline{R}_c)} \right]_{ijk} , \qquad (4.3.62)$$

and $N_{c\ell}^{ijk}$ is the number of clusters of type i,j,k.

For a system in thermal equilibrium all the different orientations j, and configurations, k, of the clusters will occur with equal probability. It is useful to define a quantity $Y_i^D(\underline{Q})$, the mean scattering per *defective* ion for a cluster of type i. If we continue to consider the case of one predominantly defective species, then

$$S_c^{D'}(\underline{Q}) = e^{-2W(\underline{Q})} \sum_{i=1}^{N_t} n_{ci} \, Y_i^D(\underline{Q}) = e^{-2W(\underline{Q})} X^{D'}(\underline{Q}) , \qquad (4.3.63)$$

where we define $X^{D'}(\underline{Q})$ as the scattering function per unit cell divided by the temperature factor. Here N_t is the number of types of cluster present, n_{ci} is the number of *defective* ions per unit cell present due to $N_{c\ell}^i$

clusters of type i. If there are N_d^i defective ions in each cluster of type i, $n_{ci} = N_d^i N_{c\ell}^i / N_c$ and

$$Y_i^D(\underline{Q}) = \frac{1}{N_d^i N_{oc}^i} \sum_{j,k}^{N_{oc}^i} \left| \mathfrak{F}_{ijk}(\underline{Q}) \right|^2 , \qquad (4.3.64)$$

where N_{oc}^i is the number of orientations and configurations of a cluster of type i. It should be noted that we have changed the notation (interchanging 'type'and 'kind') slightly from that of Hutchings et al (1983), and have also referred the scattering to that from a unit cell. It can be seen that if the scattering per unit cell is measured experimentally and the scattering per defective ion is calculated from a model of the average instantaneous disorder, the number of defective ions per unit cell may be determined.

If the dilute cluster approximation is not valid due to high concentration of defects which are correlated in position, such as occurs in doped systems, the cluster-cluster correlation must be included when evaluating equation (4.3.23). Details of this inclusion are given in the Appendix 3 of Hutchings et al (1983). The scattering function is now given by

$$S_c^{D'}(\underline{Q}) = e^{-2W(Q)} \frac{N_{c\ell}^m}{N_c} \langle b \rangle^2 \sum_{m,m'} [\delta_{mm'} + \sum_{R \neq 0} P_{mm'}(\underline{R}) e^{-i\underline{Q}.\underline{R}}] \mathfrak{F}_m(\underline{Q}) \mathfrak{F}^*_{m'}(\underline{Q}); \qquad (4.3.65)$$

here N_{cl}^m is the number of clusters of 'kind' m (type, orientation, configuration, m is used for ijk above), and $P_{mm'}(\underline{R})$ is the probability that given a cluster of kind m at the origin there will be a cluster of kind m' centred at $\underline{R}$. $\mathfrak{F}_m(\underline{Q})$ is given by expression (4.3.62).

It should be noted that if the complete expression $S_c^D(\underline{Q})$ is required, including the scattering at $\underline{Q} = \underline{\tau}$, one must include the terms $c_{\underline{d}s}$ as given in equation (4.3.23) in the structure factor of equation (4.3.65), so that instead of equation (4.3.62), $\mathfrak{F}_m(\underline{Q})$ is given by

$$\mathfrak{F}_m(\underline{Q}) = \sum_{\underline{\rho}}^{m} [(p_{\underline{\rho}} - c_{\underline{\rho}s}) \langle b_s \rangle e^{-i\underline{Q}\cdot\underline{\rho}} e^{-W_{\underline{\rho}s}(Q)}] , \qquad (4.3.66)$$

where $\underset{\sim}{\rho}$ is written for the site of any ion in the cluster relative to a local origin, and the $\langle b\rangle^2$ factor is included here instead of in equation (4.3.65). This is important when modelling extended regions of correlated defects which include a mean number of ions in excess or less than that of the rest of the lattice, since the lattice is only effectively defined over the region of correlation. If $c_{\rho s}$ is omitted the scattering in a region around $\underset{\sim}{Q} = \underset{\sim}{\tau}$, with a width dependent on the size of the region, is incorrectly calculated.

4.3.2.6 Inelastic Scattering

The cross section for one-phonon inelastic coherent scattering has been given in equation (4.3.24) . The inelastic scattering peaks when $\omega = \pm\omega_p^m(\underset{\sim}{q})$ and $\underset{\sim}{Q} = \underset{\sim}{\tau} \pm \underset{\sim}{q}$, and this enables the complete measurement of the phonon energy dispersion relation to be carried out, usually using a triple-axis spectrometer, in a now-standard manner (Dorner 1972). The zone-centre mode energies of the optic branches are usually more accurately determined using optical or infrared techniques, as are the low-energy acoustic mode energies using ultrasonic or Brillouin scattering methods. However, these techniques give measurements at very long wavelengths (low q) where relaxation effects may alter the frequencies, as discussed below, and cannot give detailed information away from $q \sim 0$. They are difficult to use at temperatures above $\sim$1200K. Measurement of the complete energy dispersion enables a searching test to be made of the models of the interatomic forces used to fit these and other data, and these forces can then be used in molecular dynamics calculations. For ionic solids the models are usually based on rigid ion models, or on shell models which include polarisabilities of the ions and the way these are affected by short range repulsive forces. Although the full set of parameters determined by fitting to the energy dispersion may not be unique nor have a readily-defined physical significance, the analytic fit can be used to calculate quantities such as the phonon energy density of states, $Z(\omega)$.

As the temperature is raised the ions vibrate more anharmonically, and the harmonic phonon modes interact, usually causing a decrease in their energy, a decrease in their lifetime and an increase in the energy width of the mode. The cross section can now be expressed in the form for a damped simple harmonic oscillator (Currat and Pynn 1979; Dorner 1982) and, if the mode energy is greater than the width, this takes the form of a Lorentzian shape function, rather than the delta function for both energy gain and loss excitations. For example, in equation (4.3.24).

$$\delta(\hbar\omega - \hbar\omega_p^m(\underline{q})) \longrightarrow \frac{1}{\pi\hbar} \frac{\Gamma}{(\omega - \omega_p^m(\underline{q}))^2 + \Gamma^2} \qquad (4.3.67)$$

The experimental lineshape may be fitted by this expression, convolved with the instrumental resolution function, to give measurement of the acoustic phonon energy. From the energies the phonon group velocities giving the adiabatic elastic constants, and hence the bulk moduli, can be determined. It is important to consider the wavevector and energy range over which the measurements are made in relation to the relaxation rates τ_{ph}^{-1} of the mode due to phonon-phonon interaction. If $\omega_p < \tau_{ph}^{-1}$ the measurement is made in the hydrodynamic, collision dominated, or *first sound* regime, whereas if $\omega_p > \tau_{ph}^{-1}$ the measurement is made in the collision free, or *zero sound*, regime (Cowley 1967; Cowley et al 1978). Ultrasonic measurements are usually made in the former regime whereas neutron scattering measurements are usually made in the latter. However in ionic solids which show dynamic disorder at high temperatures the hydrodynamic regime can extend much further in q, as is found in the case of PbF_2.

In normal ionic solids the main effect of increasing temperature is that of anharmonicity, which gives a linear fall in phonon energy as the temperature increases and a width increasing as $\sim Tq^2$. (Niklassen 1972, Garber and Granato 1975). Extensive measurements on the alkali halides have been carried out at high temperatures by Cowley and coworkers (see Cowley 1968) and the data are well accounted for by anharmonic phonon theory. However the effect of interference between one phonon and multiphonon processes must be included in the analysis (see Chapter 1 for a full discussion).

In fast-ion conducting solids the dynamic disorder can affect the phonon lifetimes in the fast-ion regime in two ways. Zeyher (1978) has considered the effect of the fluctuating ionic potentials due to the hopping motion of the ions and found an additional damping term $\sim Dq^2$, where D is proportional to the ionic conductivity. However if the disorder changes only on a time scale which is long compared with the phonon period the regularity of the lattice is effectively broken and the perfect lattice phonons are no longer good eigenmodes. One may artificially restore the regularity of the lattice and introduce a width to the phonon excitations which may be calculated by an average t-matrix approximation (Kleppmann 1978), or by the coherent potential approximation (Elliott et al 1974). Dickens et al (1979a) show that the effective static disorder gives rise to a shift in the elastic constant which is independent of q, and a width proportional to q^2. Combining the three effects gives

$$2\Gamma = [\tfrac{1}{2}(D + \eta) + b]q^2 ,$$

$$\omega_p^m(\underline{q})^2 = \omega_p^{mo}(\underline{q})^2 + aq^2. \qquad (4.3.68)$$

Here η describes the anharmonicity, D the ionic diffusion constant, and a and b the effect of disorder.

The relations (4.3.67) and (4.3.68) give a reasonable account of the experimental results on acoustic modes, as will be discussed in the next section. However, the optic modes of fast-ion conductors are found to broaden very rapidly as the temperature is increased, and to become overdamped. This precludes detailed tests of theories such as those of Boyer (1980) who predicted a soft optic mode associated with the onset of ionic diffusion in the fluorite lattice. It is clear, experimentally, that the optic modes of the perfect crystal are not good basis modes in the fast-ion state, although the long wavelength acoustic modes, which mainly depend on continuum elastic properties, do remain so.

It might be expected that a vibrational mode corresponding to the attempt frequency ν_o, (see section 4.2.2) would be a feature of fast-ion conductors. Its measurement would lead to a realistic estimate of the number of jumps per unit time of a diffusing ion. However, although a low-lying flat optic mode is observed in some cases, such as for β-AgI, the β-aluminas and $RbAg_4I_5$, it seems to be by no means universal nor is it necessarily related directly to ν_o.

Whereas thermally-induced lattice disorder and anharmonicity have a marked effect on the inelastic coherent scattering from solids at high temperatures, the scattering does not yield very direct information about the nature of the disorder. Although we are concerned here with ionic solids, we note that electronic disorder may be also thermally activated in certain compounds at high temperatures. This normally has relatively much less effect on the phonon modes. If the excitation is straightforward activation of electrons into a semiconductor conduction band one might expect screening of long-range interactions to occur as the electron density increases, giving a reduction in the splitting of longitudinal and transverse optic phonon modes (Mahan 1972). If, on the other hand dynamic valence changes are induced thermally, the resulting electrons and holes may hop through the lattice as small polarons, and these are expected to affect the longitudinal optic (LO) phonon modes (Rcik 1972). However, effects from polaron hopping have not yet been observed, although valence changes in $Sm_{0.75}Y_{0.25}S$ have been seen to broaden the LO modes at high temperature through fluctuating changes in unit cell volume (Mook et al 1978).

4.3.2.7 Density of Phonon States

Once a model has been fitted to the phonon energy dispersion relation measured in the three principal directions, the parameters may be used to calculate the mode energies at wavevectors giving a uniform distribution over the Brillouin zone. The energy density of states $Z(\omega)$, where $Z(\omega)$ is the fraction of vibrational frequencies between ω and $\omega + \Delta\omega$, is given to a good approximation by summation over a limited number of points $\underline{q}$:

$$Z(\omega) = \frac{1}{3Nd'} \sum_{m,\underline{q}} \delta[\omega - \omega_p^m(\underline{q})] , \qquad (4.3.69)$$

where d' is the number of ions in the unit cell (Stassis 1986). This method of obtaining $Z(\omega)$ is, however, a time consuming procedure, and one which becomes difficult once the phonon modes become heavily damped.

A measure of $Z(\omega)$ may be obtained more directly from powder samples by either inelastic incoherent or inelastic coherent scattering. The two techniques however are quite different. Both measurements are usually made using time-of-flight techniques with many counters detecting neutrons scattered over a range of angles. In the case of an ionic compound containing ions which scatter incoherently, the inelastic incoherent cross section gives a summation over modes weighted by the amplitude of vibration and the incoherent cross section of the ions,

$$\frac{d^2\sigma}{d\Omega dE_f} = \frac{k_f}{k_i} \sum_{s,\underline{d}} \frac{c_{\underline{ds}}}{2M_s} \frac{\sigma^2_{inc}}{4\pi} e^{-2W_{\underline{ds}}(\underline{Q})} \sum_{m,\underline{q}} \frac{|\underline{Q}\cdot\underline{\hat{e}}^m_{\underline{d}}(\underline{q})|^2}{\omega^m(\underline{q})} \times$$

$$\{n_m(\underset{\sim}{q})\ \delta(\omega + \omega_m(\underset{\sim}{q})) + (n_m(\underset{\sim}{q}) + 1)\ \delta(\omega - \omega_m(\underset{\sim}{q}))\}. \qquad (4.3.70)$$

The scattering only depends on $\underset{\sim}{Q}$ through the temperature factor, so that only the energy spectrum is measured. In the case of an incoherently scattering element with a cubic crystal structure, such as vanadium, the summation yields $Z(\omega)(n(\omega)+1)/\omega$, so that the true density of states can be deduced. The inelastic incoherent scattering from hydrogen-containing compounds is used extensively to study the lattice vibrations involving hydrogen motion through the large incoherent cross section and small mass of hydrogen.

A weighted density of states of coherently-scattering compounds may be measured using a technique in which the summation of modes of a given energy over all wavevectors throughout the Brillouin zone is made experimentally on the time-of-flight spectrometer (Breuer 1974 and references therein). The wavevector dependence of the cross section is averaged at each energy transfer by using a suitable bank of detectors to cover many $\underset{\sim}{Q}$ values uniformly distributed over a large region of reciprocal space. Use of a polycrystalline sample further ensures good averaging over phonon directions and polarisations. The excitations are observed via neutron energy loss. For the case of a single element one can show that $Z(\omega)$ is approximately given by a distribution function $g_s(\omega)$, which is determined from the measured cross section for each ω integrated over the range of scattering angles 2Θ covered by the detectors (see for example Dickens et al 1980). For a compound one must make the approximation that the cross section derives from an incoherent summation over the contributions from each element, and one can then show that one measures a weighted density of states $\overline{G}(\omega)$ given by

$$\overline{G}(\omega) = \sum_s (N_s \langle b_s \rangle^2 / M_s)\ g_s(\omega) / \sum_s (N_s \langle b_s \rangle^2 / M_s). \qquad (4.3.71)$$

Experimentally this is found to be close to $Z(\omega)$ as calculated from the dispersion relation, as is seen for PbF_2 in section 4.3.4.4. At very large scattering vectors $\underset{\sim}{Q}$, the interference effects vanish and the coherent and incoherent scattering cross sections become the same, so that the same weighted density of states is measured.

The different types of neutrons scattering experiment, and the information given, are summarised in table 4.4.

Table 4.4

The principal neutron scattering techniques and the information given

<u>Coherent Scattering</u>	
Diffraction $\frac{d\sigma}{d\Omega}$	Unit cell symmetry and size Time-averaged position of ions in cell. Mean thermal motion.
Elastic diffuse scattering $S^D(\underline{Q})$	Spatial correlation of static disordered ions.
Quasielastic diffuse scattering $S^D(\underline{Q},\omega)$	Space-time correlated dynamical behaviour of diffusing, disordered, ions.
Integrated diffuse scattering $S^D(\underline{Q})$	Snapshot picture of average spatial correlation of diffusing, disordered, ions.
Inelastic scattering $S(\underline{Q},\omega)$	Phonon energy dispersion relation. Elastic constants. Local mode energies of defects.
Small angle neutron scattering $S(\underline{Q})$	Size distribution, shape, volume fraction of inhomogeneities in scattering density. Defects 1-100 nm in size.
<u>Incoherent Scattering</u>	
Quasielastic diffuse scattering $S_i^D(\underline{Q},\omega)$	Space-time dynamical behaviour of individual diffusing ions.
Inelastic scattering $S_i(\underline{Q},\omega)$	Density of phonon states.

4.3.3 Experimental Techniques

4.3.3.1 Some Practical Considerations - Resolution, Corrections and Normalisation

The basic neutron scattering instruments are the diffractometer and the inelastic scattering spectrometer. The latter may be used to study either diffuse quasielastic or inelastic scattering and most commonly is either a triple-axis or time-of-flight spectrometer. The triple-axis instrument has the advantage that a scan of intensity may be made either of energy transfer $\hbar\omega_o$ with $\underline{Q}_o$ fixed (constant-Q), or of scattering vector $\underline{Q}_o$ with energy transfer $\hbar\omega_o$ fixed (constant-ω). The time-of-flight instrument generally performs a scan in which both $\underline{Q}_o$ and ω_o vary, and special techniques must be used to extract data with one variable held constant. These and more specialised instruments are discussed by Windsor (1986).

Resolution As in any spectroscopic measurement, when the spectrometer is set to a nominal value of $(\underline{Q}_o,\omega_o)$, the intensity measured in a neutron scattering experiment is the convolution of $\sigma(\underline{Q},\omega)$, the cross section for the specimen, with R, the instrumental resolution function:

$$I(\underline{Q}_o,\omega_o)=A\int R(\underline{Q}-\underline{Q}_o,\omega-\omega_o,\underline{Q}_o,\omega_o)\ \sigma(\underline{Q},\omega)d\underline{Q}d\omega, \qquad (4.3.72)$$

where A includes the incident neutron flux. Both triple-axis, or time-of-flight spectrometers have resolution functions which are given to a good approximation by a [4d] Gaussian function

$$R(\underline{Q}-\underline{Q}_o,\omega-\omega_o,\ \underline{Q}_o,\omega_o) = R_o(\underline{Q}_o,\omega_o)\ \exp(-\tilde{\underline{X}}\ \underline{\underline{M}}\ \underline{X}), \qquad (4.3.73)$$

where $\underline{X} = (Q^x-Q_o^x,\ Q^y-Q_o^y,\ Q^z-Q_o^z,\ \omega-\omega_o)$, and $\underline{\underline{M}}$ is the resolution matrix which depends on the spectrometer collimation angles and crystal mosaic spreads or flight-path lengths etc (Cooper and Nathans 1967; Dorner 1972). It is often discussed in terms of the ellipsoid

$$\exp\ (-\tilde{\underline{X}}\ \underline{\underline{M}}\ \underline{X}) = \text{const}\ . \qquad (4.3.74)$$

An expression similar to equation (4.3.72) holds for the diffractometer which, in the 'quasistatic' approximation, can be made to integrate over energy transfer at a given $\underline{Q}_o$ setting, so that the convolution is only over the variable $\underline{Q}$.

The cross section $\sigma(\underline{Q},\omega)$ is proportional to $S(\underline{Q},\omega)$, but contains terms such as the number of unit cells in the sample, the absorption, extinction, and, in the diffraction case, the correction for TDS. In order to place the scattering cross section on an absolute basis a comparison is made with the scattering from a sample of known cross section, making the same corrections as necessary.

Corrections In practice the effects of finite resolution are eliminated in diffraction experiments when the integral over the peak intensity is compared with experiment, and as the diffraction cross section is high, absorption, extinction and TDS are the main corrections necessary. As we have mentioned (section 4.3.2.3), these require a theoretical model for accuracy, and the data analysis is an iterative procedure. In the case of inelastic scattering, where the TDS is normally resolved in energy into well-defined excitations, usually only the absorption correction is necessary. However it is much more important to make corrections to deconvolve $\sigma(\underline{Q},\omega)$ from the measured $I(\underline{Q}_o,\omega_o)$, necessitating an accurate knowledge of the resolution function.

We first discuss the corrections to diffraction data. For a sample of well defined geometrical shape and not containing nuclei with extremely high absorption cross section, the correction for absorption may be calculated with good accuracy. The correction for extinction depends on the sample mosaic structure, and one may find that the crystal perfection improves with annealing time at high temperature increasing the extinction. There are two commonly-used formulations of the extinction correction for diffraction from well-defined sample shapes. These are given by Zachariasen (1967) and by Becker and Coppens (1974,1975). As has been noted, a small extinction correction has the same effect as an overall exponential Debye-Waller temperature factor, and so must be calculated accurately if the temperature factors are to be determined accurately. Both absorption and extinction decrease with neutron wavelength.

The principal TDS correction arises from acoustic phonon modes, the intensity of which peaks at the Bragg condition $\underline{Q} = \underline{\tau}$. The correction is discussed fully by Cooper and Rouse (1968) and by Cooper (1970). The contribution to the measured intensity at this position depends on the size of the resolution aperture in $\underline{Q}$ and the slope of the phonon modes: a knowledge of both of these is necessary to determine the corrected intensity, I^c, from that measured, I, $(I = I^c(1+\alpha))$. The phonon energy dispersion of the acoustic modes must therefore be determined by inelastic scattering at each temperature of measurement. In order for the acoustic phonon to contribute to α its velocity must be considerably less than the neutron velocity. As the acoustic modes renormalise with temperature increase, it is possible for a high-velocity phonon not to contribute at low temperature but to do so at elevated temperature. Again, the advantage of short wavelength neutrons is seen. Indeed as the use of short-wavelength neutrons allows acquisition of data to large $\underset{\sim}{Q}$, and thus yields better spatial resolution and determination of thermal parameters, the advantage is evident of a diffractometer situated on a beam from a hot-source moderator for measurement at high temperatures. In the case of x-ray diffraction similar corrections are necessary, with the velocity of the phonon always less than that of the photon.

It is important to make a number of checks on diffraction data measured at high temperature. Monitoring a standard Bragg reflection at regular time intervals at a given temperature will indicate if extinction is changing, and checks on absorption corrections may be made by comparison of intensities of symmetry equivalent peaks. At temperatures approaching the melting temperature, materials with a high vapour pressure will tend to evaporate. Encapsulation of the sample under an inert atmosphere in a tube of a material with low absorption and low chemical reactivity with the sample will help to reduce this. Correction for evaporation may be made by reference to a standard reflection of low enough intensity to be relatively insensitive to extinction changes. A further test of constancy of sample mass may be made using the overall scale factor by which the calculated intensities are fitted to those observed. If the effect of the lattice expansion on the scattering density is taken into account, this factor should be independent of temperature. If no evaporation occurs it may be used to reduce the number of parameters varied, or to help test the validity of models used in the refinement.

Analysis of inelastic scattering data necessitates corrections for the effects of the finite resolution on the intensity, the energy of excitation and the energy width of the observed profile of counts versus ω or $\underset{\sim}{Q}$, often referred to as the 'neutron group'. If we include the absorption correction in A, we can write $\sigma(\underset{\sim}{Q},\omega) = N_c S_c(\underset{\sim}{Q},\omega)$ in equation (4.3.72). It is usual to assume a form of scattering function, which contains parameters to be fitted to the data. Considering first the excitation of phonons, $S_c(\underset{\sim}{Q},\omega)$ may have the sharp excitation form given in equation (4.3.24) at low temperature, becoming broadened in the form given in equation (4.3.67) at higher temperatures and possibly becoming overdamped when the damped simple harmonic oscillator form is appropriate (Currat and Pynn 1979). The types of computer routine available for fitting the parameters in the assumed cross section to the neutron group profile have been reviewed by Hutchings et al (1986). Different integration algorithms must be used in each case. For sharp excitations, the profile is determined by the resolution function, the nature of the scan, and the slope of the excitation energy dispersion. The principal concern is to deduce the true excitation energy from the resolution-shifted peak intensity, for which the routine RESFLD is suitable. For broadened excitations the routine FITSQW may be used to deduce both the energy and the width of the excitation. Each of these routines involve a full four dimensional integration, but if the cross section varies only slowly with $\underset{\sim}{Q}$, as in the case of most quasielastic scattering, a simpler one dimensional integration over energy transfer alone maybe performed using PKFIT or GENFIT. (The routines mentioned above are described by Hutchings et al 1986).

Normalisation The normalising factor $R_o(\underset{\sim}{Q}_o,\omega_o)$ in the resolution function of a triple-axis spectrometer may be shown (Dorner 1972) to be of the form

$$R_o(\underset{\sim}{Q}_o,\omega_o) = c_1 V_I V_F / V_E \tag{4.3.75}$$

where V_E is the volume of the resolution ellipsoid defined out to a certain fraction of its peak value, c_1 is a numerical constant dependent on this fraction, and $V_I \sim k_i^3 \cot \Theta_M$ and $V_F \sim k_f^3 \cot\Theta_A$, where Θ_M and Θ_A are the monochromator and analyser Bragg angles. The proportionality factors include the reflectivity of the monochromator and analyser crystals and the counter efficiency, all of which may vary slowly with k_i or k_f, and the collimation and mosaic angles. If a constant-$\underset{\sim}{Q}$ scan of energy transfer $\hbar\omega_o$ is made by scanning k_f with k_i kept constant, it can be seen that

$$\int \frac{I(\underset{\sim}{Q}_o,\omega_o)d\omega_o}{V_F} = Ac_1 V_I \int \sigma(\underset{\sim}{Q}_o,\omega)\, d\omega \tag{4.3.76}$$

giving a determination of the total cross section integrated over ω; this is independent of the instrumental broadening and line shape effects.

In practice the integration is made either directly by a summation

$$T(\underline{Q}_o) = \hbar\int \frac{I(\underline{Q}_o,\omega_o)d\omega_o}{V_F} = \hbar\Delta\omega_o \sum_{i=1}^{n} \frac{I(\underline{Q}_o,\omega_{oi})}{V_F} \quad , \qquad (4.3.77)$$

where the scan consists of n steps $\hbar\Delta\omega_o$ of energy transfer $\hbar\omega_{oi}$, or by fitting a mathematical function to the intensity profile. We denote such an integrated phonon or diffuse scattering neutron groups by $T^P(\underline{Q}_o)$ or $T^D(\underline{Q}_o)$ respectively. The detected neutron count is usually made relative to a fixed count, C_M, of the incident beam monitor detector placed before the sample. This is a fission chamber detector which measures the incident neutron flux on the sample with a low efficiency which is proportional to k_i^{-1}. Since the flux is proportional to $k_i V_I$, we can write $C_M = c_2 V_I$, where c_2 is a constant. Thus the experimentally determined integrated intensity per monitor count is related to the theoretical cross section by

$$\frac{T(\underline{Q}_o)}{C_M} = C\,\hbar\int\sigma(\underline{Q}_o,\omega)d\omega , \qquad (4.3.78)$$

where $C = Ac_1/c_2$. It should be noted that if the scan is made with k_f kept constant, it is not necessary to divide by V_F in equation (4.3.76), and V_F can be absorbed in the constants in equation (4.3.78).

To place a measured cross section on an absolute basis, it is necessary to determine the product of the spectrometer constant, C, and the number of unit cells in the sample N_c. This is best done by using the same sample to observe scattering from a process with a cross section which may be calculated absolutely. An acoustic phonon excitation or incoherent scattering may be used for this purpose, as described below. An alternative is to use a separate sample of known cross section, such as vanadium, but careful correction for the relative amounts of the two samples in the beam must then be made.

For small acoustic phonon wavevectors the cross section may be simplified as all ions move the same amount. Using $\underline{Q} \sim \underline{\tau}$, the normalisation relations for the phonon eigenvectors, and the fact that at high temperatures $\langle n_m \rangle = [(n_m)+1] \sim k_B T/\hbar\omega_p^m(\underline{q})$, one can show that

$$\hbar\int\sigma^P(\underline{Q}_o,\omega)\,d\omega = N_c \frac{\hbar^2}{\hbar\omega_p^m(\underline{q})}\,\frac{k_B T}{\hbar\omega_p^m(\underline{q})}\,\frac{(\underline{Q}.\hat{\underline{e}}_m)^2}{2M_c} \times$$

$$\left|\sum_{\underline{d}}^{cell} \langle b_s\rangle c_{\underline{d}s}\, e^{-W_{\underline{d}s}(\underline{\tau})}\, e^{i\underline{\tau}.\underline{d}}\right|^2 . \qquad (4.3.79)$$

Here $\hat{\underline{e}}_m$ is the polarisation unit vector of the mode m, M_c is the mass of the unit cell, and N_c is the number of unit cells in the sample. It should be noted that at high temperature only the lowest wavevector modes are likely to propagate without effects from phonon-phonon interaction, and the above approximation is valid for these modes. Use of equation (4.3.79) in equation (4.3.78), where the left hand side is the experimentally measured integrated phonon intensity per monitor count $T^P(\underline{Q}_o)/C_M$, enables the unknown (CN_c) to be determined.

If the sample scatters incoherently, and the incoherent diffuse scattering can be isolated experimentally to give the integrated intensity per monitor count $I^I(\underline{Q}_o)/C_M$, then one may use the relation

$$\hbar\int\sigma^I(\underline{Q}_o,\omega)\,d\omega = CN_c\Big[\sum_s^{cell} \frac{\sigma^s_{incoh}}{4\pi} \exp(-2W_s(\underline{Q}_o)\Big] \qquad (4.3.80)$$

in equation (4.3.78) to determine the factor (CN_c).

Once (CN_c) is determined any measured scattering can be put on a quantitative basis. For example, the coherent diffuse scattering intensity is related to $S_c^D(\underline{Q}_o)$ by

$$I^D(\underline{Q}_o)/C_M = (CN_c)\, S_c^D(\underline{Q}_o). \qquad (4.3.81)$$

In the case of diffuse quasielastic scattering the variation of cross section with $\underline{Q}$ is usually not significant over the volume of the resolution function so that only the resolution effects in ω need be taken into account. The experimental intensity integrated over ω, $I^D(\underline{Q}_o)$, may then conveniently be calculated by fitting parameters in the cross section which is convolved with the resolution function. The cross section is often adequately expressed as a Lorentzian function in ω,

$$\sigma_{\alpha}^{D}(\underset{\sim}{Q}_o,\omega) = H_{L\alpha}^{D}(\underset{\sim}{Q}_o)\,\frac{\Gamma_{\alpha}(\underset{\sim}{Q}_o)}{[\Gamma_{\alpha}^{2}(\underset{\sim}{Q}_o) + \omega^{2}]}\,, \qquad (4.3.82)$$

where the temperature factor is included in $H_{L\alpha}^{D}(\underset{\sim}{Q})$, where α denotes coherent or incoherent. In this case

$$R(\underset{\sim}{Q}_o,\omega-\omega_o) = V_I V_F \exp\,[-a^2(\omega-\omega_o)^2]/\int \exp(-a^2\omega^2)\,d\omega \qquad (4.3.83)$$

where $a = 2\surd(\ln 2)/W$, and W is the FWHM of the Gaussian energy resolution window as measured by an incoherently scattering vanadium sample. The total intensity of a scan of energy transfer made at $\underset{\sim}{Q}_o$ with constant-k_i, can be written

$$I_t^D(\underset{\sim}{Q}_o,\omega_o) = B(\omega_o) + G(\omega_o) + I_{coh}^{D}(\underset{\sim}{Q}_o,\omega_o) + I_{incoh}^{D}(\underset{\sim}{Q}_o,\omega_o), \qquad (4.3.84)$$

where $B(\omega_o)$ is a background level, $G(\omega_o) = H_G\exp(-a^2\omega_o^{\,2})$ is an elastic scattering contribution, and the $I_{\alpha}^{D}(\underset{\sim}{Q}_o,\omega_o)$ terms are given by equation (4.3.72) with $\sigma_{coh}^{D}(\underset{\sim}{Q}_o,\omega)$ and $\sigma_{incoh}^{D}(\underset{\sim}{Q}_o,\omega)$ given by equation (4.3.82). In fitting to $I_t^D(\underset{\sim}{Q}_o,\omega_o)/C_M$, the parameters $\Gamma_{\alpha}(\underset{\sim}{Q})$ and $S_{L\alpha}^{D} = H_{L\alpha}^{D}C\hbar$ are determined, and $T_{\alpha}^{D}(\underset{\sim}{Q}_o)/C_M = \pi\, S_{L\alpha}^{D}$.

4.3.3.2 Separation of coherent and incoherent diffuse scattering

The experimental identification of coherent QES and incoherent QES when both contribute to the observed intensity profile requires careful data analysis. Use may be made of the different variation of integrated intensity with temperature and scattering vector. The coherent scattering intensity increases with induced thermal disorder and usually varies strongly with scattering vector, whereas for incoherent scattering the variation arises only from the temperature factor. In general the coherent scattering exhibits broader energy widths than the incoherent scattering. The change of intensity on isotopic substitution, when feasible, can be used to identify each contribution, and naturally when only one cross section dominates, such as is the case when F^- or O^{2-} (for which $\sigma_{incoh} = 0$) are the mobile ions, unambiguous data may be obtained. If it is possible to use a single isotope of the mobile ion so that the incoherent scattering arises from different nuclear spin states only, the technique of Polarisation Analysis (Moon et al 1969) may be used to separate the coherent and incoherent contributions. Recent experiments on mobile deuterium ions in the lattice gas system NbD_x illustrate this technique (Ross et al 1989).

4.3.3.3 Sample Form and Environment

The advantage of using single crystal samples to give increased intensity and details of the anisotropic nature of the scattering has been emphasised already (see section 4.3.1). If the vapour pressure of the sample is low it may be mounted directly in a vacuum or a rare-gas-filled furnace. However it is often necessary to seal the sample under vacuum in a sample tube, usually metallic, so that it is in equilibrium with its own vapour pressure. Care must be taken to ensure that there is no reaction between the sample and tube, and it may be necesssary to coat the tube to reduce any reaction. An example is the plasma-spray coating of tantalum nitride on tungsten tubes used for the encapsulation of uranium oxide samples (Clausen et al 1984b). The sample tube material and thickness depend mainly on the temperature range required, but the necessity for corrections for absorption and scattering by the tube must be borne in mind. Quartz or sapphire tubes may be used, but for increasing temperatures metallic vanadium, platinum, molybdenum and tungsten have the advantage that they can be machined into shape or fabricated by Chemical Vapour Deposition. Single crystal samples, if of sufficient size, are first orientated and then cut ultrasonically into cylindrical-shape to fit snugly into the sample tube. An appropriately-designed cap can seal the tube under vacuum by electron beam welding. If a sample is to be sealed under gas, as by argon arc welding, the rise in pressure as the temperature increases must be allowed for in the tube design.

The design of neutron-scattering furnaces depends on the conditions and performance required, such as a very uniform sample temperature and accurate control or a wide range of sample temperature. Furnaces available for both neutron and x-ray work are reviewed by Aldebert (1984). Developments at Toulouse and Grenoble (Aldebert 1984), and more recently at Harwell (Clausen et al 1984b) have enabled sample temperatures of up to 3000K to be held for periods of up to a week, sufficient to make a comprehensive neutron scattering measurement. The Harwell furnace is shown in figure 4.10. Sample temperatures of up to ∿ 2000K may be measured by thermocouples and above this by pyrometry. Careful checks are necessary to ensure that temperature gradients are minimised and that the thermometer is in equilibrium with the sample.

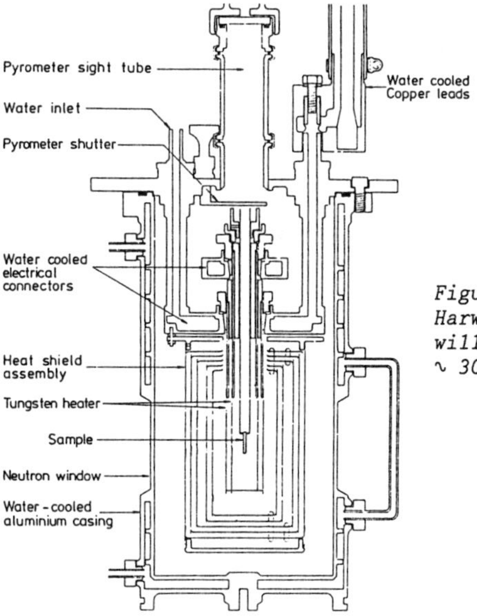

Figure 4.10 Schematic diagram of the Harwell high-temperature furnace which will attain sample temperatures up to ∿ 3000K (after Clausen et al 1984b).

4.3.4 Neutron Scattering from Pure Fluorites

In the following sections some examples will be given of the large range of measurements which have been made and of the fast-ion materials which have been studied. Emphasis will be placed on the fluorite compounds as they probably represent the most thoroughly-investigated system in which the fast-ion disorder has been studied. In building up a model of the fast-ion state and of conduction processes it is essential to utilize data from as many different types of experiment as possible.

4.3.4.1 Diffraction

The fluorite structure Fm3m may be viewed as a simple cubic array of anions with cations occupying alternate cube centres. It is a face-centred cubic structure of cube side a_o (figure 4.1), with a basis of a cation at the origin and anions at $\pm(\frac{1}{4}\frac{1}{4}\frac{1}{4})$. At low temperatures, well below T_c, it is well established that the principal thermally-induced defect in this lattice is the anion Frenkel pair, with the interstitial anion close to the centre of one of the empty cubes, i.e. the (½½½) site. Most early theories of the fast-ion phase in fluorites assumed that thermal excitation of many such defects played an important part in the onset of the fast-ion conduction. However, powder and single crystal diffraction experiments have shown that the actual cube centre site is not occupied for any appreciable length of time in the fast-ion phase (Shapiro and Reidinger 1979; Dickens et al 1979b; and Koto et al 1980), a fact supported by molecular dynamics calculations (see Chapter 3). We discuss the distribution of defective anions in more detail below. A second question, that of the actual number of defects created in the fast-ion phase, has also received considerable attention and has been the subject of some controversy. It is now clear that it is very important to define carefully the structure of the defect postulated in a given analysis or theory before comparisons are made with experiment.

The neutron diffraction from single crystals of a number of fluorites has now been measured at temperatures up to close to the melting temperature by workers at Harwell and the Clarendon Laboraory, Oxford. The measurements were made at Harwell and at the Institut Laue-Langevin (ILL) Grenoble. Crystals of PbF_2, $SrCl_2$, CaF_2, BaF_2, UO_2 and ThO_2 have now been investigated, and all the data have been interpreted in terms of a number of possible models of the defective anions. Full details of this method of analysis are given by Dickens et al (1982). At ambient temperature all the data can be well fitted by the regular fluorite structure, Fm3m, with both harmonic and anharmonic terms (equation 4.3.36) included in the potential for anion motion. The presence and amount of thermally-induced anion Frenkel disorder was determined by allowing a fraction, n_d, of anions to leave their regular lattice sites when fitting to higher temperature data. This fraction includes both anions which go to form interstitials of Frenkel pairs and anions which are relaxed from their regular site due to the presence of a nearby interstitial anion. The fraction of regular anions forming true Frenkel defects, n_f, is therefore itself only a fraction, d, of n_d, depending on the ratio of true interstitials to relaxed anions, ie $n_f=dn_d$. In these fluorite-structured compounds the cations are assumed to remain on their regular sites. The defective anions are assumed to occupy either or both of two types of defect sites, each with a harmonic temperature factor: the 12 'I' interstitial sites at positions near the mid-point of the regular anion sites in the normally empty anion cubes, such as ± (¼ + y, ¼ − y,o), and the

8 'R' relaxed anion sites at ± (x,x,x), both sets of coordinates given relative to the cation sites (see figure 4.11). The relative population of these sites was assumed to be given by one of six different possible models, described by Dickens et al (1982) and designated I to VI, with variants VII and VIII on the cluster model VI. The models are summarised in table 4.5, and were each fitted to the data at each temperature to yield values of the displacements x and y . In each case the measured intensities were first carefully corrected for thermal diffuse scattering, extinction and absorption. It was generally found that the fraction of anions which have left their lattice sites, n_d, deduced from fits to the data was relatively independent of the model used. The fraction, d, of these anions which form true Frenkel pairs depends on the choice of model, and can be deduced from the model of the disorder giving the best fit.

The principal model which was used to interpret both the diffraction data and diffuse scattering data is one of 'clusters' of defective ions , such as model VI (figure 4.11). These clusters are short-lived, as seen from the quasielastic nature of the diffuse scattering, and comprise Frenkel pairs

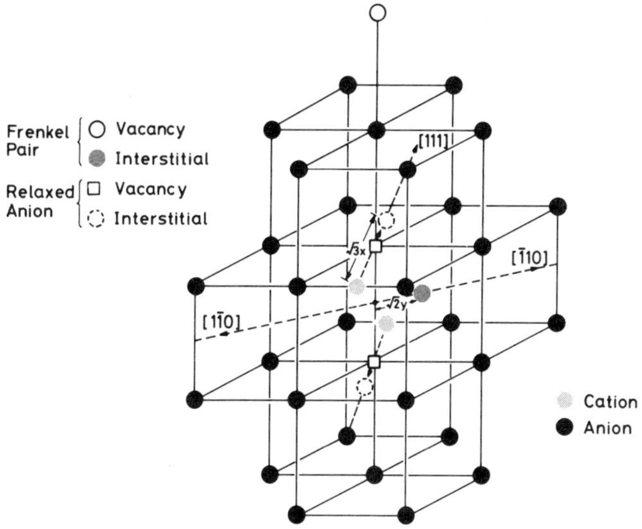

Figure 4.11 The 3:1:2 cluster showing the positions of the Frenkel interstitial anion site, the Frenkel anion vacancy site, and the relaxed anion sites (after Hutchings et al 1984).

Table 4.5

Summary of the eight models fitted to the diffraction data. n_d is the fraction of anions leaving their regular sites, and d is the fraction of these displaced anions occupying 'I' sites, n_R and n_I denote the fractional occupation of the 'R' and 'I' sites (after Dickens et al 1982)

Model	Description	Number of Parameters	n_d	d	x	y	n_R	n_I
I	Regular fluorite	5	0	-	-	-	-	0
II	Vacancies only	6	$0 \le n_d \le 1$	-	-	-	-	
III	Cube-centre-interstitial	7	$0 \le n_d \le 1$	0	0.5	-	$2n_d$	0
IV	Cube-edge interstitial	7	$0 \le n_d \le 1$	1	-	0	0	$n_d/3$
V	R-site only	8	$0 \le n_d \le 1$	0	$0.25 < x < 0.5$	-	$n_d/4$	0
VI	R&I as cluster	10	$0 \le n_d \le 1$	1/3	$0.25 < x < 0.5$	$0 < y < 0.25$	$n_d/6$	$n_d/18$
VII	R&I as cluster	10	$0 \le n_d \le 1$	1/2	$0.25 < x < 0.5$	$0 < y < 0.25$	$n_d/8$	$n_d/12$
VIII	R&I as cluster	11	$0 \le n_d \le 1$	$0 \le d \le 1$	$0.25 < x < 0.5$	$0 < y < 0.25$	$(1-d)n_d/4$	$dn_d/6$

with interstitial anions at the 'I' sites, which give rise to a region of disturbance with neighbouring anions relaxed to the 'R' sites. The location of the Frenkel vacancy was determined from static energy calculations (Catlow and Hayes 1982), and the relaxation of anions around the vacancy are neglected in the model, an approximation to the real situation. Diffraction gives only the mean occupancy of these sites in the unit cell and the relative positions of the defective anions is deduced from the diffuse scattering as discussed in the next subsection. Similar clusters of static defective ions have been postulated to explain the diffraction data from nonstoichiometric UO_{2+x} (Willis 1978), and in fluorites doped with rare earth ions (Cheetham 1973). Catlow and Hayes (1982) have made static energy calculations of the formation energy of defect clusters of some of the kinds we consider here. They found that in the static approximation stable clusters can occur, but only if the Frenkel vacancy is located in certain positions.

The different type of cluster, models VI, VII and VIII, are described in terms of the ratios of (vacancies): (interstitials): (relaxed ions) in the anion lattice, (note that different authors may use a different order in denoting the three types of defect). Model VI corresponds to a 3:1:2 cluster with d = 1/3, and model VII to a 4:2:2 cluster with d = ½. In the latter the two interstitials are in sites in opposite <110> directions relative to the same mid-ion position. Model VIII attempts to fit d, but in practice this proved to be impossible. Indeed from the diffraction data it was difficult to distinguish between models VI, VII.

The first crystal to be investigated by diffraction in detail in this manner was PbF_2 (Dickens et al 1982). The fraction of defective anions, n_d, was found to be $\sim$20% at T_c and to approach 50% at T_m. The same order of magnitude of n_d was found in each of the compounds studied, in line with the simularities from one fluorite to another noted in section 4.1. Typical values are indicated in figures 4.17a,b,c. Whereas the basic fluorite structure, model I, gave the best fit to all data at 293K and just above, the cluster models, VI to VIII were found to give the best fit to data both as T_c is approached and also in the fast-ion phase. Only in the case of BaF_2, where the data were of relatively poorer quality, was this not so.

The temperature factors for the regular sites of both ions, B_i, were found to increase smoothly, with increasing slope, as the temperature is increased; the value for the anion is larger than for the cation, indicating larger amplitude of vibration. Although it was not always possible to fit them separately, the harmonic temperature factors for the interstitial and relaxed anions were as large or slightly larger than those for the regular ions. The anharmonic parameter, c^{o}_{123} (see equation 4.3.42), for the regular anion site increased in magnitude rapidly with temperature increase. In general, the harmonic potential parameters α_i for both regular anion and cation sites, and the cubic parameter β_i for the regular anion sites (see equation 4.3.36) were found to decrease in magnitude, softening slightly with temperature increase.

In general the positions of the defective anions, given by the parameter y (for I sites) and x (for R sites), vary with the compound and the ionic radii. The fitted values of x for the relaxed anions are close to those expected from a hard sphere model of the ions but are generally a little smaller. The fitted values of y are found to be much less, implying that the interstitial ion gives a time-averaged density which lies closer to the midpoint between regular anion sites than expected. Values of x and y for six fluorites re given in table 4.6.

The model of short-lived clusters enables a visual picture to be used to describe the fast-ion disorder in fluorites. This model was developed both to account for the diffuse scattering data and to give a simple explanation of other phenomena. The cluster picture provides a description of what is observed, and it also makes contact with defect clusters in doped or very non-stoichiometric fluorites. However, the joint probability density $f_s(\underline{r})$ of the anions is perhaps the best meeting ground for molecular dynamics theory and experiment, and Walker et al (1982) have constructed this density for PbF_2 using equation (4.3.33) and the best-fitting parameters from data at 991K. Comparison of their molecular dynamics calculations of $f_s(\underline{r})$ with this experimental distribution gave good agreement; further comparisons are discussed in Chapter 3. One additional point of note is that, when defined in a way consistent with the models described above, the value of n_d calculated from molecular dynamics was of the same order as that determined by experiment.

Bachmann and Schulz (1983) have reanalysed the PbF_2 diffraction data of Dickens et al (1982) at 773K using a slightly different model, and deduced a joint probability distribution function which is shown in figure 4.12. They

Table 4.6

Displacements of interstitial and relaxed anions towards the empty cube-centre site in different types of cluster (see figure 4.11). (a) calculated using a hard sphere model at the temperature given for each compound, (b) static energy calculation (Catlow and Hayes 1982); and (c) experimental values from diffraction data (u) Hutchings et al (1984); (v) - Hutchings et al (1985); Clausen et al (1989a); (w) - Farley et al (1989).

A: Displacement from mid-anion position in <110> direction by $(y,y,0)a_o$ Å, values of y are given.

	PbF_2[u]	$SrCl_2$[u]	CaF_2[u]	UO_2[v]	ThO[v]	Li_2O[w]
T(K)	829	1078	1473	2500	2500	1273
Cluster-type						
3:1:2[a]	0.16	0.14	0.15	0.17	0.16	0.15
4:1:3[a]	0.10	0.10	0.10	-	-	-
4:2:2[a]	0.16	0.14	0.15	0.17	0.16	0.15
4:2:2[b]	-	0.17	0.17	-	-	-
9:1:8[a]	0.14	0.08	0.11	0.16	0.15	0.15
9:1:8-CC110[a]	0.07	0.11	0.09	-	-	-
3:1:2[c]	0.07±0.01	0.04±0.01	-	0.04±0.02	0.05±0.02	0.0±0.02

B: Relaxation from regular anion site in <111> direction by $(z,z,z)a_o$ Å, values of z are given (z = (x - 1/4), where x is shown in figure 4.11).

	PbF_2[u]	$SrCl_2$[u]	CaF_2[u]	UO_2[v]	ThO_2[v]	Li_2O[w]
T(K)	829	1078	1473	2500	2500	1273
3:1:2[a]	0.11	0.15	0.12	0.08	0.08	0.00
4:1:3[a]	0.10	0.16	0.14	-	-	-
4:2:2[a]	0.11	0.15	0.12	0.08	0.08	0.00
4:2:2[b]	-	~0.17	~0.16	-	-	-
9:1:8[a] n.n.	0.15	0.18	0.15	0.09	0.09	0.00
n.n.n.	0.03	0.09	0.06	0.04	0.04	0.00
9:1:1:8-CC111[a]	0.03	0.09	0.06	-	-	-
3:1:2[c]	0.12±0.01	0.13±0.01	0.08±0.03	0.09±0.01	0.09±0.02	0.04±0.01

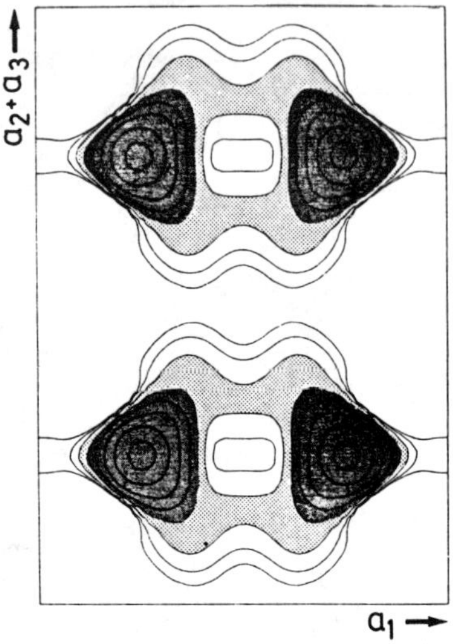

Figure 4.12 The fluorine total probability density function in the (110) plane of β-PbF_2 at T = 773K determined by Bachmann and Schulz (1983) from the neutron diffraction data of Dickens et al (1982).

then continued their analysis quite differently, and used equation (4.3.35) to deduce a potential for the F^- ion motion; they found that the potential was independent of temperature above T_c and gave a barrier height for ionic motion consistent with conductivity data. However, as was pointed out in section 4.3.2.4 there are serious doubts about the veracity of decoupling the single-ion potential in such a highly-correlated system and its use for comparison with conductivity data. The assertion that $n_f < 5\%$ also seems to be inconsistent with the model which included occupation of interstitial sites and which was used to fit the data from which the density was obtained.

Diffraction from the oxides UO_2 and ThO_2 has been measured at temperatures up to ∿ 3000K, and a similar increase in n_d to that found in the halides was observed, as shown in figure 4.13. The excitation of Frenkel defects makes a significant contribution to the heat capacity of UO_2 and is of importance in understanding the other thermodynamic properties of this nuclear fuel. (Clausen et al 1984a; Hutchings et al 1985; Hutchings 1987).

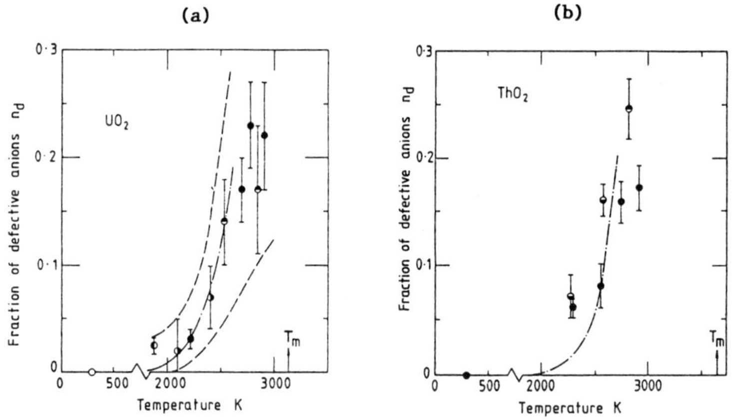

Figure 4.13 Temperature variation of the total fraction of defective anions, n_d, deduced from diffraction data from (a) UO_2, (b) ThO_2. In (a) the different symbols denote data from five different samples. Chain line is the best fit of $n_d = A \exp(-h^F/2k_BT)$ to the lower temperature data yielding h^F = 4.7 ± 0.5eV and A = 5400 ± 4000. The broken lines denote the variation deduced from the coherent diffuse QES using a 3:1:2 cluster model with values of x, y from diffraction (upper curve), or a hard sphere model (lower curve). In (b) diffraction data from several samples are denoted by different symbols. The chain line is calculated from the experimental expression for n_d with h^F scaled by the ratio of T_m for ThO_2 and UO_2 (after Clausen et al 1984a, Hutchings et al 1985).

4.3.4.2 Coherent Diffuse Scattering

As was seen in section 4.3.2.5, the diffuse scattering measures the disorder in a crystal very directly. The coherent diffuse scattering from PbF_2, $SrCl_2$, CaF_2, UO_2 and ThO_2 at temperatures from below T_c to near their melting point have now been studied in some detail at Harwell, Risϕ National Laboratory and I.L.L. Grenoble. Both fluorine and oxygen nuclei only scatter coherently whereas chlorine also exhibits a large incoherent cross section (see table 4.3). The incoherent scattering from $SrCl_2$ will be described in the next subsection. The two forms of scattering can be readily separated in this latter case because of the different energy widths of the quasielastic diffuse coherent and incoherent scattering, and also because of the experimental and theoretical observation that the coherent scattering increases with temperature increase and occurs primarily in certain regions of reciprocal space. This separation is not always so easy, as exemplified by the data on sodium compounds, and to a certain extent on AgI.

The halide fluorites were the first compounds to be investigated using single crystal samples; full details are given by Hutchings et al (1984). The weak nature of the scattering precludes a complete measurement of $S^D(\underline{Q},\omega)$ as a function of $\underline{Q},\omega$, and temperature, so that the scattering at only a few selected $\underline{Q}$ points could be investigated in detail. Initially the variation with $\underline{Q}$ of the integrated intensity, $S^D(\underline{Q})$, was determined. Details of the energy dependence and of the temperature variation of the scattering were then investigated for points of special interest, often using improved resolution. In the case of $SrCl_2$ the form of $S^D(\underline{Q})$ was determined by a number of constant-$\underline{Q}$ scans at selected $\underline{Q}$ points and the data were integrated over ω at each point. For CaF_2 and PbF_2 a series of scans at constant ω were performed over a mesh of $\underline{Q}$ points. Four or five different values of energy transfer, ω_o, were taken and the data were integrated numerically to give $S^D(\underline{Q})$ and plotted in the form of a contour diagram. In order to isolate $S^D(\underline{Q})$ great care was necessary to subtract from the total scattering the incoherent elastic and Bragg scattering from the can and furnace, and also to separate the inelastic scattering from phonons. The data were normalised and put on an absolute basis using a standard low energy acoustic phonon and also the incoherent scattering in the case of $SrCl_2$, using the method described in section 4.3.3.1.

The scattering observed from the three halide compounds at temperatures below and above T_c is illustrated in figures 4.14a,b,c where scans of energy

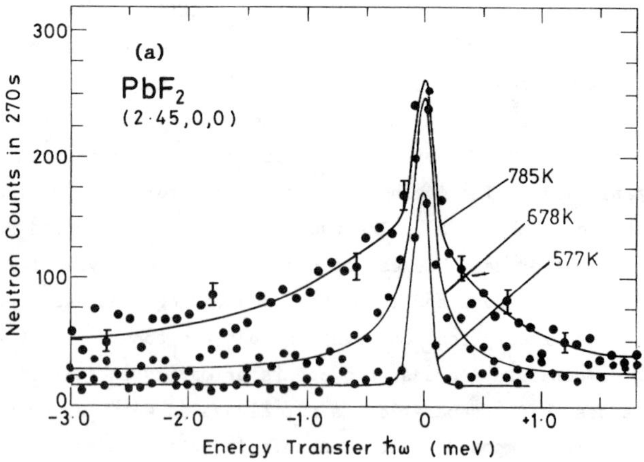

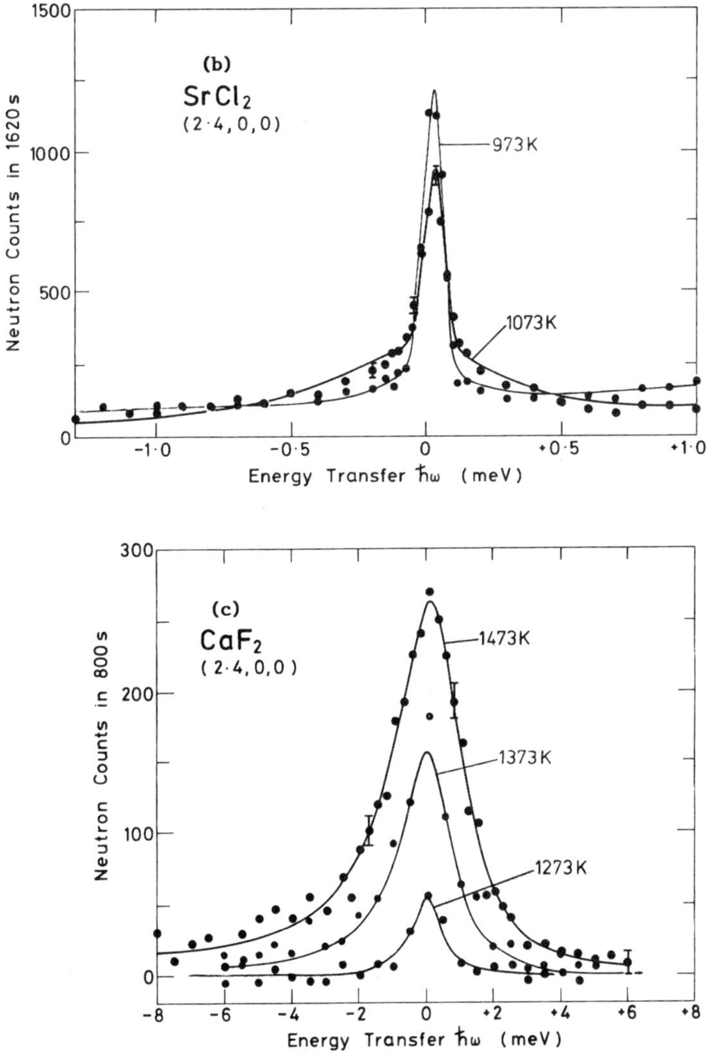

Figure 4.14 Typical coherent diffuse QES observed by constant-Q scans of energy transfer: (a) from PbF_2 (T_c = 711K), (b) from $SrCl_2$. (T_c = 1001K) and (c) from CaF_2 (T_c = 1430K) (after Hutchings et al 1984).

transfer, $\hbar\omega$, at constant $\underset{\sim}{Q}$ are shown for values of $\underset{\sim}{Q}$ close to where the maximum quasielastic intensity is observed. The total integrated scattering is seen to increase as the temperature is raised, immediately indicating its coherent origin. This coherent nature is confirmed by the marked variation with Q of the integrated intensity described below, by the fact that fluorine has no incoherent cross section, and by the results of experiments performed on $Sr^{35}Cl_2$. In the latter case a comparison of intensities of scattering from the natural and isotopic chloride confirmed that the major contribution to the scattering arises from the anion-anion correlations. In each case it is seen that there is a component of incoherent scattering, narrow in energy width, arising from the sample tube and furnace, and in the case of $SrCl_2$ from the Cl^- ions as well, superimposed upon the broader coherent scattering. The coherent QES scattering can be described quite well by a Lorentzian profile in energy transfer convolved with the resolution function, equation (4.3.84), and typical fits are shown by the solid lines in figures 4.14a,b,c. It is seen that the energy width increases with increase of temperature, and that it varies with compound, being narrowest in $SrCl_2$ and broadest in PbF_2.

The distribution of $S^D(\underset{\sim}{Q})$ with $\underset{\sim}{Q}$ at $T>T_c$ in the two principal planes of the reciprocal lattice is shown in figures 4.15a,b,c. The most detailed information is available for CaF_2, where the mesh scans gave a contour diagram for $S^D(Q)$ in both planes, figures 4.15c. Although mesh scans were also carried out on PbF_2 the scattering was rather weak and the data are best presented as areas of observed intensity and regions of no intensity (figure 4.15a). This representation is also used for $SrCl_2$ (figure 4.15b). It is seen that the scattering has the same general distribution in all three cases and its main feature is the peak in intensity along [100], just beyond the point (200). There is an anisotropic shell of intensity with peaks in certain directions and little intensity at low Q. In the case of CaF_2 and PbF_2 the scattering was examined at $\underset{\sim}{Q}$ values beyond this shell and intensity was observed beyond (400) in CaF_2 and beyond (222) in PbF_2. The wavevector and temperature variation of the intensity and width of the region of maximum intensity near (200) was investigated in most detail. At lower temperatures the diffuse scattering peaks near the (200) point but, as the temperature increases to above T_c, the intensity increases and the peak position of $S^D(\underset{\sim}{Q})$ moves away from the (200) point along [100]. The temperature variation of the scattering close to the scattering vector giving the maximum intensity is

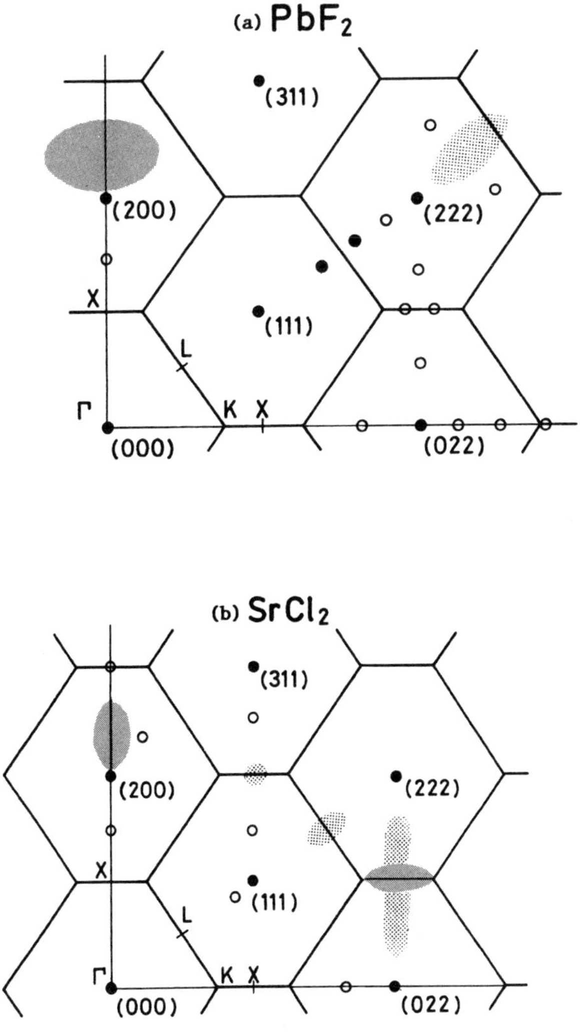
(a) PbF2
(311)
(200)
(222)
X
(111)
L
Γ
K
X
(000)
(022)
(b) SrCl2
(311)
(200)
(222)
X
(111)
L
Γ
K
X
(000)
(022)

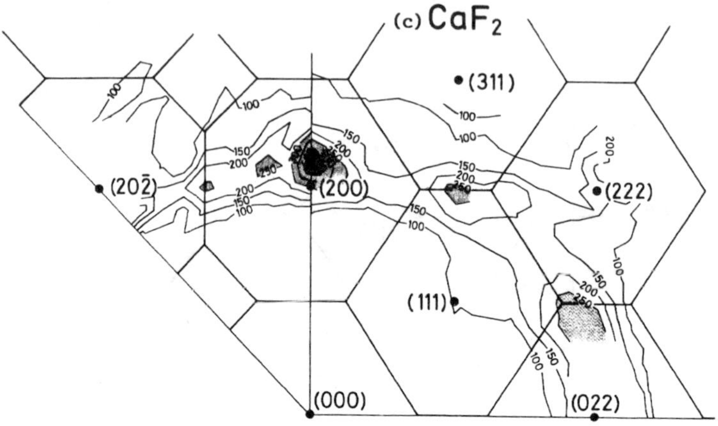

Figure 4.15 The measured distribution of $S^D(Q)$, the coherent diffuse QES intensity integrated over $\hbar\omega$, in the (01$\bar{1}$) and (010) planes of reciprocal space. (a) for PbF_2, measured at 829K, the hatched areas denote where intensity was observed and are bounded by the points where the intensity has fall to approximately half of its maximum value. The region of strongest intensity is shown in grey. Open circles denote points where no intensity was observed, and the filled circles denote points where there was a suggestion of intensity. (b) for $SrCl_2$, measured at 1078K, the symbols are as for PbF_2. (c) for CaF_2, measured at 1473K, the lines are contours of observed intensity at 1473K in the (01$\bar{1}$) plane, whereas in the (010) plane they are of the difference in intensity observed at 1473K and 1273K (after Hutchings et al 1984).

shown in figures 4.17a,b,c where $X^{D'}(\underline{Q}_o)$, as defined in equation (4.3.63), is plotted. The intensity is seen to build up just below T_c and to continue rising above T_c. In the case of $SrCl_2$ the rise of intensity with temperature has very much the same form as the increase in enthalpy due to anion disorder (Dickens et al 1978).

In order to interpret these experimental data the theoretical diffuse scattering $S^D(\underline{Q})$ was calculated by Hutchings et al (1984) using the expression (4.3.62), with the instantaneous positions of the defective anions given by a variety of clusters of one or two Frenkel pairs. Only anion-anion correlations were included since molecular dynamics calculations and $C\ell^-$ isotope substitution experiments on $SrCl_2$ show these to dominate the scattering. For the calculations a hard-sphere model was adopted to determine the position and relaxation of the anions, with the constraint that <110> relaxations (y) should be as small as possible. In this model the relaxations about a single interstitial in PbF_2 are confined essentially to the nearest neighbour (n.n.) anions, whereas for CaF_2 they extended to next nearest neighbours (n.n.n.), and for $SrCl_2$ even beyond. Relaxations about anion vacancies, and of cations, were not included. It was found that all the clusters with anions at a cube centre gave poor agreement with experiment, whereas cluster models with the interstitial in the 'I' site near the mid regular-anion position, discussed above, gave similar features to those observed i.e. an anisotropic shell of scattering in Q-space with $|(\underset{\sim}{Q})| \sim 2\text{-}3 Å^{-1}$ and subsidiary peaks, with the greatest intensity just beyond the (200) point along [100]. The best overall agreement is given by the 9:1:8 cluster, with n.n. and n.n.n. anions relaxed. The contour diagrams calculated using this cluster for the three halide compounds are shown in figures 4.16a,b,c. A Debye-Waller factor appropriate to that of the regular anion at the temperature of measurement was used in the calculation. The contours are on the same scale of intensity per anion for each diagram. There is very good general agreement with the observed $S^D(Q)$, with the relatively diffuse distribution of PbF_2 and the sharper peak along [100] for $SrCl_2$ well reproduced.

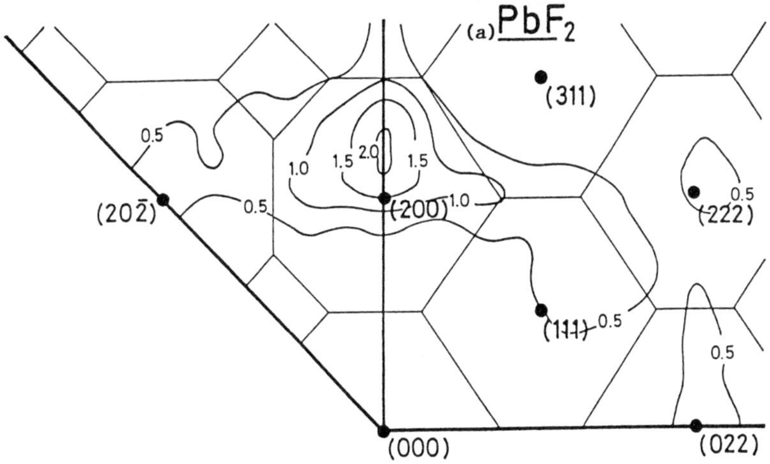

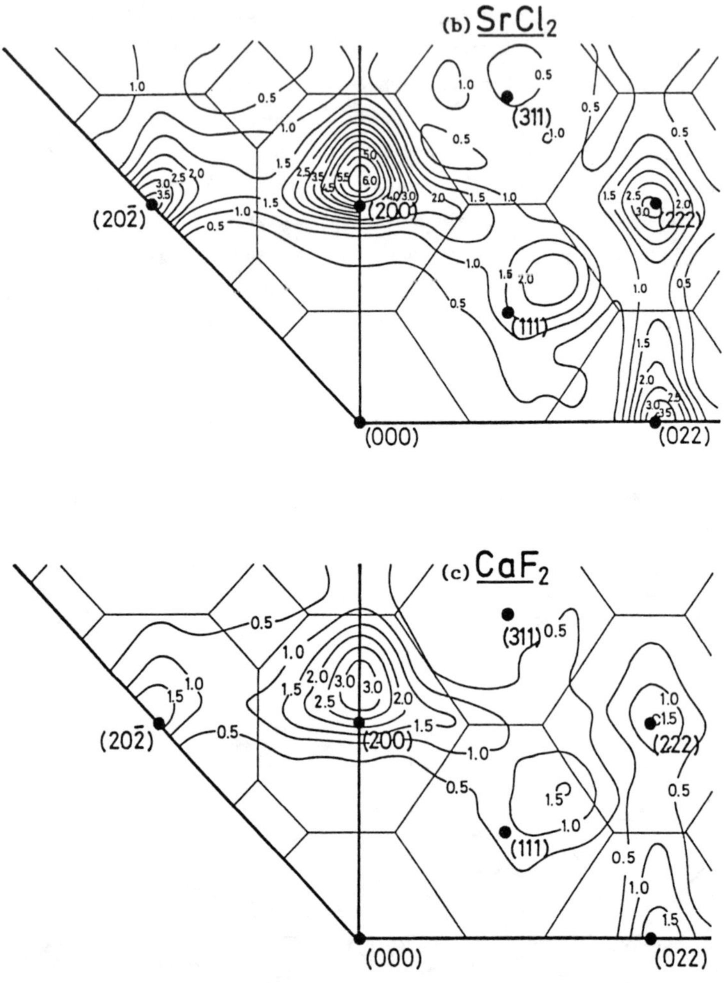

Figure 4.16 Intensity of coherent diffuse QES, $S^{D'}(\underline{Q})$, calculated using 9:1:8 clusters as a model of the anion disorder. The scale of the intensity contours is in scattering per anion and the relative normalisation between the compounds is correct. (a) PbF_2 at 829K. (b) $SrCl_2$ at 1078K. (c) CaF_2 at 1473K (after Hutchings et al 1984).

A comparison of the quantity $X^{D'}(\underline{Q}_o)$, defined as in equation (4.3.63) but in terms of the scattering per anion, determined from experiment with the calculated scattering per defective ion, $Y^D(\underline{Q}_o)$, yields another estimate of the fraction of defective anions n_d. Hutchings et al (1984) used a mean value, $\bar{Y}^D(\underline{Q}_o)$, averaged over the quite similar values calculated from four types of cluster which gave a distribution $S^D(Q)$ reasonably close to that observed. $\underline{Q}_o$ was taken as the point near (200) giving maximum intensity. The resulting n_d were used to establish the right hand scales of figures 4.17a,b,c from the values of $X^{D'}(\underline{Q}_o)$ on the left hand scale, with an estimated error of ∿15%. The shaded areas of figures 4.17a,b,c represent the estimates of n_d from diffraction including this error. They show a very good agreement between the two methods of deducing n_d, except for the case of $SrCl_2$. The discrepancy in this case could indicate that a larger extent of

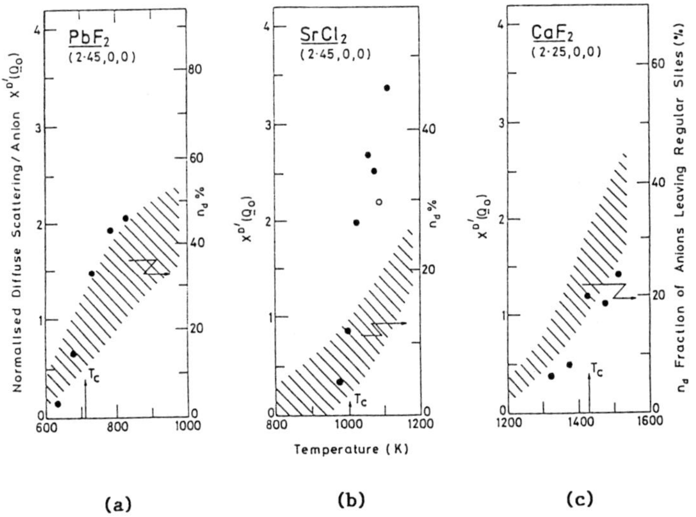

Figure 4.17 Variation of the normalised coherent diffuse QES per anion, $X^{D'}(\underline{Q}_o)$, with temperature for PbF_2, $SrCl_2$ and CaF_2, at scattering vectors $\underline{Q}_o$ near the peak intensity position. The right-hand ordinate gives the value of n_d determined using the cluster model described in the text. The shaded area represents the range of values of n_d determined by neutron diffraction including errors in the data and in the scale (after Hutchings et al 1984).

correlated defects than the 9:1:8 cluster occurs, as this would give an increase in $Y^D(\underline{Q}_o)$. The Q-width of the peaks in $S^D(\underline{Q}_o)$ near (200) confirm a correlation range $\sim a_o/2$, although the width measured for $SrCl_2$ is narrower than calculated, suggesting a larger relaxation field in this case. A crude model (Hutchings et al 1984) may be used to indicate that the movement of the peak away from (200) results from the decreasing displacement of relaxed ions as the temperature is increased.

The interstitial displacement field increases from PbF_2 to CaF_2 to $SrCl_2$, roughly in proportion to the increase in ratio r_a/r_c of the anion to cation radius. One would expect less interaction between the smaller clusters. It is interesting to note that the value of the diffuse scattering $X^{D'}(\underline{Q}_o)$ (see equation 4.3.63) at T_c is roughly the same for all three halides, in the region $\sim$ 1.0 to 1.1 (figure 4.17). The approximate values of n_d of $\sim$ 20% at T_c and $\sim$ 50% at T_m suggest that even at T_c every one of the empty cubes is involved in fluctuating clusters whereas at T_m each empty cube contains one defective anion. Estimates of the number of Frenkel pairs n_f deduced from the cluster model are $\sim$ 3-8%, 1-5% and 2-7% for PbF_2, $SrCl_2$ and CaF_2 respectively.

It was not possible to obtain as much data on the $\underline{Q}$ variation of $S^D(\underline{Q})$ for the oxides UO_2 and ThO_2 at the much higher temperatures, between 2000-3000K, required to observe oxygen disorder. However with reasonable assumptions regarding the cluster type, the values of n_d estimated from diffuse scattering are found to be in fair agreement with those from diffraction, and from the temperature variation of n_d an estimate of the activation energy of Frenkel defect formation could be made (Clausen et al 1984a; Hutchings 1987; Clausen et al 1989a). In the case of the antifluorite Mg_2Si (Hutchings et al 1988) only the region near (200) was investigated; the results indicate longer-range correlations between the disordered ions since the diffuse peak lies very close to (200) and is narrower in $\underline{Q}$ than in the halides.

The temporal behaviour of the clusters gives rise to the energy width of the coherent QES, as discussed in section 4.3.2.5. A simple exponential decay of the correlations between ions in a cluster in time τ_{coh} ps gives a Lorentzian of FWHM $2\Gamma_{coh}(\underline{Q})$ in meV, these being related by

$$\tau_{coh} = \frac{4.14}{\pi[2\Gamma_{coh}(Q)]} \qquad (4.3.85)$$

The clusters' spatial variation with time is given by the $\underset{\sim}{Q}$ dependence of $\Gamma_{coh}(\underset{\sim}{Q})$. In the cases of $SrCl_2$ and CaF_2 these coherent energy widths $\Gamma_{coh}(\underset{\sim}{Q})$ tend to very low values at the $\underline{\tau}$= (200) point and increase with $\underset{\sim}{q} = \underset{\sim}{Q} - \underline{\tau}$ along [100]. For CaF_2 this increase is approximately linear at lower temperature for low q, becoming quadratic at higher temperatures (see figure 4.18). However there appears to be little variation of $\Gamma_{coh}(\underset{\sim}{Q})$ with $\underset{\sim}{Q}$ for PbF_2. The case of $SrCl_2$ has been extensively investigated using a time-of-flight spectrometer to yield a comprehensive variation of $\Gamma_{coh}(\underset{\sim}{Q})$ with $\underset{\sim}{Q}$ in the $(0\bar{1}1)$ plane (Osborn et al 1989). Although the data analysis involved some approximations, the results show that the coherent width $\Gamma_{coh}(\underset{\sim}{Q})$ has the periodicity of the reciprocal lattice of the cubic chlorine ion lattice, as is found for the incoherent widths described in the next section. These data strongly suggest that the cluster is intimately involved in the anion hopping mechanism, with the cluster essentially dissociating and reforming at a later time a distance $a_o/2$ away with the same or different ions in the same relative positions. There appears to be a correlation between the lifetime and size of the cluster, the longer lifetime being associated with the larger cluster; this is exemplified by both $SrCl_2$ and the covalent antifluorite Mg_2Si. The lack of coherence in the scattering from different clusters is most probably due to their dynamic nature.

4.3.4.3 Incoherent Diffuse Scattering

The fluorite $SrCl_2$ and the antifluorite Li_2O exhibit a fast ion phase in which the mobile ion scatters incoherently. Details of the geometry of the individual ionic motion may therefore be obtained by measurement of the self pair-correlation function as given by equation (4.3.4). Detailed measurements have been made on $SrCl_2$ and to date provide the cleanest example of scattering from the single particle motion in a fast ion conductor. (Dickens et al 1983; Schnabel et al 1983). Lithium ion diffusion in the antifluorite $^7Li_2O$ has also been investigated (Farley et al 1988, 1989).

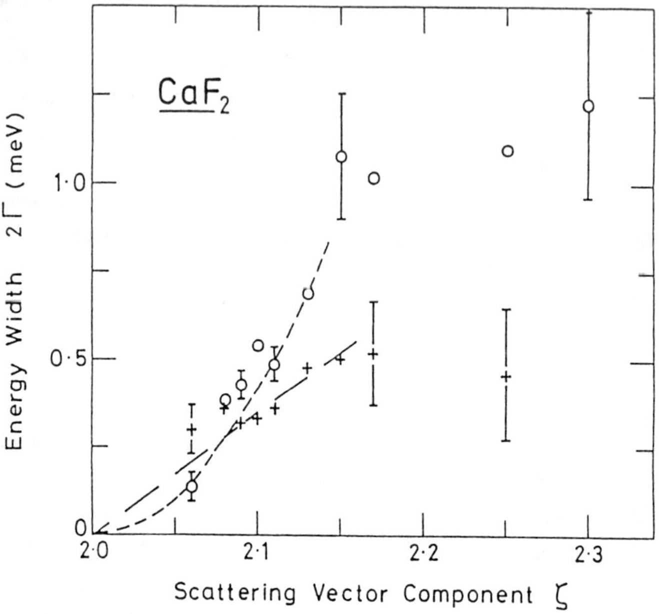

Fig 4.18 Variation of energy linewidth, $2\Gamma_{coh}(Q)$, of coherent diffuse QES from CaF_2 with scattering vector along [ζ,0,0] at 1323K(+) and 1473K(O). Typical error bars are given where clarity permits. The lines are the best fit of a linear (+) and quadratic (o) variation of width with q=(ζ-2) (after Hutchings et al 1984).

Measurement of the incoherent quasielastic scattering from $SrCl_2$ has been made at three temperatures 1001, 1053 and 1087K using the high resolution triple-axis spectrometer IN12 at ILL Grenoble. Data were taken in the region of low |Q|, where the coherent scattering is mainly weak, for Q along the three principal crystallographic directions. The Lorentzian HWHM's, $\Gamma_{incoh}(Q)$, at 1053K are shown in figure 4.19. They were determined by subtraction of the small amount of scattering from the can, and then fitting equation (4.3.84) to the data. The periodicity of the widths with Q shows immediately that the Cl^- ions are predominantly hopping between n.n. regular anion sites. The data were fitted either by using the Chudley-Elliott expressions given by equation (4.3.55) for uncorrelated, unblocked, random diffusion over Bravais lattice sites, allowing for both n.n. and n.n.n. hops, or by using the encounter model developed by Wolf (1977) and Göltz et al

(1980), which takes account of spatial correlations between successive ionic jumps. Both expressions give a good account of the data. The cross section may be written in the form of equation (4.3.82) with

$$\Gamma_{incoh}(\underset{\sim}{Q}) = \frac{1}{\tau_{enc}} \sum_i \frac{W(i)}{n_i} \sum_j [1 - \exp(-\underset{\sim}{Q}.\underset{\sim}{\rho}_{ij})] , \qquad (4.3.86)$$

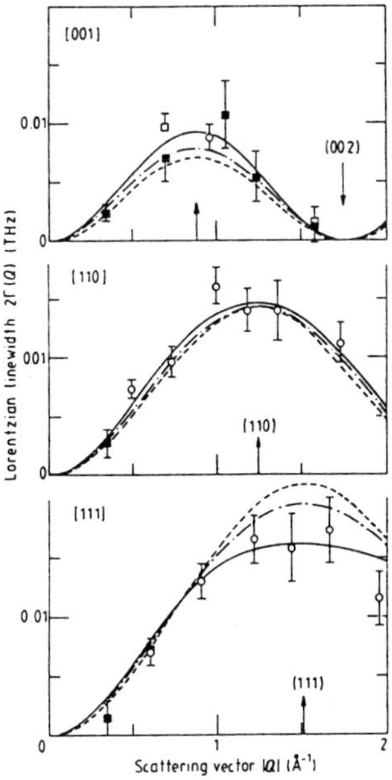

Figure 4.19 Variation of the FWHM of the incoherent QES from $SrCl_2$ at 1053K with scattering vector in three principal directions. The symbols ■ and • denote data taken with $k_i = 1.047Å^{-1}$ and $1.160Å^{-1}$ respectively, and full and open symbols denote data taken from different samples. The broken (n.n hops) and full (n.n. and n.n.n. hops) curves are the best fit of the Chudley-Elliott expression (4.3.86), and the chain curve that of Wolf's encounter model, as described in the text (after Dickens et al 1983).

where j denotes one of the n_i neighbouring sites of type i (n.n., n.n.n. etc) at $\underset{\sim}{\rho}_{ij}$, and W(i) is the probability of reaching the site of type i. At low $|Q|$, $\Gamma_{incoh}(\underset{\sim}{Q}) = DQ^2$, where D = D* as discussed in section 4.3.2.5.

In the Chudley-Elliott model the W(i) are treated as parameters, W(0) is zero, and τ_{enc} is the mean anion residence time, τ_d, denoted τ_p in equation (4.3.55). The data were fitted by varying τ_d alone with W(1) = 1 for hopping to n.n. sites only, at $d = a_o/2$, giving the broken lines in figure 4.19, and by varying both τ_d and $W(1) = p_1$, the probability for n.n. hops, giving the full line. The probability for n.n.n. hops at $a_o/\surd 2$, is given by $W(2) = p_2 = 1-p_1$. The inclusion of hops to both types of sites fits the data better, and gives $p_1 = 0.73\pm0.05$, and $\tau_d = 29.0 \times 10^{-12}s$. The diffusion constant is given by $D = (p_1 + 2p_2)d^2/6\tau_d = 9.32 \times 10^{-6} cm^2 s^{-1}$ at 1053K. It is interesting to note that a molecular dynamics simulation of CaF_2 gave p_1 = 0.79 and p_2 = 0.21 when analysed in terms of this model (Jacucci and Rahman, 1978), in very reasonable agreement with the experimental result for $SrCl_2$.

Wolf (1977) has attempted to treat the effect of spatial correlation of the hopping anions in concentrated lattices by considering the ionic motion as resulting from encounters with a defect, such as a single vacancy, which itself undergoes truly stochastic motion. As a result of each encounter, the anion may make a number of essentially instantaneous n.n. hops to nearby lattice sites $\underline{\rho}_{ij}$, or back to its original site i = 0 at $\underline{\rho}$ = 0. Correlations are included because the successive very rapid jumps of an anion due to an encounter with the same defect are not independent of each other. The probabilities W(i) of reaching each type of site, i, by such an encounter with mono-vacancies have been calculated by Wolf, so that the only unknown parameter is the mean time between encounters, τ_{enc}. The best fit of this model is given by the chain curve in figure 4.19. It does not represent the data quite as well as the n.n. plus n.n.n. Chudley-Elliott expression, which has one more parameter, but gives $\tau_{enc} = 24.8 \times 10^{-12}s$ and $D = 8.28 \times 10^{-6} cm^2 s^{-1}$. The accuracy of D determined from the models is estimated to be $\sim$ 10%. Data at 1001K, where the widths are much narrower, give $D \sim 2.3 \times 10^{-6} cm^2 s^{-1}$, and the Chudley-Elliott model gives $p_1 \sim 0.91$. At 1087K $D \sim 13.8 \times 10^{-6} cm^2 s^{-1}$ and $p_1 \sim 0.78$. Further analysis in terms of theories which allow for non-zero flight times (Gissler and Stump 1973), or analytical treatment of correlations and site blocking effects (Tahir-Kheli and Elliott 1983) did not appear justified in this case. The latter theory predicts small changes in spectral shape of the QES which were not detectable from the available data.

The variation of the integrated intensity of the incoherent scattering generally follows that expected from the Debye-Waller factor. In some regions of Q-space both coherent and incoherent scattering can be resolved in one triple-axis scan of ω, and this affords a normalisation of the coherent scattering, as discussed in section 4.3.3.1. As mentioned above, a recent analysis of coherent scattering data taken at larger scattering vectors in terms of a Lorentzian spectral shape shows that the coherent widths roughly follow those of the incoherent scattering in $SrCl_2$, having the same periodicity. It therefore appears that equation (4.3.52) is a good first approximation to the coherent scattering (Osborn et al 1989). This periodicity in Q of both the coherent and incoherent scattering reflects the hopping of both the mobile anion and its distortion field between regular anion sites. It has not been possible to detect any contribution from hopping to interstitial sites or between sites in a cluster. Conventionally, hopping between regular sites is termed a 'vacancy' diffusion mechanism, but the results are fully consistent with a correlated defect hopping or a crowdion mechanism, as in the interpretation of the molecular dynamics calculations given in Chapter 3.

The Chudley-Elliott and encounter approaches described above assume that the ion vibrates about its site for a mean time τ_d, which we shall here call τ_{incoh} (which is much longer than the phonon period τ_{ph}) before hopping to a nearby site. The coherent scattering is characterised by the lifetime of the cluster, τ_{coh}, which reflects the correlated transit time. The clusters appear static to the majority of the phonon modes as discussed in the next section, so that in general $\tau_{incoh} > \tau_{coh} > \tau_{ph}$.

The ratio of the coherent and incoherent widths affords another estimate of the fraction of defective ions n_d. From equation (4.3.58)

$$\frac{N_{v'}}{N_{a'}} = \frac{\Gamma_{incoh}}{\Gamma_{coh}} = \frac{\tau_{coh}}{\tau_{incoh}} \simeq \frac{n_d}{(1 - n_d)} , \qquad (4.3.87)$$

where $N_{a'}$ is the number of anions on the regular lattice and $N_{v'}$ the number of vacancies due to Frenkel pairs and relaxed anions. Using estimates of τ_{coh} for $SrCl_2$ from data at (2.4,0,0), Hutchings et al (1984) find values of n_d ranging from 0.01 at 1001K to 0.10 at 1087K, in fair agreement with those obtained from diffraction. Estimates of an effective τ_{incoh} from diffusion data and nmr for PbF_2 and CaF_2 give similarly reasonable values for n_d when combined with τ_{coh}.

4.3.4.4 Inelastic Scattering

The fluorites were one of the first series of ionic compounds to have their lattice dynamics studied by neutron scattering, and early studies of UO_2 (Dolling et al 1965) and CaF_2 (Elcombe and Pryor 1970) were followed by the investigation of $SrCl_2$, SrF_2 and BaF_2 (Hayes and Stoneham 1974). In the course of the study of fast-ion behaviour of the fluorites measurements have been made of the phonon energy dispersion curves of PbF_2 (Dickens and Hutchings 1978), ThO_2 and CeO_2 (Clausen et al 1987), and the antifluorites Mg_2Si (Hutchings et al 1988) and Li_2O (Farley et al 1988). These energy dispersion relations have been used to establish inter-atomic potentials used in the molecular dynamics simulations of the crystals. The temperature variation of the lattice modes in PbF_2 have been studied to temperatures close to T_m, and this will be discussed in most detail here. Because of its importance as a nuclear fuel UO_2 has been similarly investigated (Clausen et al 1985) and most recently Li_2O (Hull et al 1988a). Less comprehensive data on the temperature variation have been obtained for $SrCl_2$, ThO_2, CeO_2 and Mg_2Si (Hutchings M T, private communication).

One of the early motivations for studying the temperature variation of the lattice dynamics was to search for a possible soft mode associated with the onset of fast-ion conduction. Boyer (1980) has suggested possible modes which will soften. However, none have been observed in the fluorites, and the rapid broadening of the optic modes even below T_c would appear to make such an observation very difficult. In the case of PbF_2 the optic modes are broadened even at 295K and the dispersion relation had to be determined at 10K (Dickens and Hutchings 1978b). One interesting feature is the *softening* in the energy of the transverse optic (TO) mode in PbF_2 as the temperature is reduced below 295K, due to a weak tendency towards ferrolectricity (Dickens and Hutchings 1978a).

The rapid broadening of optic phonon modes at high temperature due to anion disorder is illustrated by the Raman mode in $SrCl_2$ (Dickens et al 1976) and the Raman TO and LO modes in UO_2 (Clausen et al 1983, 1985, and 1989b; Hutchings 1987) (see also section 4.2.5). The broadening of the LO mode in UO_2 is found to be similar to that in CaF_2 and is thought to be due largely to defect creation (Clausen et al 1989b). The temperature variation of the density of phonon states in PbF_2 has been measured using the powder time-of-flight technique discussed in section 4.3.2.7, and illustrates the rapid smoothing of features as the temperature is increased (Dickens et al 1980). At 10K the measured density of states shown in figure 4.20a closely resembles that calculated from shell-model parameters fitted to the energy dispersion relation. At 910K only a peak at $\sim$ 5meV and a broad peak at $\sim$ 32 meV are observed, as seen in figure 4.20b.

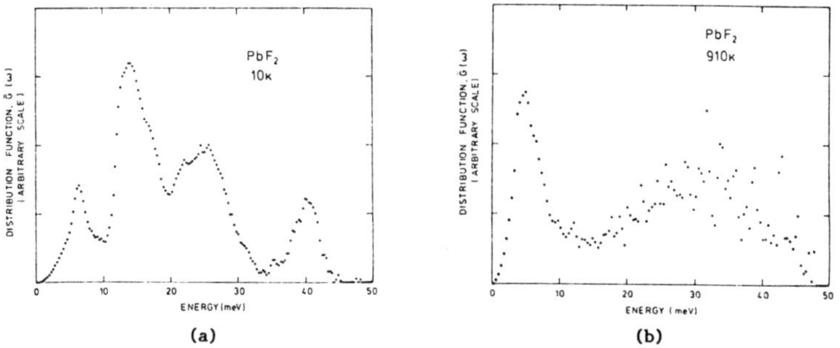

Figure 4.20 Weighted density of phonon states $\bar{G}(\omega)$ for PbF_2 measured at : (a) 10K; (b) 910K (after Dickens et al 1980).

The temperature variation up to 898K of the low lying acoustic modes in PbF_2 was studied in detail using a single crystal sample and triple axis spectrometer in order to determine both the mode energies, and hence the elastic constants, and the energy widths. Both necessitate careful

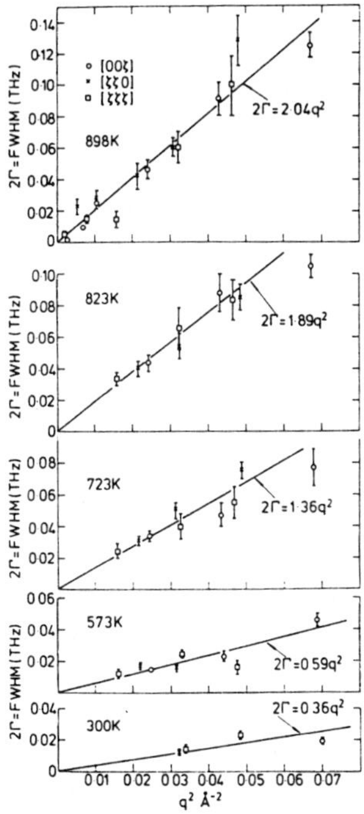

Figure 4.21 Transverse acoustic phonon line widths, 2Γ, in PbF_2, showing a q^2 dependence and rapid increase near T_c (after Dickens and Hutchings 1978c).

deconvolution of the instrumental resolution effects (Dickens and Hutchings, 1978c). The neutron groups observed below 300K were resolution-limited but, at 300K and above, a broadening of the TA modes was observed which was proportional to $|q|^2$ and approximately independent of the direction of q. The average broadening is shown in figure 4.21, and suggests first sound (or hydrodynamic) regime behaviour. The widths increase approximately linearly with temperature at lower temperatures indicating anharmonic behaviour, but then broaden more rapidly near T_c. The variation of the elastic constants with temperature is shown in figure 4.22. They decrease linearly with temperature increase to near T_c and more rapidly above T_c, particularly in the case of C_{11} (see also section 4.2.5 and figure 4.8). Dickens et al (1979a) have shown that the q^2 dependence of the line widths is most likely to arise from the creation of defects which are effectively static on the phonon time scale. The increase in width of the TA modes is consistent with ∿ 10% Frenkel defect creation above T_c. As mentioned in section 4.2.5, Catlow et al (1978) and Kleppmann (1978) have suggested that the drop in C_{11} above T_c is due to changes in both short-range and long-range forces due to defect creation. For C_{12} and C_{44} these effects nearly cancel. Measurement of the detailed behaviour of the acoustic modes in UO_2 at temperatures up to ∿ 2930K have also been made (Clausen et al 1985). The widths vary linearly with q at temperature ∿ 2,500K, but above they vary more rapidly with q. All three elastic constants decrease linearly with temperature increase up to ∿ 2100K, and then more rapidly, with C_{44} showing the most marked decrease. Clearly, an explanation of this behaviour will require a different change in short- and long-range forces from that in PbF_2.

4.3.5 Neutron Scattering from Doped and Nonstoichiometric Fluorites

The addition in solid solution of monovalent, trivalent or tetravalent cations to alkaline earth halides with the fluorite structure is possible for a range of concentrations, depending on the ions involved. The change in valency of the cations is charge-compensated by the creation of vacancies in the anion lattice or by the addition of interstitial anions. In general, either type of charge compensation doping increases the ionic conductivity at lower temperatures, but as the temperature is raised the conductivity tends to saturate at the level appropriate to the pure compound. Doping also affects the specific heat anomaly at T_c, causing it to broaden and weaken in magnitude. Andersen et al (1983) suggest that this indicates an increased

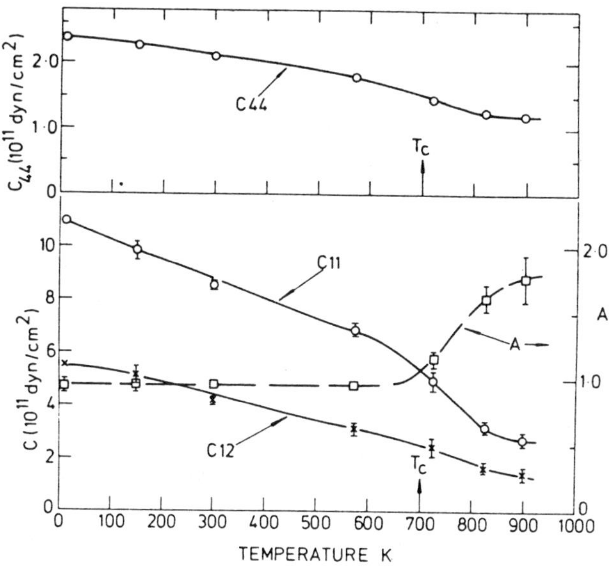

Figure 4.22 Temperature variation of the elastic constants and anisotropy parameter $A = 2C_{44}/(C_{11}-C_{12})$ *for* PbF_2 *(after Dickens and Hutchings 1978c).*

concentration of thermally-generated defects at lower temperatures in anion-excess doped fluorites.

At low dopant levels, less than 1 mole %, ESR techniques show that the additional anions occupy the empty cube-centre position (Bleaney et al 1956; Baker et al 1968). However at higher concentrations of excess anions or vacancies, diffraction data suggests that the defects form static clusters. Most of these defect structures have been studied at ambient temperature and below, but recently measurements have been extended to high temperatures. One of the earliest anion-excess compounds to be studied was hyperstoichiometric $UO_{2.13}$. Neutron diffraction measurements made by Willis (1978) at 1073K, above the temperature at which the excess oxygen precipitates as U_4O_9, were interpreted in terms of static 2:2:2 clusters of the type discussed in section 4.3.4.1. In the absence of detailed diffuse scattering measurements the presence of clustering must be inferred, using steric arguments and static energy calculations, from the presence of vacancies on the regular sites, and from the population of the I and R sites (sometimes called F' and F" sites).

The system YF_3/CaF_2 has been studied by Bragg neutron diffraction by Cheetham et al (1971). Single crystal diffraction from $(Ca/Y)F_{2.06}$, and powder diffraction from $(Ca/Y)F_{2+x}$ with x = 0.1, 0.15, 0.25 and 0.32, was investigated at 293K and the concentration of vacancies and population of I and R sites were determined. At the lower concentrations 2:2:2 and 3:4:2 clusters were proposed, with more extended clusters at the higher levels of doping. Measurements at 773K on $(Ca/Y)F_{2.10}$ and at 1173K on $(Ca/Y)F_{2.15}$ showed no major structural differences from 293K. A similar study of the $PrCl_3/SrCl_2$ and $LaCl_3/SrCl_2$ systems was made by Bendall et al (1984), but only at293K. They made measurements on a single crystal of $(Sr/Pr)Cl_{2.10}$, and on powders of $(Sr/La)Cl_{2.1}$ and $(Sr/La)Cl_{2.18}$. Although evidence for population of the I sites was found, that of the R sites was very small. Diffuse scattering was also observed from the powder samples. The absence of R site population led the authors to suggest a model of a larger cluster in these compounds, of M_6X_{12} form, consisting of an octahedron of trivalent cations with I type interstitials along each edge. Such a cubo-octahedral 'super' cluster was also proposed by Catlow et al (1983) to account for neutron diffraction studies of single crystals of $(Ca/La)F_{2.05}$, $(Ca/Er)F_{2.05}$ and $(Ca/Er)F_{2.10}$ at 80K and, in the case of $(Ca/Er)F_{2.05}$, at 1125K. They represent their data in the form of partial Fourier maps, where the reference structure factors are determined by the best fit of the data by a perfect fluorite structure in which only the anion regular site occupancy was allowed to vary in addition to the usual parameters. Presented in this way, R site interstitials were only found to be present in the case of $(Ca/La)F_{2.05}$. Furthermore, there was a marked difference in the $(Ca/Er)F_{2.05}$ maps at 80K and at 1125K which they suggest is possibly due to the dissociation of the interstitials from the octahedral aggregate. In the examples given above the static cluster models proposed were based on Bragg diffraction as the only experimental data, so that the relative positions of the defective anions had to be inferred from other considerations. Diffuse scattering can give more direct information on the cluster form, as we have seen for the pure fluorites. This diffuse scattering will be elastic for static defects, but is expected to become quasielastic as some of the ions involved become mobile. Furthermore, any correlations between the position of different clusters will give rise to additional diffuse peaks in the scattering (equation 4.3.65). These points are seen in the examples we give next, which rely primarily on diffuse scattering as the experimental technique.

Coherent diffuse neutron scattering has been used to examine cluster formation in anion excess fluorites at Risø National Laboratory (Andersen et al 1983). A single crystal of $(Ba/U)F_{2.2}$ has been studied at 293K, with the diffuse intensity mapped in the (100) and $(0\bar{1}1)$ planes. Broad peaks were observed, centred on (2.6,0,0) and (1.5,1.5,1.5), and these are quite well accounted for by calculations similar to those described in section 4.3.2.5, equation (4.3.65), based on uncorrelated 2:2:2 clusters and averaging over possible configurations and orientations. In these clusters the two fluorine interstitials are charge-compensated by a nearby U^{4+} ion, rather than by vacancies as in the pure compound. Within the energy resolution of the instrument the clusters were found to be static, probably pinned by the U^{4+} ions. No correlations between the clusters were indicated in these data. However the diffuse scattering observed from single crystals of $(Ba/La)F_{2.209}$ and $(Ba/La)F_{2.492}$ shown in figure 4.23, takes the form of a shell of intense scattering with $2Å^{-1} < Q < 3Å^{-1}$, containing local maxima and minima (Kjems et al 1983; Andersen et al 1986a). The more highly-doped sample exhibits relatively sharp elliptical peaks at positions ±(2/3,0,0) relative to the fluorite peaks; the overall intensity is enveloped by the (2:2:2) cluster scattering pattern as given by equations (4.3.65 and 4.3.66). These diffuse satellite peaks indicate a short-range correlation between the 2:2:2 clusters, with spacing $(3a_o/2,0,0)$ along the <100> directions. The extent of the correlation was estimated from the width of these peaks in Q as 30Å to 40Å along [100] but only <10Å in the transverse direction, suggesting one-dimensional aggregates. Measurement was made of the satellite peak intensities along [100] and these could be well accounted for by a model of 2:2:2 clusters with long correlation length, (corresponding to a linear crystal of cell size $3a_o/2,a_o,a_o$), the positions of the interstitials being treated as parameters. The less heavily-doped sample showed weaker satellite peaks and more closely resembled the scattering from isolated 2:2:2 clusters. In these cases the 2:2:2 cluster is expected to be charge compensated by two La^{3+} ions on adjacent cation sites. As the temperature is raised the satellite peak intensity at (2.69,0,0) in $(Ba/La)F_{2.492}$ decreases rapidly between 673K and 1073K; the steepest rate of change of intensity occurs at 923K, the temperature at which the ionic conductivity shows the most rapid rate of increase. The Q-width of the satellites also broadened in this temperature range and the results were interpreted as indicating a loss of

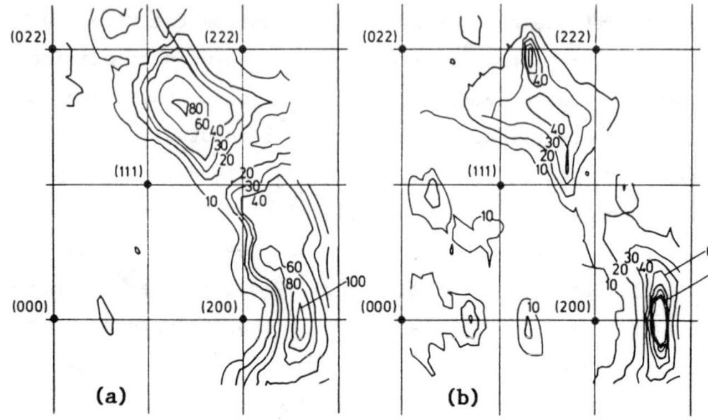

Figure 4.23 Contour maps (arbitrary scales) of measured coherent diffuse elastic intensities at 293K in the ($0\bar{1}1$) plane of (a) $Ba_{0.791}$, $La_{0.209}$ $F_{2.209}$ and (b) $Ba_{0.508}$ $La_{0.492}$ $F_{2.492}$ at 293K (after Andersen et al 1986a).

correlation between the clusters due to thermal fluctuations and simultaneous generation of mobile F^- ions. The clusters themselves, however, are thought to remain largely intact. It was not possible to observe any energy broadening of the diffuse scattering, which remained elastic within the experimental resolution, indicating a lifetime of the clusters greater than 10^{-10}s. Despite the new information provided by these measurements the exact conduction mechanism at high temperatures remains to be fully explained (Andersen et al 1986a).

The best example of a doped fluorite system with anion vacancy compensation is that of yttria doped zirconia and this has attracted a great deal of attention over the past ten years on account of its technological importance as an oxygen conductor in fuel cells and oxygen sensors and as a as a high-temperature ceramic. But it is due to its use as an artificial gemstone that a ready supply of a range of dopant levels in single crystal form has been available from J F Wenkus of Ceres Corporation, USA. At 293K pure ZrO_2 has a monoclinic structure, transforming to a tetragonal phase at 1273K and to the cubic fluorite phase between 2643K and its melting point of 2988K. The fluorite phase can be stabilised by the addition of various divalent (CaO, MgO etc) or trivalent (Y_2O_3,La_2O_3) cationic oxides in a wide range of concentrations. Electrical neutrality is maintained by vacancies in the oxygen sublattice. The ionic conductivity arises from this high concentration of oxygen vacancies and increases with temperature increase. At fixed temperature the conductivity goes through a maximum as the dopant level increases. The very high vacancy concentrations lead to vacancy-vacancy and vacancy-dopant ion correlations and these have been probed by diffuse electron and x-ray, as well as diffuse neutron scattering. The time-average structure of the unit cell has been studied by diffraction using electrons, x-rays and neutrons. Most of these measurements have been made at 293K or on samples rapidly quenched from high temperatures. However the temperature dependence of coherent diffuse neutron scattering has now been studied in some detail at Harwell , Risϕ and ILL Grenoble and this has enabled different contributions to the scattering to be identified (Hull et al 1988; Andersen et al 1985, and 1986b; Osborn et al 1986; see also Liu at al 1987; where references to previous neutron scattering and the other work can be found).

At 293K all the diffuse neutron scattering from ZrO_2 doped with Y_2O_3 is elastic within the instrumental resolution but, as the temperature is raised, part of the scattering becomes quasielastic with a Lorentzian profile and the intensity of the elastic component varies. Intensity maps at 293K in the (100) and $(0\bar{1}1)$ planes have been measured for samples of y mole% Y_2O_3 in ZrO_2 with y = 9.4, 12, 15, 18, 21 and 24. These show features which evolve as the concentration increases and a typical contour map for 12 mole % is shown in figure 4.24. The diffuse scattering may be identified as having two main origins. One is from regions of the crystal which are relatively vacancy-free and have a tetragonally distorted fluorite structure, giving rise to quite sharp peaks at the (112) and (114) points. These decrease in intensity as y increases and show that the tetragonal regions become insignificant for y above 18. Their intensity is in fact somewhat dependent on the thermal history of the sample and the peaks rapidly become quasielastic as the temperature is raised; this indicates an increasingly shorter lifetime due either to fluctuations between cubic and tetragonal symmetry or to the diffusion of vacancies. The rest of the diffuse scattering can be identified as arising from a vacancy defective fluorite structure in which vacancies, vacancy pairs and aggregrates of vacancy pairs with a range of sizes occur. A vacancy pair with associated relaxed anions and cations is shown in figure 4.25.

Owing to the similarity of the scattering lengths of Y and Zr, the location of the yttrium ions can not be determined easily by neutron scattering. The aggregate vacancy pairs give rise to satellite peaks which

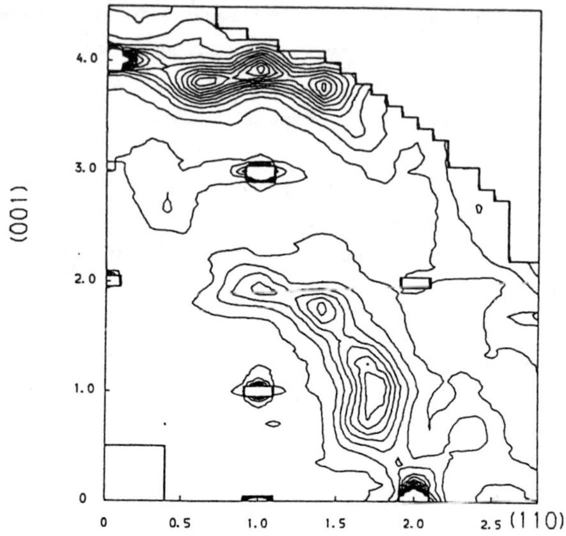

Figure 4.24 Contour map of coherent elastic diffuse scattering intensity in the ($1\bar{1}0$) plane of 12 mol% Y_2O_3 in ZrO_2 at 293K (after Osborn et al 1986).

occur near points at ±(0.4,0.4,±0.8) relative to the regular fluorite recriprocal lattice points. These arise from the separation of vacancy pairs by (1,-0.5,0.5) lattice vectors, roughly corresponding to the Pr_7O_{12} structure (von Dreele et al 1975). Not all the satellites are observed because of the enveloping structure factor of the vacancy pair in equation (4.3.65). However the correlated regions are not large, extending only over a few lattice spacings, and as y increases they tend to increase in number rather than size. Using equations (4.3.65 and 66) contour maps of the diffuse scattering can be reproduced quite well by such a combination of the two types of region (Andersen et al 1985).

The temperature variation of the scattering from y = 12 and y = 18 crystals has been investigated up to 2780K (Hull et al 1988b) and constant-Q scans of $\hbar\omega$ generally show a combination of a resolution limited Gaussian profile with a quasielastic Lorentzian profile. The intensity of the two components varies with $\underline{Q}$. The scattering is predominantly Gaussian at the diffuse satellite peak positions indicating that the strong correlated regions of vacancy pairs remain relatively stable with temperature increase, and the intensity can be well accounted for by a Debye-Waller factor. At a

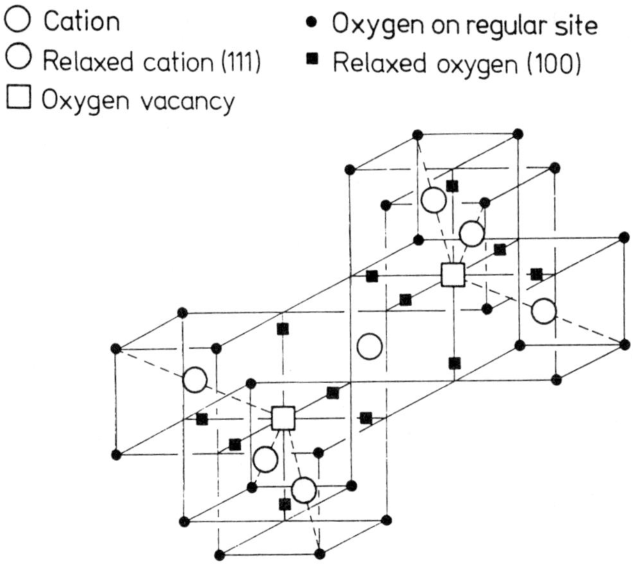

Figure 4.25 Defect cluster consisting of a vacancy pair, separation (½,½,½)a_o and centred on a cation. n.n. oxygens relax towards the vacancy along ⟨100⟩ directions and n.n. cations relax away from the vacancy along ⟨111⟩ directions (after Andersen et al 1986).

point such as(1.7,1.7,1.0), where the scattering is expected to have a strongest contribution from single vacancies and their relaxation field, the Lorentzian component increases in intensity and energy width as the temperature increases, while the Gaussian component decreases in intensity, as shown in figure 4.26. This suggests that such small defects are becoming mobile as the temperature increases. A component with a Lorentzian profile component, broadening as the temperature increases, is also observed along the [100] axis.

It therefore appears that the maximum in the dopant-dependent conductivity, at a given temperature, is due to the trapping of the vacancies in the aggregates, as suggested by Catlow (1984) and others, and that as the temperature rises there is an increasing diffusion of vacancies near the periphery of the larger aggregates.

The mixed conductors Cu_2Se and $Cu_{1.8}Se$, single crystals of which have been investigated by Cava et al (1986), provide another example of a vacancy-defective antifluorite system. Cu_2Se is nominally stoichiometric above 140°C, but $Cu_{1.8}Se$ contains Cu vacancies. The coherent diffuse scattering, S($\underline{Q}$), from Cu_2Se was investigated at 453K, and that from $Cu_{1.8}Se$

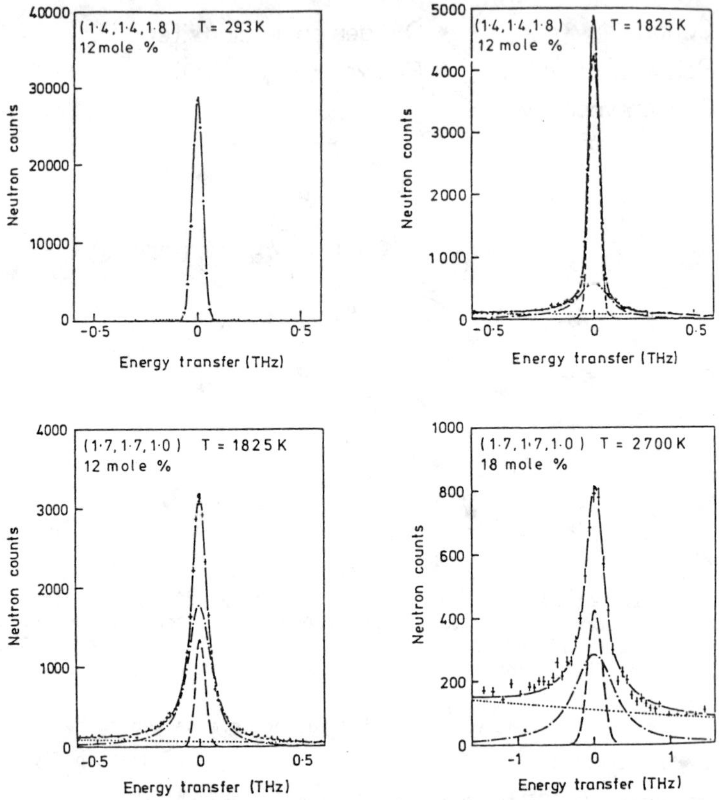

Figure 4.26 Energy scans at constant-Q through the coherent diffuse QES from 12 mole% Y_2O_3 and 18 mole% Y_2O_3 in ZrO_2 (after Hull et al 1988b).

at 324K. Both samples showed a similar characteristic pattern of S($\underline{Q}$) which differs from that of the other fluorite systems studied, even in the stoichiometric case. That for $Cu_{1.8}Se$ is quasielastic, indicating its dynamic origin.

4.3.6 <u>Neutron Scattering from Silver Compounds</u>

Several silver compounds exhibit fast-ion conduction of the silver ions at elevated temperature and they form an important class of materials which have now been quite extensively studied by neutron scattering techniques. As mentioned in section 4.1, AgI exhibits a very high conductivity above its first-order transition at 420 K to the α structure, in which there is a b.c.c. array of relatively immobile I ions (see figure 4.2). α-AgI was one of the first fast-ion conductors to be investigated by diffraction (Hoshino 1957) and by inelastic scattering (Funke et al 1974), and indeed is still the object of both theoretical and experimental work. The nature of the lattice of sites for mobile ions differs from that in the fluorites in that in α-AgI there are many possible Ag sites per b.c.c. (Im3m) unit cell, only a few of which are occupied by the two Ag ions per cell. The same b.c.c. anion lattice occurs in the related high temperature form of the compounds α-Ag_3SI above 515K, α-Ag_2Se above ∿400K and β-Ag_2S above 450K. The latter two compounds are mixed conductors. The number of Ag^+ ions per b.c.c. cell increases in α-AgI, α-Ag_3SI, and α-Ag_2Se or α-Ag_2S by 2:3:4 respectively, as the average anion charge increases from 1:3/2:2. This increased density of Ag ions is expected to lead to increased correlation in their motion and the neutron scattering data confirm this. It is now established that the Ag ions occupy principally the 12(d), tetrahedrally cooordinated sites in the b.c.c. cell (see figure 4.2) particularly at high temperature, and that the conduction involves translational motion between these sites by hopping. Comparison of the scattering from each member of this series of compounds is proving useful in identifying the conduction mechanism.

Extensive diffraction experiments on these compounds have led to time-averaged density distributions of the Ag ions being determined for each compound above and below the fast-ion transition, as described briefly below. Investigation of quasielastic diffuse neutron scattering has been made difficult by the problem of experimental separation of the coherent and incoherent contribution. The former has a cross section an order of magnitude larger (see table 4.4), and probably dominates the observed scattering in most Ag compounds. This, combined with the apparent complex nature of the motion, makes interpretation of the data in terms of models of the diffusion quite difficult (Funke 1988). In several Ag compounds, AgI (Funke et al 1974), $RbAg_4I_5$ (Shapiro et al 1977) and Ag_2S (Grier et al 1984), a low-lying excitation is observed by inelastic scattering below the fast-ion transition. This excitation disappears at different rates as the temperature is raised into the fast-ion phase, merging into the increasing quasielastic diffuse scattering. The excitation is strongly $\underline{Q}$ dependent in Ag_2S and has been identified by Ebbsjö et al (1987), using molecular dynamics, as an Einstein-like mode of the Ag ions at their residence sites. They suggest that the potential well in which the Ag ions reside becomes very shallow, leading to anharmonic motion and diffusion as the temperature is increased.

Powder diffraction experiments on α-AgI by Hoshino et al (1977) and Wright and Fender (1977) showed the predominant occupation of the 12(d) sites by the Ag ions, and this was confirmed by Cava et al (1977) using single crystal samples. Cava et al found high anharmonicity in the Ag vibrations, needing up to 4th degree thermal parameters in the analysis. They used partial Fourier synthesis to obtain the distribution of Ag ion density in the (001) plane, illustrating the sensitivity of this technique to show up low densities of highly disordered ions. Their density distribution at 433K from α-AgI is given in figure 4.27a, and shows the occupation of the 12(d) sites at (½½0) etc. A corresponding map has been obtained for a single crystal of α-Ag_3SI by Didisheim et al (1986) and is shown in figure 4.27b. They found that the low temperature phase, β-Ag_3SI, resembled α-AgI in its scattering, and suggest that the delocalised density in (100) bands is an average of positional disorder over closely spaced sites rather than a time average of dynamic disorder. Oliveria et al (1988) and Cava et al (1980) have also studied α-Ag_2Se and α-Ag_2S respectively by single crystal neutron diffraction in its fast-ion phase. The Ag ion distributions are shown in figure 4.27c and d. They also suggest that the smeared distribution arises from considerable positional disorder of the Ag ions and from Ag-Ag interaction effects. The data suggest a more continuous diffusion in the latter case, rather than the hopping between reasonably well-defined sites, as occurs in α-AgI.

Quasieleastic diffuse neutron scattering has been investigated in several of the compounds discussed above. The most extensive experimental and theoretical work has been performed on α-AgI. Early powder diffuse scattering indicated two contributions to the quasielastic energy spectra (Eckold et al 1976; Funke et al 1978; Lechner et al 1978), one narrow and one broad, which were attributed respectively to the superposition of a translational jump diffusion and a large-amplitude local motion of the Ag ions. The anisotropy in the scattering was investigated using a 16 cc single crystal of α-AgI by Funke et al (1980), and the narrower component was found to exhibit an anisotropic width whereas the broader component was more isotropic. The broad component is interpreted as arising from local motion of individual Ag ions within a cage of neighbours which has a lifetime τ_d. This cage opens to allow a diffusion of ions between 12(d) sites along <100> with a characteristic jump time τ_f. In contrast to the simple jump diffusion model where $\tau_f \ll \tau_d$, they used the theory of Gissler and Stump (1973) and found $\tau_f \sim 3\tau_d$ with a characteristic jump distance of $\sim a_o$ (see also section

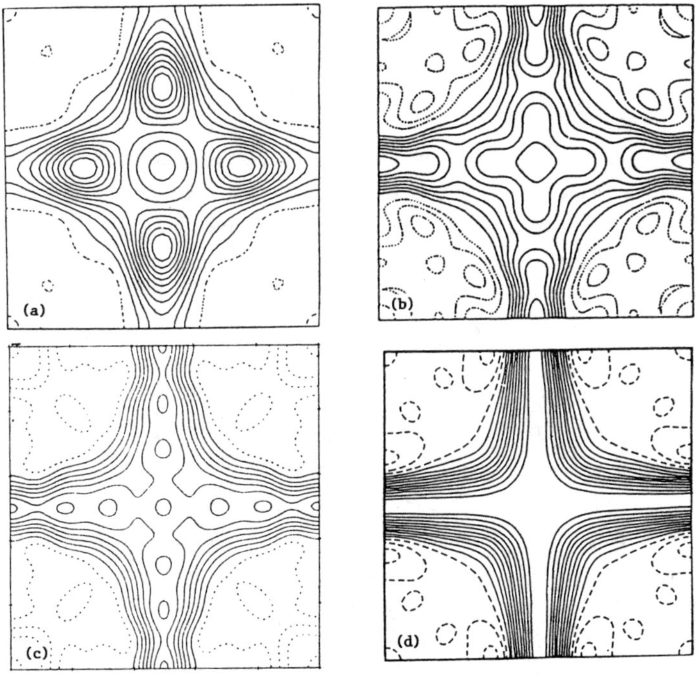

Figure 4.27 Partial Fourier synthesis of the Ag scattering density distribution $\Delta\rho_c(\underline{r})$, in the (100) plane of:- (a) α-AgI at 433K; (b) α-Ag_3SI at 748K; (c) α-Ag_2Se at 573K; and (d) β-Ag_2S at 598K. The scattering contributions from I, Se and S ions have been subtracted. Contour levels are (a) 3.8×10^{-15} cm/Å^3, (b) 7.5×10^{-15} cm/Å^3, (c) 30×10^{-15}cm/Å^3 and (d) 4.8×10^{-15} cm/Å^3. ((a) (d) after Cava et al 1977, 1980, (b) after Didisheim et al 1986, and (c) after Oliveria et al 1988).

4.2.5). Funke (1988) has recently considered more complex motion of ions in crystals, involving non-periodic local motion, non-hopping translational motion and non-statistical hopping motion, particularly with regard to forward-backward hopping which maintains the ion in its cage, and has discussed the scattering from α-AgI in terms of these ideas. However the difficulty of isolating the incoherent from the dominant coherent scattering contributions in this compound, the large number of assumptions necessary and their use of the Sköld approximation (Sköld 1967) to discuss the coherent scattering make a unique analysis difficult. More experimental data are required, and the contribution from correlated motion needs to be considered. Use of a triple-axis spectrometer rather than a time-of-flight spectrometer would be an advantage in obtaining accurate $\underline{Q}$ dependence of the scattering.

The effects of correlations among mobile Ag ions has been seen in β-Ag_2S by Grier et al (1984) who used inelastic neutron scattering to investigate the nature of the diffuse peak observed with x-ray scattering near $\underline{Q}$ = (1.6,1,0) by Cava and McWhan (1980). A single crystal sample and triple-axis spectrometer were used. They found an inelastic mode at $\sim$ 2meV in the low temperature phase (< 450K) near (1.6,1,0) which merges into the diffuse quasielastic scattering when the temperature is raised to give the fast-ion β-phase. The diffuse distribution S($\underline{Q}$) of coherent scattering increases in intensity rapidly from zero at the transition temperature but then decreases slowly with temperature increase. It can be accounted for quite well by a model of randomly orientated cigar-shaped clusters or microdomains of correlated Ag ions, with a length of four unit cells along <100> and with Ag ions in n.n.n. tetrahedral sites. The energy width of the quasielastic scattering has a characteristic quadratic variation with scattering vector relative to the (1.6,1,0) point and increases with temperature increase. This variation has yet to be explained. Vashishta et al (1985) have made molecular dynamics calculations of S($\underline{Q}$) and show that the diffuse intensity does indeed arise from highly correlated Ag motion due to the strong constraints of charge neutrality within the unit cell.

Surprisingly the reported neutron scattering from a single crystal of α-Ag_2Se (Höch et al 1983) appears to differ from that from β-Ag_2S. α-Ag_2Se is a mixed conductor with almost metallic electronic conductivity. The time-of-flight energy spectra showed two main components to the quasielastic scattering, as observed from α-AgI, but in this case both are isotropic. A broad component is essentially ascribed to localised irregular movements of the Ag ions, and the narrow component is consistent with continuous diffusion of the Ag ions within the <100> channels. A further broader, structureless, component to the scattering was also observed.

4.3.7 Neutron Scattering from other Fast-Ion Conductors

Although the fluorites and the silver compounds represent the two main classes of fast-ion conductors which have been extensively studied, several other compounds have been investigated by neutron scattering at elevated temperatures. Some of these will be briefly mentioned in this subsection, which is by no means comprehensive.

Lithium compounds have attracted attention because of the interest in lithium batteries. ^{7}Li isotopic substitution is preferred, to avoid high absorption (see table 4.3). However, the coherent and incoherent cross sections are of the same magnitude (table 4.3). The antifluorite Li_2O has been mentioned in section 4.3.4.4. Li_3N has been investigated by Rabenau (1982), but early work was confused by the presence of H in the samples. LiAl is a defective lattice and Li jump diffusion was inferred from the Q dependence of the narrower incoherent quasielastic component, superimposed on a broader coherent component, consistent with the Chudley-Elliott model (Brun et al 1981). Another disordered lattice is that of the one-dimensional fast-ion conductor, β- eucryptite $LiAlSiO_4$, in which coherent and incoherent quasielastic scattering were observed at temperatures up to 1073K using a high resolution triple-axis spectrometer (Renker et al 1983). The coherent scattering indicated the presence of Li ion correlations and the incoherent scattering, interpreted in terms of the Chudley-Elliott formalism, reflected the one dimensional diffusion along the c axis, with predominant jumps of $c_o/3$ between Li sites with the same symmetry. Very narrow incoherent quasielastic energy widths (∿2μeV) have been observed using a back-scattering spectrometer from in-plane diffusion of Li ions in the intercalation compound LiC_6 below the disordering temperature of 715K, (Magerl et al 1985). Above this temperature more rapid diffusion gives rise to broader widths observable by time-of-flight techniques . In a study at temperatures between 630-730K different jump vectors were found to dominate in the two phases. Other materials which have been studied by neutron diffraction and diffuse scattering over a range of temperature are Li_2SO_4 (Aronnson et al 1983), Cu_2S (Cava et al 1981), and K-Hollandite (Rosshirt et al 1988).

The β-alumina system has attracted the most attention as a solid electrolyte as far as practical application is concerned, but relatively little inelastic or quasielastic neutron scattering has been performed on it. This is due to in part the complexity of the crystal structure and in part to the complex nature of the diffusion process being investigated, as well as to the difficulty in obtaining large single crystals. In contrast a large amount of neutron diffraction work has been carried out on the β-aluminas at elevated temperature and has helped to elucidate the time-averaged distribution of ions in the conducting plane (see for example Roth et al 1976). Quasielastic energy widths have been observed in scattering from a polycrystalline sample of silver β-alumina, $Al_{11}O_{17.125}Ag_{1.25}$ at 473K by Gavarri et al (1983) which led to an estimate of the translational diffusion constant for the Ag^+ ion. Lucazeau et al (1987) have observed quasielastic diffuse scattering at temperatures of up to 673K from Na^+ motion in a sodium β-alumina, $Al_{11}O_{17.125}Na_{1.25}$, powder sample. More recently Lucazeau et al

(1988) have obtained improved data from a single crystal sample with Q in the conducting plane. Both experiments used the IN6 time-of-flight spectrometer at ILL Grenoble. Analysis is made difficult by the problem of the unambiguous separation of coherent and incoherent contributions to the scattering because of the very similar cross sections for Na. The proportion of Na^+ ions mobile on a picosecond time scale was deduced to be less than 0.2 at 573K. Two components to the quasielastic scattering suggest that these mobile ions undergo both fast localised motions, restricted to a given cell, and long range motion with a longer time scale. Although the data are as yet insufficient to uniquely fit the details of the complex motion of the Na^+ions in the plane, they suggest that by use of a triple-axis spectrometer further refinement will be possible, giving unique information on this important system.

4.4 The Nature of the Fast-Ion State: Conclusions

In this Chapter we have reviewed the properties of ionic crystals at high temperature when thermally induced lattice disorder leads to fast-ion behaviour with high ionic diffusion and conductivity. Emphasis has been given to crystals with the fluorite structure which, because of their relative simplicity, have received a great deal of experimental and theoretical attention and at present are probably the best understood fast-ion system. Even so this understanding lacks a complete pictorial model of what is a complex high temperature phase.

An acceptable model of the high temperature disordered phase must be able to account for, or predict, the complete range of physical properties of the crystal. These range from bulk physical properties such as specific heat and electrical conductivity, properties measured by more specific probes such as nmr, diffusion measurements and light scattering, to the comprehensive information given by x-ray diffraction and neutron scattering. Such a model describes the complex state in terms of parameters and must be built up using all available data. Ancillary information, and an alternative more fundamental theoretical approach, is provided by molecular dynamics calculations as described by Gillan (1985, 1986, and in Chapter 3 of this volume). Although such calculations can provide the position in space and time of every ion in the system, it can be difficult to extract a detailed picture of the behaviour of the average ion and its environment due to the wealth of information. In some ways it provides another type of experiment which requires its own interpretation. The connection between the data from a practical experiment, the analytic model or the molecular dynamics approach to interpretation of these data, and the 'true' situation is outlined diagrammatically in figure 4.28. This emphasises the close inter-relationships needed for an understanding of the properties of a material.

Neutrons have a distinct advantage over other probes of solids at high temperatures in their penetration of furnace walls and heat shields, and this has enabled measurements at sample temperatures of up to 3000K to be made. There are several types of neutron scattering experiment which can be employed to study ionic disorder at high temperature. Diffraction gives information on the time-averaged location of ions in the unit cell, and their mean thermal vibrational displacements, whereas coherent diffuse scattering gives a 'snapshot' picture of the average instantaneous relative positions of the dynamically disordered ions. Incoherent, and coherent, diffuse quasielastic scattering gives information on the motion of individual disordered ions, and of their correlated behaviour, respectively (see table 4.4). These types of neutron scattering experiment have been used to build up a picture of the disordered phase in fluorites in which there are short-lived ($\sim 10^{-12}$s), dynamically fluctuating or mobile, clusters of disordered anions, comprising anion Frenkel pairs with their surrounding distortion field. Their dynamical nature and independent motion means that the correlation between different clusters is weak and so the scattering from each may be added incoherently. Static clusters of a similar nature are found at low temperatures in intrinsically doped fluorites, but these tend to be correlated and thus scatter coherently with each other. As the temperature increases some of the anions become mobile and the correlation decreases.

Although in principle a time-averaged distribution of scattering length density, the so called probability scattering density function, can be obtained by Fourier transform of diffraction data, in practice a parameterised model is usually needed for their interpretation. The Partial Fourier synthesis method may be used to give the probability scattering density function of the disordered, or mobile, ions if a model for the positions of the static ions is assumed. A model which fits the data well may be used to determine the total ionic probability density function, or

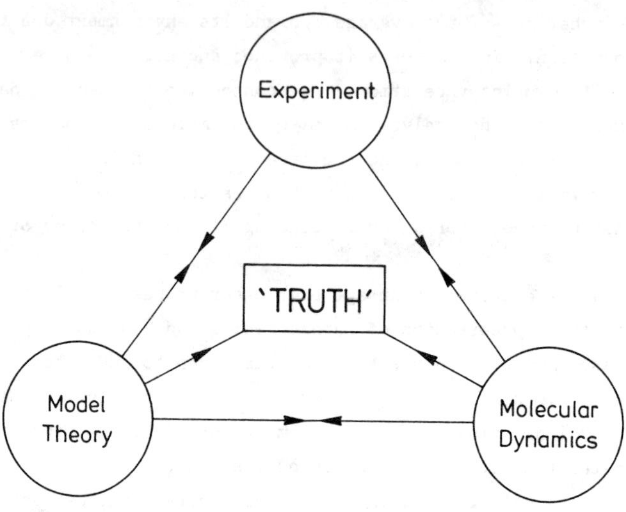

Figure 4.28 Schematic relationship between experimental data, molecular dynamics data, analytical theory and 'truth'.

distribution, for one disordered species of ion, and this is a good point for comparison with molecular dynamics results to be made as has been demonstrated for PbF_2 (Walker et al 1982). The ionic probability density function has also been used to infer the potential in which the ions move, but this procedure gives rise to a number of points of concern.

The dynamic cluster model which has successfully accounted for the main features of the neutron scattering data from fluorites at high temperatures has been derived largely by trial and error, and, although necessarily an oversimplification of the true complex state, does provide a working picture. The data may be interpreted in terms of clusters with one or two Frenkel pairs, with the best results obtained from the 9:1:8 cluster, in which a Frenkel pair has its interstitial displaced in a <110> direction from the mid regular-anion site into an empty cube in the lattice, and causes n.n. and n.n.n. anion relaxations in <111> directions into the empty cubes. A hard sphere model of displacements gives a good estimate of these relaxations, and these are found to be largest for $SrCl_2$ and the antifluorite Mg_2Si, and least for the antifluorite Li_2O. In general the larger the extent of the distortion field, the longer the lifetime of the cluster. As the temperature increases the extent of the correlated distortion field decreases.

A number of quantitative results may be deduced from the cluster model for fluorites. The intensity of the coherent QES is directly related to the number of defective anions, and gives an estimate of this number and of their activation energy. Values for the fraction of defective anions, n_d , found in this way are consistent with those determined from diffraction data. The relationship between n_d and n_f, the fraction of Frenkel defects, involves an estimate of the size of the distortion field. Although this estimate depends on some assumptions, values obtained for n_f are of the same order as those estimated from other data and from molecular dynamics if the same definitions are used. Great care is necessary for example when estimating the configurational entropy in deducing n_f from specific heat data. The temperature variation of the incoherent QES energy width yields an activation energy for the anion diffusion, and the $\underline{Q}$ dependence of the width yields the mean anion residence time and a diffusion constant equivalent to that given by tracer methods. The ratio of coherent to incoherent energy widths also yields a further estimate of n_d which is consistent within a factor of two with other estimates. Values of activation energies which are in principle more accurate than those obtained from neutron scattering data may be obtained from conductivity, nmr and diffusion measurements as discussed section 4.2 and in Chapter 2. However care must be taken to ensure that the correct temperature region is used in the analysis. Light scattering and inelastic neutron scattering show that at high temperature the increasing lattice disorder destroys the phonon as an eigenmode of the system for all but the long wavelength acoustic modes. The defects are found to be effectively static as regards scattering of phonons.

The fluorites do not give any indication of undergoing a true phase transition at T_c. Often the term 'sublattice melting' is used to describe the fast-ion phase, but care must be taken over the sense in which such a term is used. In the case of the fluorites this does not seem to be directly appropriate. Neither $S^D(\underline{Q})$ nor $g(\underline{r})$, for example, closely resemble those of a liquid.

The relationship between the molecular dynamics calculations, and the neutron scattering experimental data on fluorites now seems to be very satisfactory, as described in Chapter 3. Early calculations of Jacucci and Raman on CaF_2 gave results very consistent with the incoherent QES data from $SrCl_2$, and this is confirmed by later calculations by Gillan and Dixon (1980) and Dixon and Gillan (1980) on $SrCl_2$. The wavevector dependence of the energy width of the incoherent QES from $SrCl_2$ shows that the anions do hop between regular lattice sites, predominantly between n.n. sites. This type of motion is conventionally called 'vacancy' diffusion but the geometry of hops would be the same for the correlated, or caterpillar, type diffusion. Gillian and Dixon show that the coherent QES arises predominantly from the disordered anions, and their work and that of Gillan (1985, 1986) suggests that the anions move between regular lattice sites by a correlated diffusional motion, rather than by the simple vacancy mechanism. Gillan shows that it is this motion which gives rise to the coherent QES, but provides no visual model of the distortion field of the mobile ions. The wavevector dependence of the energy width of the coherent QES from $SrCl_2$ and from CaF_2 suggests that the anion interstitial and its correlated distortion field move in a similar manner to the anion alone. From the relative energy widths of the QES it seems probable that the mobile anions vibrate for several phonon periods about their regular site before hopping, on a time scale of correlated motion reflected by the coherent QES width, to the neighbouring site. The total coherent QES, $S^D(\underset{\sim}{Q})$, thus provides an average 'snapshot' picture of the mobile anions and their distortion field, which is interpreted in the form of the short-lived cluster. The alternative possibility of an essentially static Frenkel cluster providing a vacancy into which anions can hop, used by Mościński and Jacobs (1985), seems less appropriate.

Silver compounds which undergo a first order phase transition to a fast-ion phase were the first to be studied by neutron scattering, and have attracted a great deal of attention; however a fully detailed picture of their more complex conduction process, in which there are many more sites available per mobile cation than in the case of the fluorites, has yet to emerge. Molecular dynamics methods have been used to give a good account of the coherent neutron scattering from disorder in β-Ag_2S, which has been studied in single crystal form using a triple-axis spectrometer, in terms of the correlated motion of silver ions. It would seem that further studies of the other silver compound using the same methods are necessary in order to fully understand the nature of the disorder. As the complexity increases, as in the case of the β-aluminas, more detailed and extensive data from neutron scattering is necessary in order to cover the range of time scales of the motions involved. Nevertheless, by a combination of all sources of experimental data with molecular dynamics simulation it should be possible to develop a working model which can give a satisfactory account of the data. Once fully satisfactory models are established for these and the other technologically important materials, it should be possible to engineer their properties more successfully for specific applications.

Acknowledgements

We are indebted to our many colleagues, especially to K N Clausen, M H Dickens and J Kjems, for close collaboration on aspects of much of the work described in this Chapter. The work was supported in part by the Underlying Research Programme of the UKAEA, and by the UK SERC.

References

Aboagye J K and Friauf R J 1975 Phys. Rev. B11 1654

Ackerman R J, Gilles P W and Thorn R J 1956 J. Chem. Phys. 25 1089

Aldebert P 1984 Revue Phys. Appl. 19 649

Allnatt A R, Chadwick A V and Jacobs P W M 1986 Proc. Roy. Soc. A410 385

Andersen N H, Clausen K and Kjems J K 1983 Solid State Ionics 9/10 543

Andersen N H, Clausen K, Hackett M A, Hayes W, Hutchings M T, Macdonald J E and Osborn R 1985 Transport-Structure Relations in Fast Ion and Mixed Conductors eds. Poulson F W et al (Risø National Laboratory, Denmark) p279

Andersen N H, Clausen K N, Kjems J K and Schoonman J 1986a J. Phys. C 19 2377

Andersen N H, Clausen K, Hackett M A, Hayes W, Hutchings M T, Macdonald J E and Osborn R 1986b Physica 136B 315

Andersen N H, Clausen K N and Kjems J K 1987 Methods of Experimental Physics, Vol 23B, Neutron Scattering eds. Sköld K and Price D L (Academic Press, New York) p187

Aronsson R, Knape H, Lunden A, Nilsson L, Torell L M, Andersen N and Kjems J 1983 Radiation Effects 75 79

Azemi A, Carr V M, Chadwick A V, Kirkwood F G and Saghafian R 1984 J. Phys. Chem. Sol. 45 23

Bachmann R and Schulz H 1983 Solid State Ionics 9/10 521

Baker J M, Davies E R and Hurrell J P 1968 Proc. Roy. Soc. A308 403

Becker P and Coppens P 1974 Acta Cryst. A30 129 and 148

Becker P and Coppens P 1975 Acta Cryst. A31 417

Belosludov V R, Efremova R I and Matizen E V 1974 Sov. Phys. St. 16 847

Bendall P J and Catlow C R A and Fender B E F 1984 J. Phys. C 17 797

Benière M, Chemla M and Benière F 1979 J. Phys. Chem. Solids 40 729

Beyeler H U, Brüesch P, Pietronero L, Schneider W R, Strässler S and Zeller H R 1979 Physics of Superionic Conductors, Topics in Current Physics 15 ed. Salamon M B (Springer-Verlag, Berlin)p77

Bleaney B, LLewellyn P M and Jones D A 1956 Proc. Phys. Soc. B69 858

Boyce J B, Mikkelsen J C and O'Keefe M 1977 Solid State Commun. 21 955

Boyce J B and Huberman B A 1979 Phys. Rev. 51 189

Boyer L L 1980 Phys. Rev. Letters 45 1858

Breuer N 1974 Z. Phys. 271 289

Browning P, Hyland G J and Ralph J 1983 High Temp.-High Pressures 15 169

Brun T O, Susman S, Rowe J M and Rush J J 1981 Solid State Ionics 5 417

Catlow C R A 1984 Solid State Ionics 12 67

Catlow C R A, Comins J D, Germano F A, Harley R T and Hayes W 1978 J. Phys. C 11 3197

Catlow C R A, Comins J D, Germano F A, Harley R T, Hayes W and Owen I B 1981 J. Phys. C 14 329

Catlow C R A and Hayes W 1982 J. Phys. C 15 L9

Catlow C R A, Chadwick A V and Corish J 1983 Radiation Effects 75 61

Cava R J, Reidinger F and Wuensch B J 1977 Solid State Commun. 24 411

Cava R J and McWhan D B 1980 Phys. Rev. Letters 45 2046

Cava R J, Reidinger F and Wuensch B J 1980 J. Solid State Chem. 31 69

Cava R J, Reidinger F and Wuensch B J 1981 Solid State Ionics 5 501

Cava R J, Andersen N H and Clausen K 1986 Solid State Ionics 18/19 1184

Chadwick A V 1983 Radiation Effects 74 17; and 1983 Solid State Ionics 8 209

Cheetham A K, Fender B E F and Cooper M J 1971 J. Phys. C 4 3107

Cheetham A K 1973 Chemical Applications of Neutron Scattering ed. B T M Willis (O.U.P., Oxford)p225

Christy R W and Lawson A W 1951 J. Chem. Phys. 19 517

Chudley C T and Elliott R J 1961 Proc. Phys. Soc. (London) 77 353

Clausen K, Hayes W, Macdonald J E, Schnabel P G, Hutchings M T and Kjems J K 1983 High Temp.-High Pressures 15 383

Clausen K, Hayes W, Macdonald J E, Osborn R and Hutchings M T 1984a Phys. Rev. Letters 52 1238

Clausen K, Hayes W, Hutchings M T, Macdonald J E, Osborn R and Schnabel P 1984b Revue Phys. Appl. 19 *719*

Clausen K, Hayes W, Hutchings M T, Kjems J K, Macdonald J E and Osborn R 1985 High Temp. Science 19 *189*

Clausen K, Hayes W, Macdonald J E, Osborn R, Schnabel P G, Hutchings M T and Magerl A 1987 J. Chem. Soc., Faraday Trans. 2 83 *1109*

Clausen K N, Hackett M A, Hayes W, Hull S, Hutchings M T, Macdonald J E, McEwen K A, Osborn R and Steigenberger U 1989a Physica B 156/157 *103*

Clausen K, Hackett M A, Hayes W, Hull S, Hutchings M T and Steigenberger U 1989b (to be published)

Cooper M J 1970 Thermal Neutron Diffraction *ed. Willis B T M (O.U.P, Oxford)p51*

Cooper M J and Nathans R 1967 Acta Cryst. 23 *357*

*Cooper M J and Rouse J D 1968 Acta Cryst. A*24 *405*

*Cooper M J, Rouse K D and Willis B T M 1968 Acta Cryst. A*24 *484*

Cowley R A 1967 Proc. Phys. Soc. 90 *1127*

Cowley R A 1968 Rep. Prog. Phys. 31 *123*

Cowley R A, Buyers W J L, Svensson E C and Paul G L 1978 Neutron Inelastic Scattering *Vol I (IAEA, Vienna)p281*

Cummins H Z and Levanyuk A P 1983 Light Scattering Near Phase Transitions *(North Holland, Amsterdam)*

Currat R and Pynn R 1979 Treatise on Materials Science and Technology, Vol 15, Neutron Scattering *ed. Kostorz G (Academic Press, New York)p131*

Dawson B 1970 Thermal Neutron Diffraction *ed. Willis B T M (O.U.P, Oxford)p101*

*Dawson B, Hurley A C and Maslen V W 1967 Proc. Roy. Soc. (London) A*298 *289*

Dickens M H, Hayes W and Hutchings M T 1976 J. de Physique Colloque C7 37 *C7-353*

Dickens M H, Hutchings M T, Kjems J and Lechner R E 1978 J. Phys. C 11 *L583*

Dickens M H and Hutchings M T 1978a Lattice Dynamics *ed. Balkanski M (Flammarion, Paris)p540*

Dickens M H and Hutchings M T 1978b J. Phys. C 11 *461*

Dickens M H and Hutchings M T 1978c Neutron Inelastic Scattering 1977 *Vol II (IAEA, Vienna)p285*

Dickens M H, Hayes W, Hutchings M T and Kleppmann W G 1979a J. Phys. C 12 *17*

Dickens M H, Hayes W, Smith C and Hutchings M T 1979b Fast Ion Transport in Solids *eds. Vashishta P et al (Elsevier North Holland, Amsterdam)p225*

Dickens M H, Hutchings M T and Suck J-B 1980 Solid State Commun. 34 559

Dickens M H, Hayes W, Hutchings M T and Smith C 1982 J. Phys. C 15 4043

Dickens M H, Hayes W, Schnabel P G, Hutchings M T, Lechner R E and Renker B 1983 J. Phys. C 16 L1

Didisheim J-J, McMullan R K and Wuensch B J 1986 Solid State Ionics 18/19 1150

Dixon J M, La Grange L D, Merten U, Miller C F and Porter J T 1963 J. Electrochem. Soc. 110 276

Dixon M and Gillan M J 1980 J. Phys. C 13 1919

Dolling G, Cowley R A and Woods A D B 1965 Can. J. Phys. 43 1397

Dorner B 1972 Acta Cryst. A28 319

Dorner B 1982 Coherent Inelastic Neutron Scattering in Lattice Dynamics (Springer-Verlag, Berlin)

Dworkin A S and Bredig M A 1963 J. Chem. Eng. Data 8 416

Dworkin A S and Bredig M A 1968 J. Chem. Physics 72 1277

Ebbsjö I, Vashishta P, Dejus R and Sköld K 1987 J. Phys. C 20 L441

Eckold G, Funke K, Kalus J and Lechner R E 1976 J. Phys. Chem. Solids 37 1097

Elcombe M M and Pryor A M 1970 J. Phys. C 3 492

Elliott R J, Krumhansl J A and Leath P L 1974 Rev. Mod. Phys. 46 465

Elliott R J, Hayes W, Kleppmann W G, Rushworth A J and Ryan J F 1978 Proc. Roy Soc. A360 317

Farley T W D, Hayes W, Hull S, Ward R, Hutchings M T and Alba M 1988 Solid State Ionics 28/30 189

Farley T W D, Hayes W, Hull S, Hutchings M T, Alba M and Vrtis M 1989 Physica B 156/157 99

Faux D A and Ross D K 1987 J. Phys. C 20 1441

Figueroa D R, Chadwick A V and Strange J H 1978 J. Phys. C 11 55

Funke K 1988 Solid State Ionics 28/30 100

Funke K, Kalus J and Lechner R E 1974 Solid State Commun. 14 1021

Funke K, Eckold G and Lechner R E 1978 in Microscopic Structure and Dynamics of Liquids eds. Dupuy J and Dianoux A J (Plenum, New York)

Funke K, Höch A and Lechner R E 1980 J. de Physique Colloque C6 41 C6-17

Garber J A and Granato A V 1975 Phys. Rev. B11 3390

Gavarri J R, Lucazeau G, Dianoux A J and Colomban Ph 1983 Solid State Commun. 45 203

Gillan M J 1985 Physica 131B 157

Gillan M J 1986 J.Phys. C 19 3391 and 3517

Gillan M J 1989 This Volume Ch 3

Gillan M J and Dixon M 1980 J. Phys. C 13 1901

Gillan M J and Wolf D 1985 Phys. Rev. Letters 55 1299

Gissler W and Rother H 1970 Physica 50 380

Gissler W and Stump N 1973 Physica 65 109

Göltz G, Heidemann A, Mehrer H, Seeger A and Wolf D 1980 Phil. Mag. A41 723

Gordon R E and Strange J H 1978 J. Phys. C 11 3213

Grier B H, Shapiro S M and Cava R J 1984 Phys. Rev. B29 3810

Harding J H 1989 This Volume Ch 2

Harding J H, Masri P and Stoneham A M 1980 J. Nucl. Mat. 92 73

Hayes W 1982 Light Scattering in Solids III, Topics in Applied Physics, 51 eds. Cardona M and Güntherodt G (Springer-Verlag, Berlin)p93

Hayes W 1986 Contemp. Phys. 27, 519

Hayes W and Loudon R 1978 Scattering of Light by Crystals (Wiley, New York)

Hayes W and Stoneham A M 1974 Crystals with the Fluorite Structure ed. Hayes W (Clarendon, Oxford)Ch 2

Hayes W and Stoneham A M 1985 Defects and Defect Processes in Non-metallic Solids (Wiley, New York)

Hempelmann R, Richter D, Faux D A and Ross D K, 1988 Z. Physik Chemie NF 159 175

Höch A, Funke K, Lechner R E and Ohachi T 1983 Solid State Ionics 9/10 1353

Hoshino S 1957 J. Phys. Soc. Japan 12 315

Hoshino S, Sakuma T and Fujii Y 1977 Solid State Commun. 22 763

Howard R E and Lidiard A B 1964 Rep. Prog. Phys. 27 161

Hull S, Farley T W D, Hayes W and Hutchings M T 1988a J. Nucl. Mat. 160 125

Hull S, Farley T W D, Hackett M A, Hayes W, Osborn R, Andersen N H, Clausen K, Hutchings M T and Stirling W G 1988b Solid State Ionics 28/30 488

Hutchings M T 1987 J. Chem. Soc., Faraday Trans. 2 83 1083

Hutchings M T, Clausen K, Dickens M H, Hayes W, Kjems J K, Schnabel P G and Smith C 1983 Harwell Report AERE-R11127

Hutchings M T, Clausen K, Dickens M H, Hayes W, Kjems J K, Schnabel P G and Smith C 1984 J. Phys. C 17 3903

Hutchings M T, Clausen K, Hayes W, Macdonald J E, Osborn R and Schnabel P G 1985 High Temp. Science 20 97

Hutchings M T, Lowde R D and Tindle G 1986 Inst. Phys. (London) Conf. Proc. 81 p151

Hutchings M T, Farley T W D, Hackett M A, Hayes W, Hull S and Steigenberger U 1988 Solid State Ionics 28/30 1208

Jacobs P W M and Ong S H 1980 Crystal Lattice Defects 8 177

Jacucci G and Rahman A 1978 J. Chem. Phys. 69 4117

James R W 1948 Acta Cryst. 1 132

Johnson C K 1970 Thermal Neutron Diffraction ed. Willis B T M (O.U.P, Oxford)p132

Kerrisk J F and Clifton D G 1972 Nucl. Technol. 16 531

Kjems J K, Andersen N H, Schoonman J and Clausen K 1983 Physica 120B 357

Kleppmann W G 1978 J. Phys. C 11 L91

Koto K, Schulz H and Huggins R A 1980 Solid State Ionics 1 355

Lawson A W 1950 Phys. Rev. B78 185

Le Claire A D 1970 Physical Chemistry, An Advanced Treatise eds. Eyring H, Henderson D and Jost W (Academic Press, New York) Vol 10 p261

Lechner R E, Eckold G and Funke K 1978 in Microscopic Structure and Dynamics of Liquids eds. Dupuy J and Dianoux A J (Plenum, New York)

Lidiard A B 1974 Crystals with Fluorite Structure ed. Hayes W (Clarendon Press, Oxford)p101

Liu D W, Perry C H, Feinberg A A and Currat R 1987 Phys. Rev. B36 9212

Lovesey S W 1984 Theory of Neutron Scattering from Condensed Matter (Clarendon Press, Oxford) Vols I and II

Lucazeau G, Gavarri J R and Dianoux A J 1987 J. Phys. Chem. Solids 48 57

Lucazeau G, Dohy D, Fanjat N and Dianoux A J 1988 Solid State Ionics 28/30 1611

MacInnes D A and Catlow C R A 1980 J. Nuc. Mat. 89 354

Magerl A, Zabel H and Anderson I S 1985 Phys. Rev. Letters 55 222

Mahan G 1972 Polarons in Ionic Crystals and Polar Semiconductors ed. Devresse J T (North-Holland, Amsterdam)p553

Mook H A, Nicklow R M, Penney T, Holtzberg F and Shafer M W 1978 Phys. Rev. B18 2925

Moon R M, Riste T and Koehler W C 1969 Phys. Rev. 181 920

Mościński J and Jacobs P W M 1985 Proc. R. Soc. A398 173

Naylor B F 1945 J. Amer. Chem. Soc. 67 150

Niklasson G 1972 Phys. Condens. Matter 14 138

Oberschmidt J 1981 Phys. Rev. B23 5038

Oliveria M, McMullan R K and Wuensch B J 1988 Solid State Ionics 28/30 1332

Osborn R, Andersen N H, Clausen K, Hackett M A, Hayes W, Hutchings M T and Macdonald J E 1986 Mat. Science Forum 7 55

Osborn R, Boland B C, Bowen Z A, Taylor A D, Hackett M A, Hayes W and Hutchings M T 1987 J. Chem. Soc., Faraday Trans. 2 83 1105

Osborn R, Taylor R D, Bowden Z A, Hackett M A, Hayes W, Hutchings M T, Amoretti G, Caciuffo R, Blaise A and Fournier J M 1988 J. Phys. C 26 L931

Osborn R. Dickens M H, Hutchings M T and Lechner R E 1989 (in course of preparation)

Perenthaler E, Schulz H and Beyeler H U 1981 Solid State Ionics 5 493

Perry C H and Feinberg A 1980 Solid State Commun. 36 519

Price D L and Sköld K 1986 Methods of Experimental Physics, Vol 23A, Neutron Scattering eds. Sköld K and Price D L (Academic Press, New York)p1

Rabenau A 1982 Solid State Ionics 6 277 (and references therein)

Ralph J and Gillan M J 1984 J. Nucl. Mat. 126 111

Ralph J and Hyland G J 1985 J. Nucl. Mat. 132 76

Reik H G 1972 Polarons in Ionics Crystals and Polar Semiconductors ed. Devresse J T (North-Holland, Amsterdam)p679

Renker B, Bernotat H, Heger G, Lehner N and Press W 1983 Solid State Ionics 9/10 1341

Roberts R B and White G K 1986 J. Phys. C 19 7167

Roth W L, Reidinger F and Laplaca S 1976 Superionic Conductors eds. Mahan G D and Roth W L (Plenum, New York)p223

Ross D K, Faux D A, McKergow M W, Wilson D L T and Sinha S K 1986 Atomic Transport and Defects in Metals by Neutron Scattering eds. Janot C et al Proc. in Phys. Vol 10 (Springer-Verlag, Berlin)p116

Ross D K, Faux D A, Benham N J, Sinha S K, Hempelmann R, Cook J C, Richter D, Schaerpf O and Anderson I S 1989 Physica B 156/157 121

Rosshirt E, Frey F, Boysen H, Eckold G and Steigenberger U 1988 Mat. Science Forum 27/28 129

Salamon M B (ed.) 1979 Physics of Superionic Conductors, Topics in Current Physics 15 (Springer-Verlag, Berlin)

Schnabel P, Hayes W, Hutchings M T, Lechner R E and Renker B 1983 Radiation Effects 75 73

Schoonman J 1980 Solid State Ionics 1 121

Schröter W and Nölting J 1980 J. de Physique Colloque C6 41 C6-20

Schulz H and Zucker U H 1981 Solid State Ionics 5 41

Sears V F 1986 Methods of Experimental Physics, Vol 23A, Neutron Scattering eds. Sköld K and Price D L (Academic Press, New York)p521

Shapiro S M, Semmingsen D and Salamon M 1978 Lattice Dynamics ed. Balkanski M (Flammarion, Paris)p538

Shapiro S M and Reidinger F 1979 Physics of Superionic Conductors, Topics in Current Physics 15 ed. Salamon M B (Springer-Verlag, Berlin)p45

Sköld K 1967 Phys. Rev. Letters 19 1023

Slagle O D 1983 J. American Ceram. Soc. 67 169

Springer T 1972 Quasielastic Neutron Scattering for the Investigation of Diffusive Motions in Solids and Liquids (Springer-Verlag, Berlin) Tract 64

Squires G L 1978 Introduction to the Theory of Thermal Neutron Scattering (Cambridge, England)p61

Stassis C 1986 Methods of Experimental Physics, Vol 23A, Neutron Scattering eds. Sköld K and Price D L (Academic Press, New York)p369

Strock L W 1934 Z. Phys. Chem., Abt. B25 411; and 1936 ibid 31 132

Suemoto T and Ishigama M 1983 Solid State Commun. 45 641

Tahir-Kheli R A and Elliott R J 1983 Phys. Rev. B27 844

Tallon J L and Roberts R 1985 Transport-Structure Relations in Fast Ion and Mixed Conductors eds. Poulsen F W et al (Risø National Laboratory, Denmark)p210

Tannhauser D S, Bruner L J and Lawson A W 1956 Phys. Rev. 102 1276

Ubbelohde A R 1978 The Molten State of Matter (Wiley, New York)

Van Hove L 1954 Phys. Rev. 95 249

Vashishta P, Ebbsjö I, Dejus R and Sköld K 1985 J. Phys. C 18 L291

Vineyard G H 1958 Phys. Rev. 110 999

von Dreele R B, Eyring L, Bowman A L and Yarnell J L 1975 Acta Cryst. B31 971

Walker A B, Dixon M and Gillan M J 1982 J. Phys. C 15 4061

Willis B T M 1969 Acta Cryst. A25 277

Willis B T M 1970 Thermal Neutron Diffraction ed. Willis B T M (O.U.P, Oxford)p124

Willis B T M 1978 Acta Cryst. A34 88

Windsor C G 1986 Methods of Experimental Physics, Vol 23A, Neutron Scattering eds. Sköld K and Price D L (Academic Press, New York)p197

Wolf D 1977 Solid State Commun. 23 853

Wright A F and Fender B E F 1977 J. Phys. C 10 2261

Zachariasen W H 1967 Acta Cryst. 23 558

Zeyher R 1978 Z. Phys. B31 127

Chapter 5

HIGH TEMPERATURE EXPERIMENTS WITH THE LASER-HEATED DIAMOND CELL

R. Jeanloz

Department of Geology and Geophysics, University of California
Berkeley, CA 94720 USA

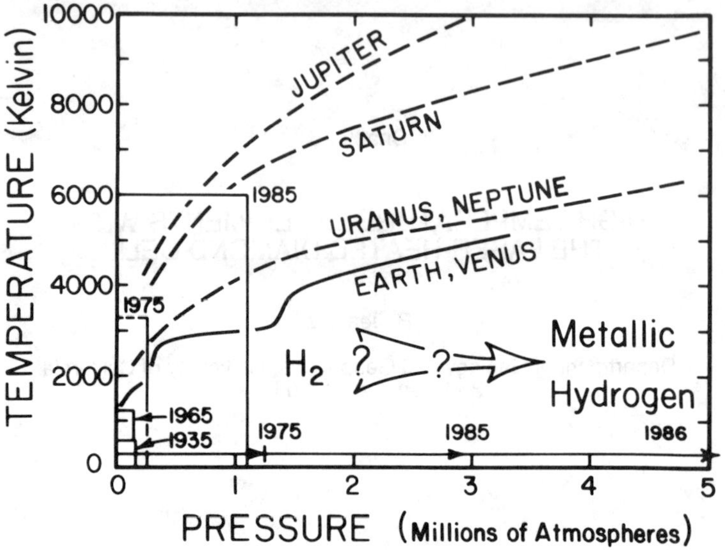

Figure 1. Sustained pressure-temperature conditions that have been achieved experimentally are shown as functions of time. For comparison, the average temperature-pressure (or depth) profiles estimated for some of the terrestrial and giant planets are indicated, as is the range of pressures at which hydrogen is expected to metallize. By 1975, temperatures in excess of 3000 K and pressures in excess of 100 GPa (1 million atmospheres) had been achieved in the diamond cell.

INTRODUCTION

The diamond cell is usually thought of as an instrument for high-pressure research. Indeed, the highest sustained pressures (over 500 GPa) have been experimentally achieved using the diamond cell (Figure 1). With the development of laser-heating techniques, however, the diamond cell can also be considered an instrument for studying condensed matter at extremely high temperatures. The diamonds act as windows through which one can examine a hot and reactive sample. In addition, they allow pressure to be used as an independent variable in the experiments. This can be advantageous in that even modest pressures suppress vaporization or surface contamination which often make it difficult to observe the bulk of the sample. As the melting temperature of most solids increases with pressure, compression of the sample also allows one to investigate solid-state phenomena at much higher temperatures than would otherwise be possible.

THE DIAMOND CELL

There are several recent reviews of the diamond cell, including descriptions of a variety of designs and their applications in high-pressure research (e.g., Bassett[2]; Hazen and Finger[8]; Jayaraman[13,14]; Jephcoat et al.[17]). The basic design, as illustrated in Figure 2, involves two, brilliant-cut diamonds that are oriented to contain a sample between the culets. These must be gem-quality diamonds, typically between 1/5 and 1/3 carat in size. Although synthetic diamonds have been used (e.g., Ruoff et al.[34]), natural diamonds of this size and quality are more readily available. Depending on the experiment to be carried out, diamonds ranging from colored (high impurity) Type I to clear, low-fluorescence (low impurity) Type II are used; the former appear to be stronger, but the latter are usually preferable for spectroscopic studies (Xu et al.[44]).

The advantage to using diamonds is threefold: (i) diamond is the hardest and strongest material known, (ii) it is transparent to a broad range of electromagnetic radiation, and (iii) it is extremely refractory, having a melting temperature above 4000 K and a zero-pressure stability limit above 2000 K (under vacuum) (Evans[6]; Shaner et al.[37]). The first point is usually emphasized for high-pressure experimentation, but the second and third are more significant for studies emphasizing high temperature. In particular, the low atomic number of diamond allows the sample to be probed with X-rays above 5-10 keV in energy, especially if the high intensity available from a synchrotron source is used. This energy range is ideal for diffraction and at least some spectroscopic investigations: e.g.,

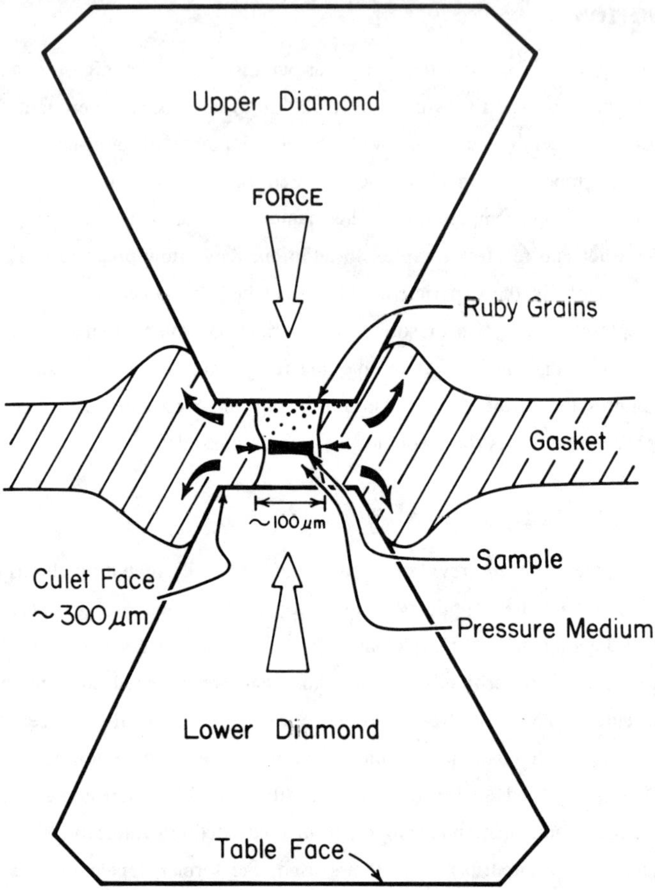

Figure 2. Schematic cross-section of the diamond cell showing the location of the sample (green) and of the ruby grains used for pressure calibration (red). As force is exerted onto the diamonds, the metal foil acting as a gasket is extruded outward — providing lateral support for the diamonds — and inward into the sample area.

EXAFS and XANES on elements beyond, and possibly including, the first transition series. In contrast, most window materials such as sapphire absorb hard X-rays much more severely. Also, the high thermal conductivity and refractory nature of diamond are advantageous for containing hot samples, as described below.

The sample is held between the culet faces with a metal gasket. This is simply a metal foil (for example stainless steel or Inconel) 75 μm to 250 μm in original thickness. Prior to the experiment, the foil is indented between the diamonds and a hole is drilled out for the sample in the center of the indentation. Typically, culets of 200 μm to 1 mm are used and the sample area in the gasket is roughly 50-800 μm in diameter. After indentation, the gasket is between 10 μm and 200 μm thick, leaving a sample volume of about 20 nl to 100 μl. This volume must contain the sample, a medium to surround the sample and any calibration standard that might be required. The medium varies depending on the experiment: powdered sample material can be used as a medium to avoid contaminating the sample itself; ruby, diamond or other refractory materials can be used to minimize reactivity with the medium at high temperatures; and alcohol mixtures, salts or noble gasses are used to achieve hydrostatic or nearly hydrostatic pressures (Piermarini et al.[31,32]; Knittle and Jeanloz[19,22]; Mao et al.[27]; Williams et al.[42]; Heinz and Jeanloz[10]). Among the calibration standards that are employed, fine-grained ruby powder is common because it can be used to monitor pressure, and at least moderate temperatures, by spectroscopic means (Barnett et al.[1]; Mao et al.[26]; Shimomura et al.[38]; Sato-Sorensen[35]).

The critical factor in using the diamond cell at high pressures is to ensure that the culets are as close to parallel as possible. As the force is applied to the tables of the two diamonds, the culets must be precisely cut parallel to the table faces. Also, considerable attention is paid in the design of diamond cells so that the tables of the two diamonds are kept parallel when the sample is taken to high pressures (e.g., Jayaraman[13,14]; Jephcoat et al.[17]). Optical techniques, including the examination of interference fringes between the culets and of stress-induced birefringence within the diamonds, are used to ensure parallelism to within ~ 0.1 μm across the entire culet face.

TABLE I

TECHNIQUES FOR HEATING SAMPLES IN THE DIAMOND CELL

	Advantages	*Disadvantages*	*Maximum Temperature*	*Reference*
RESISTANCE HEATING	Stability and Control of of Temperature	Limited Temperature Range		
Internal	High Efficiency	Difficult Fabrication and Calibration	1700 K	Boehler[4] Mao et al.[25]
External	Low Temperature Gradients	Diamonds and Cell are Heated	1500 K	Schiferl et al.[36]
LASER HEATING	High Temperatures Optical Access to Sample	Large Temperature Gradients Laser-Sample Coupling		
Pulsed	High Temperatures	Difficulty in Achieving Equilibrium Difficult Calibration	>5000 K	Ming and Bassett[30] Weathers and Bassett[41] Bassett and Weathers[3]
Continuous	Stability Achieving Equilibrium	High Power is Necessary	5000 -7000 K	Heinz and Jeanloz[9] Williams et al.[42] Knittle and Jeanloz[24]

LASER HEATING AND SPECTRORADIOMETRY

Techniques that have been used for heating samples inside the diamond cell are summarized in Table 1. These are of two types, resistance heating and laser heating. The advantage to resistance heating is that temperature can be accurately controlled and stabilized. Thus, resistance heating has been primarily used for examining thermal expansion and phase transformations at elevated temperatures and pressures by means of synchrotron-based X-ray diffraction (Skelton et al.[39]; Furnish and Bassett[7]; Huang and Bassett[12]; Schiferl et al.[36]). With a heater that is external to the sample volume, temperature gradients across the sample are also minimized. This is achieved at the cost of heating the diamonds and gasket, however, and it is usually these components which set an upper limit on the temperatures that can be achieved with a resistance heater. In order to avoid this limitation, microfurnaces have been placed inside the sample volume itself, but these are difficult to fabricate. Also, they generate large temperature gradients which can complicate the determination of the sample temperature (Williams et al.[42]).

The highest temperatures so far achieved in the diamond cell have been with laser heating. Although heating has been attempted with pulsed lasers, no quantitative measurements have so far been made (Ming and Bassett[30]; Bassett and Weathers[3]). Also, it is difficult to determine whether or not thermodynamic equilibrium is reached during pulsed laser heating. Therefore, because of the large temperature range and the relative stability of heating that are achieved by continuous laser heating, the following discussion concentrates almost exclusively on this technique.

Both CO_2 and Nd:YAG continuous (cw) lasers have been used with the diamond cell. The latter is more common because the 10.6 μm CO_2-laser radiation is subject to absorption by all but the highest-purity (hence costliest) diamonds. As the technique depends on the laser radiation being absorbed by the sample, there must either be an inherent absorption mechanism (e.g., the 1 μm crystal-field absorption of Fe^{2+} in oxides), or an absorbing material such as platinum black must be mixed into the sample in small proportions. Along with this requirement that there be a mechanism for the sample to couple with the laser beam, a primary disadvantage of laser heating is that large temperature gradients are produced across the sample (Table 1).

A laser-heating system is schematically illustrated in Figure 3, with a detailed description being provided elsewhere (Jeanloz and Heinz[15,16]; Heinz and Jeanloz[9]). Simply,

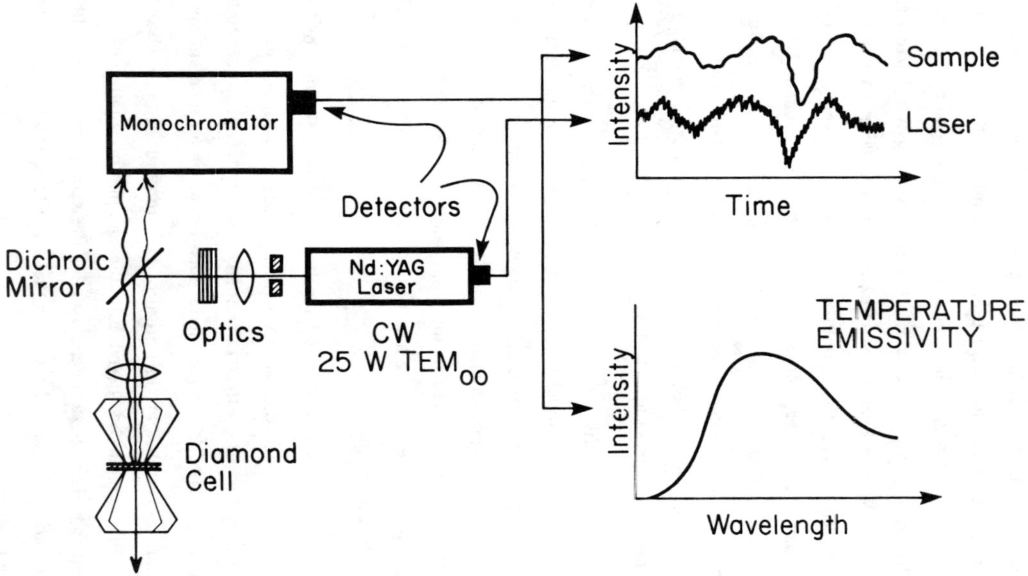

Figure 3. Laser-heating and spectroradiometer system that is described in the text. The thermal radiation emitted from the sample (green) is collected through the same microscope optics that are used to focus the beam from the cw Nd:YAG laser (red). The temperature and emissivity of the sample are determined from a greybody fit to the thermal radiation collected at visible and near-infrared wavelengths. Fluctuations in laser output and sample temperature are monitored with time (after Jeanloz and Heinz[15]).

the 1064 nm radiation from a 100 W cw Nd:YAG laser is focused into the sample volume in the diamond cell by way of a microscope with a long (15 mm) working-distance objective. We use only the TEM_{00} mode, which has a maximum output of 25 W, passing it through a filter to avoid that light from the pump lamp contaminates the thermal radiation emitted from the sample. The microscope contains a dichroic mirror to reflect the laser beam and transmit the visible thermal radiation emanating from the sample.

The sample temperature is best determined by spectroradiometry. This technique takes advantage of the optical access which the diamonds provide to the sample, and it allows the spatial distribution of temperature to be mapped out. We analyze the thermal radiation emitted from the sample by means of a holographic-grating monochromator attached to the microscope. The image of the hot spot in the sample area is focused onto the entrance slit of the monochromator after filtering out scattered laser radiation. The remaining thermal radiation, between wavelengths of 400 and 850 nm, is measured with 10 nm spectral resolution using either a silicon photovoltaic detector or a photomultiplier tube. The output is collected and reduced by computer to determine the temperature and emissivity of the sample according to a greybody model; that is, a model with emissivity being constant over the spectral bandwidth of interest (Jeanloz and Heinz[15]).

The temperature of the sample is changed by varying the output of the laser. This is best done by means of a polarizer that is external to the laser cavity, because any change inside the cavity (e.g., changing power to the pump lamp) causes the beam character to change. In fact, even after the laser is warmed up and running with a relatively steady power, the output varies sufficiently with time that it must be continuously monitored. We find that there are 1 to 3 percent fluctuations in laser output which cause temperature variations of hundreds of Kelvin in the sample on time scales of $\lesssim$ 200 msec. These fluctuations are problematical because they must be averaged out in any measurement of sample temperature.

Correlations of fluctuations in sample temperature with the fluctuations in laser output can in principle yield much information about the thermal properties of materials at high temperatures. As shown schematically in Figure 3, the temperature fluctuations are commonly found to be smoothed and slightly lagged in time relative to the fluctuations in laser power. Thus, quantitative modeling of the temperature distribution inside the diamond cell may make it possible to derive the thermal diffusivity or heat capacity of the sample from such comparative measurements of laser output and sample temperature. Also, it is observed that the correlation between the laser power and sample temperature breaks down when melting or a reaction is taking place; presumably the heat of reaction (or fusion) and kinetic hindrances play an important role in buffering the temperature fluctuations in the sample. Correlations in sample temperature and laser-output fluctuations can be used in this way to identify the occurrence of phase transitions at extreme conditions.

A schematic view of the heated region illustrates the axial (*z*-direction) and radial (*r*-dependent) temperature profiles that are achieved around the focal spot of the laser beam (Figure 4). Typical dimensions are indicated, and with peak temperatures of 5000 K to 7000 K gradients as large as 10^8 to 10^9 K m^{-1} are obtained. As noted above, this spatial variation of temperature results in the hot part of the sample being entirely enclosed by cold, relatively unreactive sample material. Consequently, ionic crystals that are heated in this manner, with cold sample material as the surrounding medium, rarely have the opportunity to react with and damage the diamonds. Metals, semiconductors and other strongly absorbing samples must be surrounded by refractory oxides (e.g., Al_2O_3) to achieve the same chemical and thermal isolation of the laser-heated spot from the diamond windows (e.g., Williams et al.[42]).

The temperature distribution is measured by a tomographic technique that involves scanning a narrow slit across the focused image of the hot spot, prior to that image entering the monochromator shown in Figure 3 (Jeanloz and Heinz[16]). Alternatively, a small aperture (circular or square) can be used to sample the image of the hot spot, and hence obtain the temperature as a function of position inside the sample. This tends to yield less reproducible results than the slit, however, because there is more severe diffraction and lower throughput with the aperture than with the slit. A smaller (narrower) slit than aperture can often be used in practice, and since spatial resolution trades off directly against aperture or slit size, it is generally advantageous to use a slit.

As the slit is scanned across the image of the hot spot, the spectrum of the thermal radiation emitted from the sample is observed as a function of slit position. From this information it is possible to derive the vertically (*z*-)averaged temperature as a function of radial distance from the center of the image of the hot spot by way of a Radon transform. Moreover, we have found that the third dimension (the *z*-dependence) of the temperature distribution can be modeled with little uncertainty once a single length scale, the *r*-dependence, has been determined for the temperature field inside the diamond cell (Heinz and Jeanloz[9]).

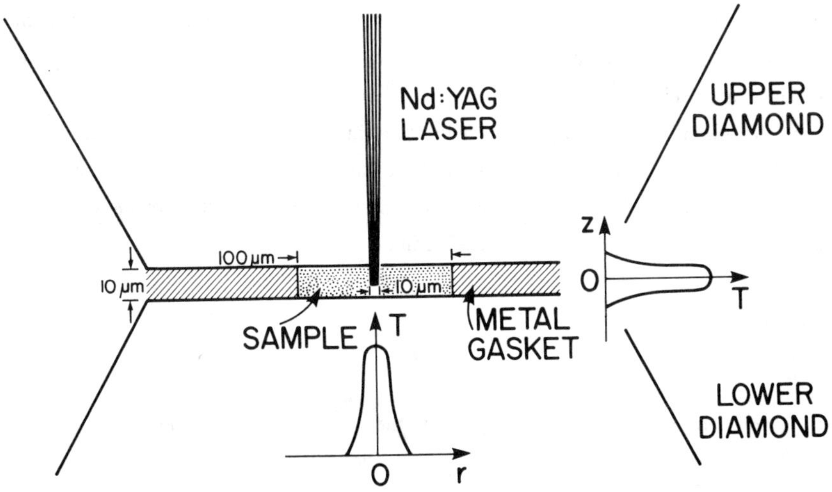

Figure 4. Illustration of the axial (z) and radial (r) temperature dependencies near the focal spot of the laser beam inside the diamond cell.

It is worth emphasizing that the strong variation of temperature across the sample is largely due to the presence of the diamonds acting as heat sinks, rather than to the Gaussian intensity distribution of the TEM_{00} mode of the laser beam. The details of the thermal distribution inside the diamond cell are not fully understood at present, but simple modeling and experimental observations support this conclusion. For example, spherical volumes ~ 100 to 200 μm in diameter are readily heated with the focused laser when the diamonds are not present. Such heating is used to measure the zero-pressure melting point of calibration materials in order to confirm the accuracy of the spectroradiometric temperatures (Jeanloz and Heinz[15]; Heinz and Jeanloz[9]). Also, it is found that heating becomes limited by the diminishing sample thickness in ultra-high pressure experiments (e.g., Williams et al.[42]). Thus, thermal insulation from the diamonds, as well as chemical isolation, becomes a consideration in choosing a medium to surround the sample. For ionic materials, the sample itself often provides the best medium.

APPLICATIONS

Spectroscopy

Optical properties are readily measured in the laser-heated diamond cell. Among these, the emissivity is obtained the most directly because it is determined through the spectroradiometric measurement of the sample temperature. Measuring the emissivity is especially important in studying ionic materials because their transparency at visible wavelengths results in extremely low values: emissivities of 10^{-1} to 10^{-4} have been found for the ~ 10 μm thick samples that are typical of diamond-cell experiments (Heinz and Jeanloz[9]). As a result, single-wavelength optical pyrometry yields severely biased estimates of temperature as compared with spectroradiometry. Also, the wavelength dependence of the emissivity is used to check the validity of the greybody model from which the temperatures are calculated. Any spectral range over which the emissivity varies significantly with wavelength is excluded from the temperature measurement because the variation in emissivity is evidence for a strong absorption band (non-greybody spectrum) occurring at those wavelengths.

The absorption and reflectivity of the sample inside the diamond cell are best measured with laser sources in the case of high temperature experiments. Although conventional lamps are adequate as sources at near-infrared and shorter wavelengths for temperatures less than about 1000 K (Figure 5), the thermal radiation of the sample is usually too

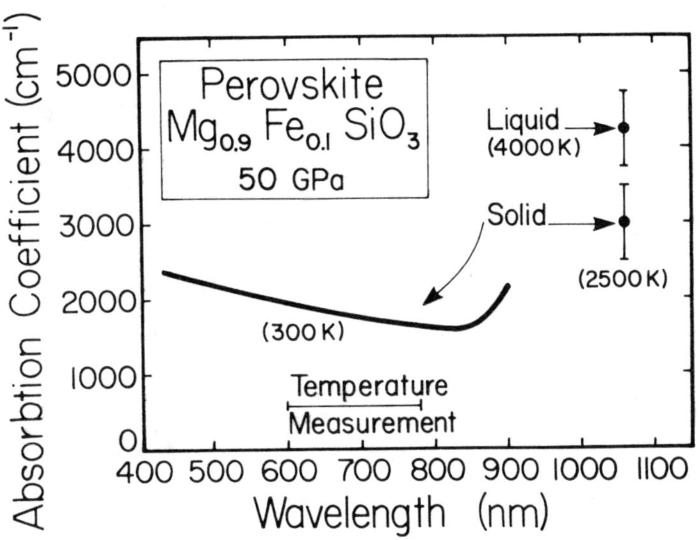

Figure 5. Absorption coefficient of silicate perovskite at visible and near-infrared wavelengths as measured in the diamond cell at elevated pressure. High-temperature measurements of the solid and melt are obtained from the attenuation of the Nd:YAG laser beam that is transmitted through the sample, whereas room-temperature measurements are obtained with a Xe lamp as source. The spectral range over which temperature is determined is indicated at the bottom (from Heinz and Jeanloz[10]).

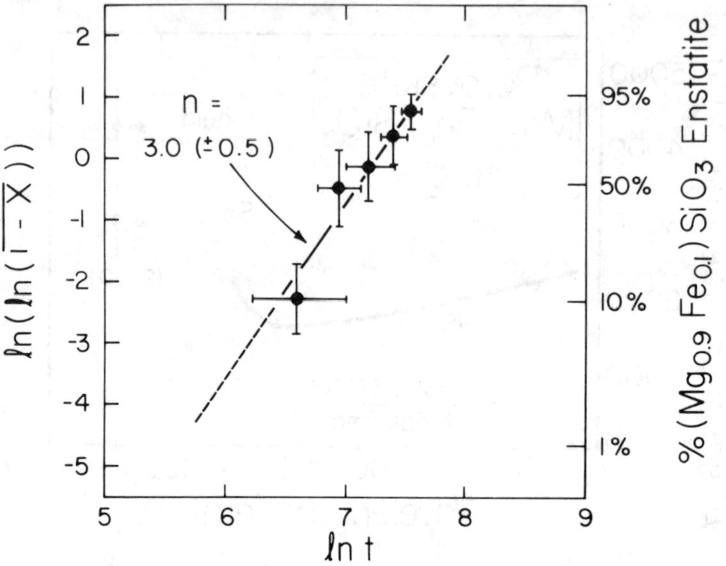

Figure 6. Extent of transformation (X) determined spectroscopically as a function of time (t) for the conversion of silicate perovskite to pyroxene (enstatite) at 850 (± 10) K and zero pressure. The data are plotted according to the Avrami relation, $X = 1 - \exp(-kt^n)$, so that the slope yields the exponent $n = 3.0$ (± 0.5) (from Knittle and Jeanloz[20]).

intense to allow reliable measurements at higher temperatures. As the reflectivity of ionic materials is low, the absorption spectrum of the sample can be directly compared with the emissivity that is derived spectroradiometrically. We have found that the greatest difficulty in making quantitative comparisons between emissivity and absorption coefficients is in measuring the optical geometry with adequate accuracy (e.g., sample thickness; surface area and solid angle of emittance). Nevertheless, results with no more than a few percent uncertainty in absorption coefficient and up to a few tens of percent uncertainty in emissivity have been obtained in the diamond cell (Heinz and Jeanloz[10]).

As an illustrative example, the visible absorption spectrum shown in Figure 5 was recently used to monitor the kinetics of the high-temperature transformation from the perovskite to the pyroxene phases of $(Mg,Fe)SiO_3$ (Knittle and Jeanloz[23]). The spectra of these polymorphs differ sufficiently at wavelengths of 400 to 800 nm that the amount of transformation can be quantitatively monitored as a function of time. The result, obtained on about 1.3 μg of sample that had been synthesized at high pressures in the diamond cell, clearly documents the time dependence of the transformation kinetics as given by the Avrami relation (see Figure 6). In this case, a resistance heater rather than the high-power laser was used to heat the sample because temperatures less than 1000 K are sufficient to overcome the kinetic barrier for the transformation of silicate perovskite to pyroxene.

There is much promise in the further development of laser-spectroscopic techniques with the diamond cell. Spatial resolution and sensitivity are not problematical with laser sources (e.g., Hemley et al.[11]), and with the availability of broadband turnable lasers, absorption and reflectivity spectra can be accurately measured over a wide range of wavelengths. Intrinsically weak signals, such as those from Raman scattering, can also be obtained at high temperatures through the use of laser-beam modulation and frequency lock-in techniques. Although well established (Demtröder[5]), these techniques have so far not been used with the diamond cell.

Electrical Conductivity

One of the most successful applications of the laser-heated diamond cell has been in obtaining the electrical conductivity of solids and melts at temperatures as high as 4500 K (Knittle and Jeanloz[21]). A pseudo four-lead geometry is used, with the contacts at the sample being separated by less than 150 μm (Figure 7). For an insulating sample the separation L between the leads must be less than the diameter of the hot spot, $D \sim 30$ μm, but for semiconducting samples the separation can be larger. The sample geometry is sufficiently well determined by direct observation into the diamond cell that absolute resistivity values are derived with an uncertainty of less than 10 to 20 percent; this uncertainty derives almost entirely from uncertainties in measuring the lead geometry. Comparisons with resistivities that have been measured independently confirm this level of accuracy for values obtained in the diamond cell (e.g., Knittle and Jeanloz[20]).

Results on $Fe_{0.94}O$, which is a poor semiconductor at ambient conditions, illustrate the qualitative change in the temperature dependence of the resistivity as the sample is converted to its metallic phase (Figure 8). Due to competing magnetic interactions, the metallization transition occurs only at simultaneously high pressures and temperatures: above 1000 K and 70 GPa in the solid (Figure 9). An important technical result of this work was to document that there is good agreement between high-temperature data collected on micro-samples in the diamond cell and data collected on ~ 1g samples under dynamic conditions (Figure 10). The critical step in this comparison is to model the sample geometry and temperature distribution in the laser-heated diamond cell sufficiently well to obtain reliable values of resistivity for the hot portion of the sample alone (Knittle and Jeanloz[21]). That this can be done successfully opens a new range of pressure-temperature conditions at which ionic mobilities and electronic transitions can be studied in condensed matter.

Soret Diffusion

The large temperature gradients that are achieved in the diamond cell are useful in many ways. They provide a chemical isolation of the hot part of the sample, and they can be used to reverse high-temperature phase transitions during a single *in-situ* measurement (Heinz and Jeanloz[10]). Another application is in studying phenomena that are driven by spatial variations of temperature, such as Soret diffusion. An example of such chemical diffusion being induced in an oxide by a strong temperature gradient is summarized here.

Figure 11 shows the temperature distribution in a crystalline sample of $(Mg,Fe)SiO_3$ perovskite under modest laser irradiation. This particular experiment was carried out at a pressure of 50 GPa, with the temperature profile being measured by the slit-sampling technique. In response to the nonuniform temperature, chemical diffusion is observed after approximately 45 minutes to 1 hour of continuous laser heating. An X-ray emission map of one of the samples confirms that the effect of the temperature gradient is to enrich the outer, cooler regions in iron at the expense of the inner, hottest zone of the sample (Figure

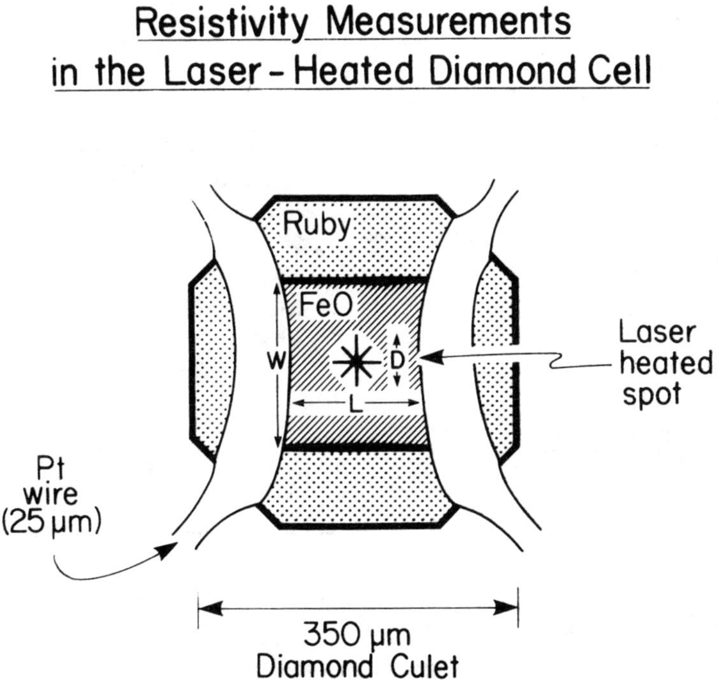

Figure 7. Geometry of electrical leads (two wires with 4 external leads) as observed in the laser-heated diamond cell. The hot spot at the focus of the laser beam (diameter D) is located between the leads, which are metal wires 5 to 25 μm in initial diameter (Pt, W, Fe, Au or Cu have been used successfully). The leads are spaced a distance L apart, and the resistivity of the sample is determined from the measured geometry (L, sample thickness T and width W) and temperature distribution. The temperature dependence of the resistivity is modeled in terms of a grid of resistors, in parallel across the width W and in series across the length L, which simulate the variation of resistivity with position (Knittle and Jeanloz[24]). This model is confirmed by measuring the resistance at several values of wire-spacing, L, which corresponds to extrapolating the measurements to obtain the resistivity as $L \rightarrow 0$.

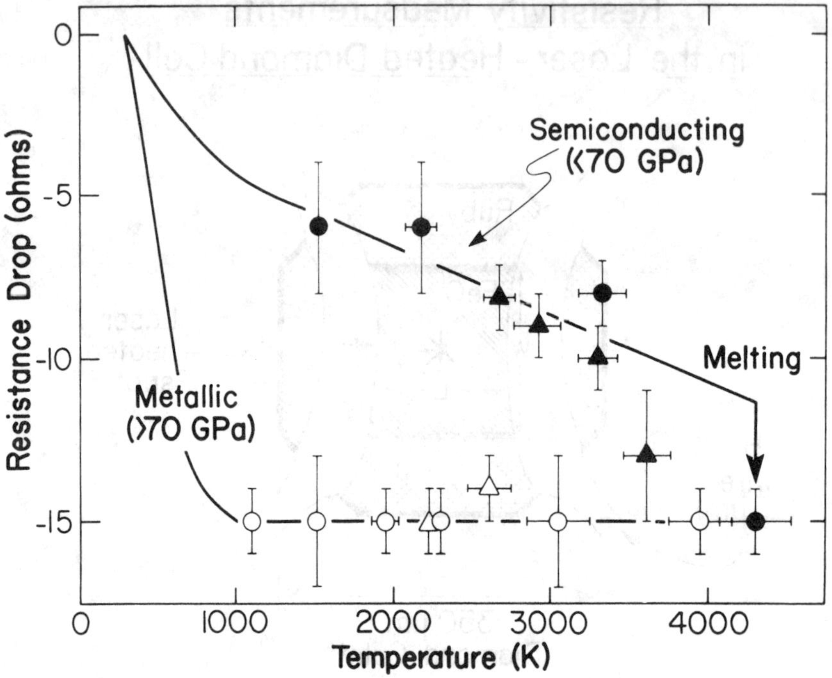

Figure 8. Electrical resistance of $Fe_{0.94}O$ as a function of temperature, and at pressures above and below the metallization transition at 70 GPa (Figure 9). A strong temperature dependence of resistance is evident for the semiconducting phase, from which an activation energy E_a = 0.06 (± 0.02) eV is obtained. In contrast, the resistance of the metallic phase does not vary measurably with temperature. The absolute resistivity is determined by modeling the measured resistance as indicated in Figure 7; its value in the metallic phase is in quantitative agreement with shock-wave measurements at simultaneously high temperatures and pressures: see Figure 10. Note that the melt is metallic below 70 GPa (after Knittle and Jeanloz[24]).

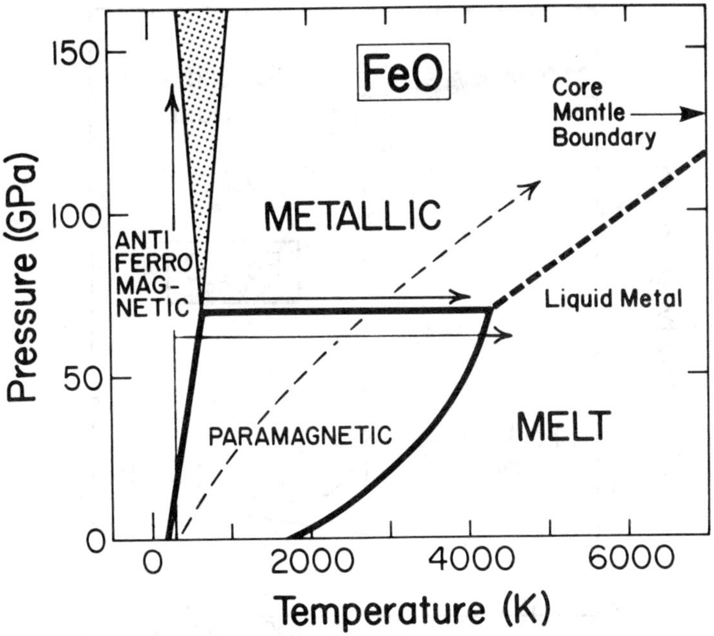

Figure 9. Pressure-temperature phase diagram of $Fe_{0.94}0$ based on: 1) diamond-cell experiments at room temperature (electrical resistivity and X-ray diffraction along the blue path); 2) high-temperature experiments with the laser-heated diamond cell (electrical resistivity along the green and red paths); 3) equation of state and electrical resistivity measurements under shock-wave loading (Hugoniot). The green and red paths correspond to the two curves in Figure 8 (below and above 70 GPa, respectively).

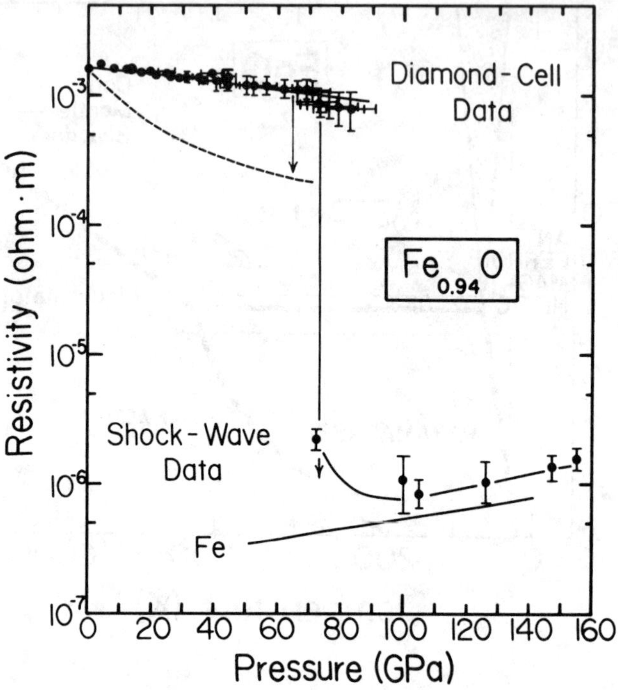

Figure 10. The electrical resistivity of $Fe_{0.94}O$ measured at room temperature in the diamond cell is compared with the resistivities of $Fe_{0.94}O$ and Fe measured at simultaneously high pressures and temperatures under shock loading. The green and red arrows indicate the resistivities obtained at high temperatures using the laser-heated diamond cell: see Figures 8 and 9. Both shock-wave and laser-heated diamond cell measurements indicate that the electrical resistivity of FeO at pressures above 70 GPa and temperatures above 1000 K is comparable to that of Fe (modified from Knittle and Jeanloz[21]).

12). This is as would be expected, in that the diffusing species of highest mass are selectively enriched by Soret diffusion toward the cold end of the sample (e.g., Jost[18]). The effect is especially visible here as an "enrichment front" of iron surrounding the spot on which the laser had been focused. The reason that this front appears so clearly is that as the iron diffuses down the temperature gradient the chemical diffusivity decreases rapidly with decreasing temperature. This causes the iron to pile up roughly 15 μm away from the center of the hot spot. Although the piling up makes the Soret effect readily visible, it complicates the quantitative analysis of the diffusivity. So far, detailed modeling of the temperature and compositional variations (paralleling the models of electrical resistance described above) have not been attempted for the chemical diffusion observed in the laser-heated diamond cell.

Yield Strength

Variations in pressure across a sample that is loaded between the diamonds have been used to investigate phase transitions and mass transfer in the same way that strong temperature variations in the diamond cell are used. For example, the pressure at which a particular reaction or transformation occurs can be rapidly determined by tracing out the pressure contour between product and reactant phases after a given period of laser heating (Yagi et al.[45]). Similarly, the yield strength of a solid can be determined by contouring the pressure distribution inside the diamond cell.

The most common technique used to measure pressure in the diamond cell is to observe the fluorescence of fine particles of ruby that are included in the sample area (Figure 2). The R_1 and R_2 ("laser") lines of the ruby are excited by means of a low-power laser beam that is focused through a microscope (Figure 13). As the wavelengths of the fluorescence lines shift systematically with pressure (Barnett et al.[1]; Mao et al.[26]; Xu et al.[44]), fluorescence spectroscopy of the ruby grains allows pressure to be accurately measured with a ~ 1 to 10 μm spatial resolution across the sample. No more than a few percent of the total sample volume needs to contain ruby in order to have a good calibration of the pressure distribution.

In the yielding experiments, polycrystalline samples are loaded into the diamond cell without a pressure medium. Under the partly uniaxial loading that is achieved, the samples first undergo a transient deformation that is dominated by extrusion. This is followed by a more reproducible regime of deformation in which the maximum pressure gradients

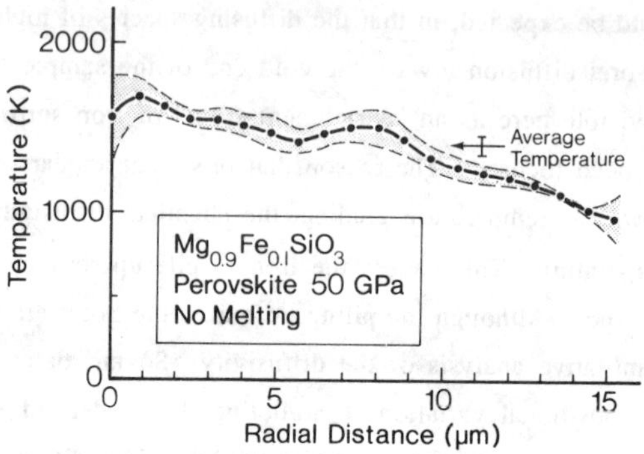

Figure 11. Sample temperature as a function of distance from the center of the laser-beam focus. The shaded band indicates the tomographically computed profile and its uncertainty. For comparison, the average temperature of the hot spot and its uncertainty are shown on the right. The sample is silicate perovskite at 50 GPa in the diamond cell: with modest laser power no melting occurs in the sample (Heinz and Jeanloz [10]).

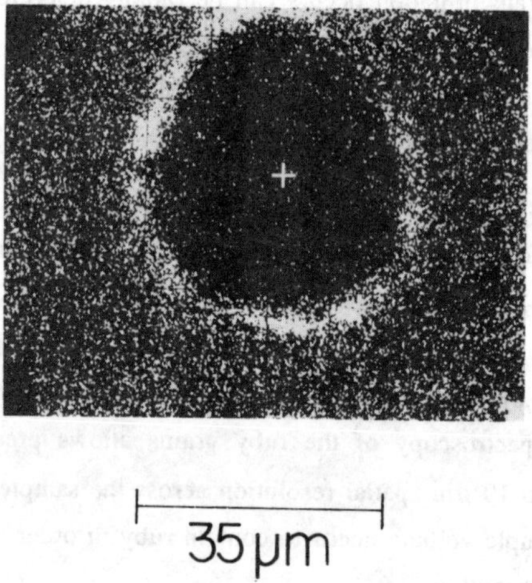

Figure 12. Fe K_α X-ray emission map across a silicate perovskite sample held at conditions similar to those shown in Figure 11 for 45 minutes. The density of white points is proportional to the X-ray emission induced under a rastered electron-microprobe beam, and hence represents the iron concentration. Depletion of iron in the hot zone around the laser-beam focus and concentration of iron in the cooler regions away from the focal spot are indicative of Soret diffusion. The cross indicates the center of the focal spot (Heinz and Jeanloz [10]).

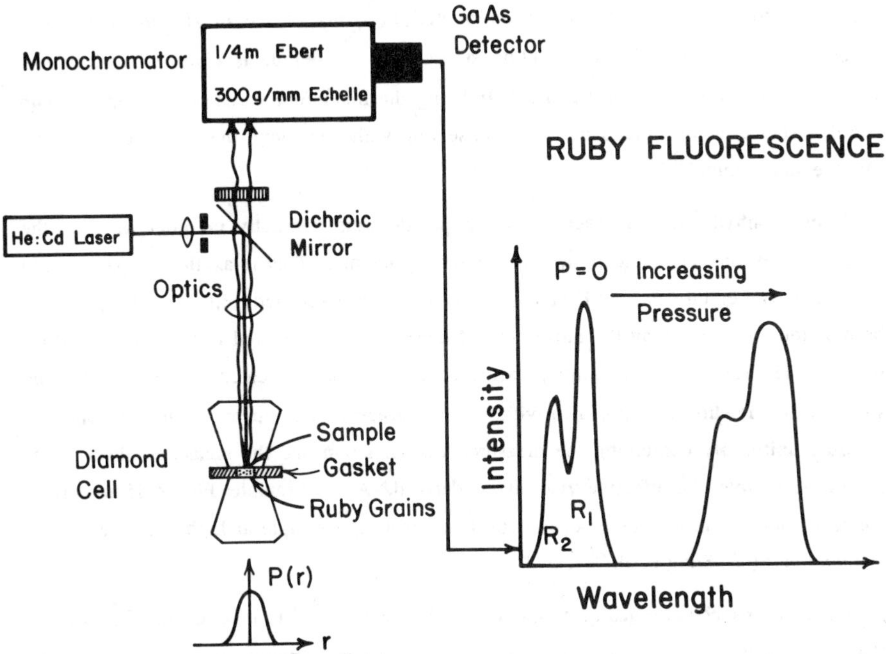

Figure 13. Microspectrometer system for measuring the pressure distribution in the diamond cell *(P(r))* using the ruby fluorescence technique. The output from a 14 W cw He:Cd laser (blue) is focused into the sample area that contains ruby grains (Figure 2), thus exciting the ruby fluorescence (red) that is measured by a high-resolution monochromator and photomultiplier tube. The shift of the fluorescence wavelength is used to determined the pressure. Changes in the width of the fluorescence peaks and in the fluorescence lifetime have also been used to calibrate temperature inside the diamond cell.

observed across the sample can be interpreted in terms of the maximum supportable differential stress. That is, the yielding strength of the sample is given by $(h/2)\partial P/\partial r$, with h being sample thickness and $\partial P/\partial r$ being the maximum variation of pressure with radial distance across the sample, as observed with the ruby fluorescence technique (Meade and Jeanloz[28]).

Under nonhydrostatic pressures, the sample deforms internally in order to relax the shear stresses that are created by the pressure gradients. This relaxation is particularly evident if temperature is raised, and the viscous strain associated with the relaxation can be monitored by following the movement of marker particles mixed into the sample (Figure 14). The ruby grains used for pressure calibration can be treated as markers, and the strain is then readily determined to within the accuracy of a geometric factor, β, which is of order unity. By combining the measured rate of strain and the measured shear stress across the sample, the effective viscosity is derived. As an example, Figure 15 illustrates the relaxation of shear stresses caused by laser heating a sample at high pressures in the diamond cell (cf., Sung et al.[40]).

The marker technique has been applied by Poirier et al.[33] to measure the flow rate of H_2O ice under modest pressures (1.2 GPa) and room temperature. This is a high-temperature measurement in the sense that it is carried out very close to the melting temperature of ice. At these conditions, Poirier et al.[33] documented a relatively low viscosity for the high-pressure polymorph Ice VI. Although viscosity and yield strength normally increase with increasing pressure, this is counteracted in H_2O ice by the polymorphic transformations that effectively weaken the structure. Similarly, Meade and Jeanloz[29] have found from measurements of pressure gradients across the sample that the strength of SiO_2 glass decreases as pressure is increased above 30 GPa at room temperature. This weakening appears to be induced by a coordination change that occurs in the glass structure at elevated pressures (Williams and Jeanloz[43]). Because it is reversible, the coordination change and its associated weakening are expected to occur at high temperatures in both glassy and molten SiO_2. That is, although measured at room temperature, the weakening of SiO_2 glass at high pressures is thought to reflect a weakening that must occur in molten SiO_2 at ultrahigh pressures and temperatures.

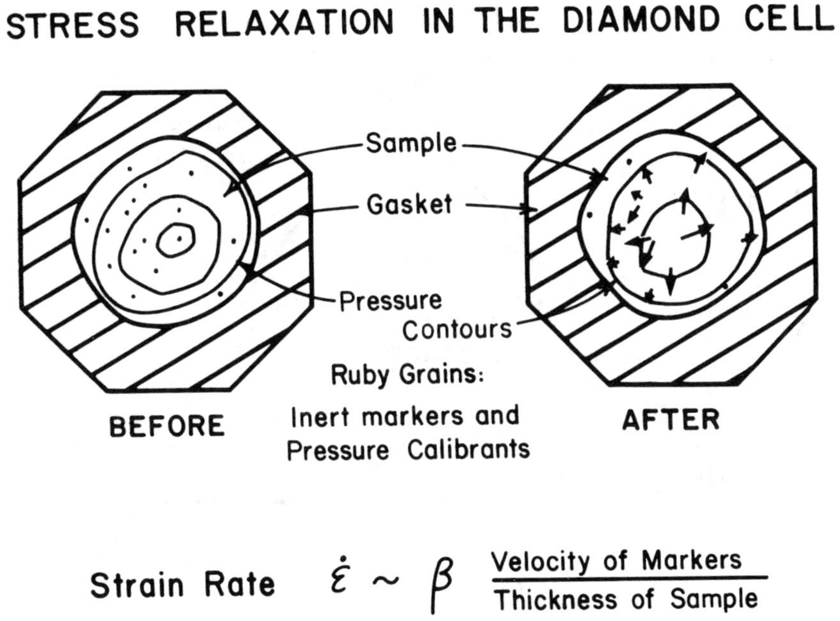

Figure 14. Schematic view of a nonhydrostatically loaded polycrystalline sample inside the diamond cell: immediately upon loading (Before) and after sufficient time for creep deformation (After). As a result of viscous deformation the pressure distribution (marked by green contours, maximum at the center) relaxes and material moves down the pressure gradient. Ruby grains (red) are used both to determine pressure and as markers in the sample. The rate of creep deformation is given by the ratio of marker velocity to sample thickness, with β being a factor to correct for geometric effects and friction against the diamonds.

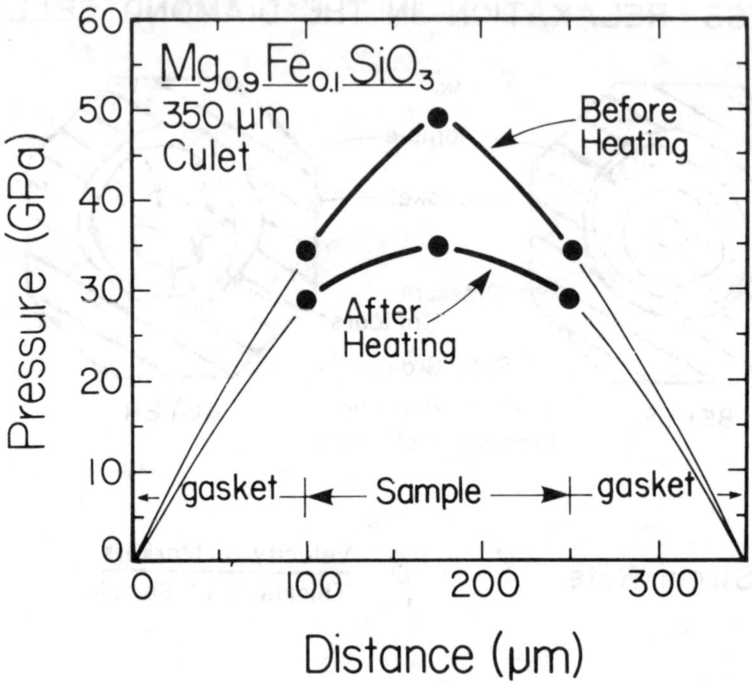

Figure 15. Relaxation in the pressure distribution *P(r)* caused by laser heating a sample of (Mg,Fe)SiO_3 to 3000 K. Creep deformation at high temperatures reduces the shear stresses and hence the maximum pressure gradients (Heinz and Jeanloz[10]).

CONCLUSIONS

The laser-heated diamond cell provides new opportunities for studying the properties of materials over a broad range of temperatures and pressures. Although mainly thought of as anvils for high-pressure research, the diamonds act as windows that provide direct optical access to the sample while it is at high temperatures. The spatial variations of temperature within the diamond cell raise some experimental difficulties, but they are an advantage in isolating the sample thermally and chemically. The problem of working with small ($\sim 10^{-6}$ g) samples is largely overcome by the application of laser sources and synchrotron radiation as probes. Thus, despite its relatively brief history as a research tool, the laser-heated diamond cell is yielding new insights into the properties of matter at extreme conditions.

Acknowledgments—Work supported by the National Science Foundation, by NASA, and by the U.S. Department of Energy through Lawrence Berkeley Laboratory. I thank Q. Williams, E. Knittle, and C. Meade for helpful discussions.

REFERENCES

1. Barnett, J. D., S. Block and G. J. Piermarini (1973) An optical fluorescence system for quantitative pressure measurements in the diamond-anvil cell. *Rev. Sci. Instrum.*, *44*, 1-9.

2. Bassett, W. A. (1979) The diamond cell and the nature of the Earth's mantle. *Ann. Rev. Earth Planet. Sci.*, *7*, 357-384.

3. Bassett, W. A., and M. L. Weathers (1987) Temperature measurement in a laser-heated diamond cell, in: *High-Pressure Research in Mineral Physics* (M. H. Manghnani and Y. Syono, editors) Am. Geophys. Union, Washington, D.C., pp. 129-33.

4. Boehler, R. (1986) The phase diagram of iron to 430 kbar. *Geophys. Res. Lett.*, *13*, 1153-1156.

5. Demtröder, W. (1982) *Laser Spectroscopy*, Springer-Verlag, New York, 696 pp.

6. Evans, T. (1979) Changes produced by high temperature treatment of diamond, in: *The Properties of Diamond* (J. E. Field, editor), Academic, Orlando, Fla., pp. 403-424.

7. Furnish, M. D., and W. A. Bassett (1983) Investigation of mechanism of the olivine-spinel transition in fayalite by synchrotron radiation. *J. Geophys. Res.*, *88*, 10333-10341.

8. Hazen, R. M., and L. W. Finger (1982) *Comparative Crystal Chemistry*, J. Wiley, New York, 231 pp.

9. Heinz, D. L., and R. Jeanloz (1987a) Temperature measurements in the laser-heated diamond cell, in: *High-Pressure Research in Mineral Physics* (M. H. Manghnani and Y. Syono, editors), Am. Geophys. Union, Washington, D.C., pp. 113-127.

10. Heinz, D. L., and R. Jeanloz (1987b) Melting of (Mg,Fe)SiO_3 perovskite measurement at lower mantle conditions and geophysical implications. *J. Geophys. Res.*, *92*, 11437-11444.

11. Hemley, R. J., P. M. Bell and H. K. Mao (1987) Laser techniques in high-pressure geophysics. *Science*, *237*, 605-612.

12. Huang, E., and W. A. Bassett (1986) Rapid determination of Fe_3O_4 phase diagram by synchrotron radiation. *J. Geophys. Res., 91,* 4697-4703.

13. Jayaraman, A. (1983) Diamond anvil cell and high-pressure physical investigations. *Rev. Modern Physics, 55,* 65-108.

14. Jayaraman, A. (1986) Ultrahigh pressures. *Rev. Sci. Instrum., 57,* 1013-1031.

15. Jeanloz, R., and D. L. Heinz (1984) Experiments at high temperature and pressure: Laser heating through the diamond cell. *J. Physique, 45,* (C8) 83-92.

16. Jeanloz, R., and D. L. Heinz (1986) Measurement of the temperature distribution in cw-laser heated materials, in: *Laser Welding, Machining and Materials Processing* (C. Albright, editor), Springer-Verlag, New York, pp. 239-243.

17. Jephcoat, A. P., H. K. Mao and P. M. Bell (1987) Operation of the Megabar diamond-anvil cell, in: *Hydrothermal Experimental Techniques* (G. C. Ulmer and H. L. Barnes, editors), Wiley Interscience, New York, pp. 469-506.

18. Jost, W. (1960) *Diffusion in Solids, Liquids, Gasses,* Academic, Orlando, Fla.

19. Knittle, E., and R. Jeanloz (1985) X-ray diffraction and optical absorption studies of CsI at high pressures. *J. Phys. Chem. Solids, 46,* 1179-1184.

20. Knittle, E., and R. Jeanloz (1986a) High-pressure electrical resistivity measurements of Fe_2O_3: Comparison of static-compression and shock-wave experiments to 61 GPa. *Solid State Commun., 58,* 129-131.

21. Knittle, E., and R. Jeanloz (1986b) High-pressure metallization of FeO and implication for the Earth's core. *Geophys. Res. Lett., 13,* 1541-1544.

22. Knittle, E., and R. Jeanloz (1987a) Synthesis and equation of state of $(Mg,Fe)SiO_3$ perovskite to over 100 GPa. *Science, 235,* 236-244.

23. Knittle, E., and R. Jeanloz (1987b) The activation energy of the back transformation of silicate perovskite to enstatite, in: *High-Pressure Research in Mineral Physics* (M. H. Manghnani and Y. Syono, editors), Am. Geophys. Union, Washington, D.C., pp. 243-250.

24. Knittle, E., and R. Jeanloz (1988) The high-pressure phase diagram of $Fe_{0.94}O$: A possible constituent of the Earth's core. *J. Geophys. Res.* (in press).

25. Mao, H. K., P. M. Bell and C. Hadidiacos (1987) Experimental phase relations of iron to 360 kbar, 1400°C, determined in an internally heated diamond-anvil apparatus, in: *High-Pressure Research in Mineral Physics* (M. H. Manghnani and Y. Syono, editors), Am. Geophys. Union, Washington, D.C., pp. 135-138.

26. Mao, H. K., P. M. Bell, J. W. Shaner and D. J. Steinberg (1978) Specific volume measurements of Cu, Mo, Pd and Ag and calibration of the ruby R_1 fluorescence pressure gauge from 0.06 to 1 Mbar. *J. Appl. Phys.*, *49*, 3276-3283.

27. Mao, H. K., J. Xu and P. M. Bell (1986) Calibration of the ruby pressure gauge to 800 kbar under quasi-hydrostatic conditions. *J. Geophys. Res.*, *91*, 4673-4676.

28. Meade, C., and R. Jeanloz (1988a) Yield strength of MgO to 40 GPa. *J. Geophys. Res.*, *93*, 3261-3269.

29. Meade, C., and R. Jeanloz (1988b) The effect of coordination change on the strength of amorphous SiO_2. *Science*, *241*, 1072-1074.

30. Ming, L. C., and W. A. Bassett (1974) Laser heating in the diamond anvil press up to 2000°C sustained and 3000°C pulsed at pressures up to 260 kilobars. *Rev. Sci. Instrum.*, *45*, 1115-1118.

31. Piermarini, G. J., S. Block and J. D. Barnett (1973) Hydrostatic limits in liquids and solids to 100 kbar. *J. Appl. Phys.*, *44*, 5377-5382.

32. Piermarini, G. J., R. A. Forman and S. Block (1978) Viscosity measurement in the diamond anvil pressure cell. *Rev. Sci. Instrum.*, *49*, 1061-1066.

33. Poirier, J. P., C. Sotin and J. Peyronneau (1981) Viscosity of high-pressure ice VI and evolution and dynamics of Ganymede. *Nature*, *292*, 225-227.

34. Ruoff, A. L., S. T. Weir, K. E. Brister and Y. K. Vohra (1987) Synthetic diamonds produce pressure of 125 GPa (1.25 Mbar). *J. Mater. Res.*, *2*, 614-618.

35. Sato-Sorensen, Y. (1987) Measurements of the lifetime of the ruby R_1 line and its application to high-temperature and high-pressure calibration in the diamond-anvil cell, in: *High-Pressure Research in Mineral Physics* (M. H. Manghnani and Y. Syono, editors), Am. Geophys. Union, Washington, D.C., pp. 53-59.

36. Schiferl, D., J. N. Fritz, A. I. Katz, M. Schaefer, E. F. Skelton, S. B. Qadri, L. C. Ming and M. H. Manghnani (1987) Very high temperature diamond-anvil cell for

X-ray diffraction: Application to the comparison of the gold and tungsten high-temperature—high-pressure internal standards, in: *High-Pressure Research in Mineral Physics* (M. H. Manghnani and Y. Syono, editors), Am. Geophys. Union, Washington, D.C., pp. 75-83.

37. Shaner, J. W., J. M. Brown, C. A. Swenson and R. G. McQueen (1984) Sound velocity of carbon at high pressures. *J. Physique, 45,* (C8) 235-237.

38. Shimomura, O., S. Yamaoka, H. Nakazawa and O. Fukunaga (1982) Application of a diamond-anvil cell to high-temperature and high-pressure experiments, in: *High-Pressure Research in Geophysics* (S. Akimoto and M. H. Manghnani, editors), Center Acad. Pub. Japan, Tokyo, pp. 49-60.

39. Skelton, E. F., J. Kirkland and S. B. Qadri (1982) Energy-dispersive measurements of diffracted synchrotron radiation as a function of pressure: Applications to phase transitions in KCl and KI. *J. Appl. Cryst., 15,* 82-88.

40. Sung, C. M., C. Goetze and H. K. Mao (1977) Pressure distribution in the diamond anvil press and the shear strength of fayalite. *Rev. Sci. Instrum., 48,* 1386-1391.

41. Weathers, M. S., and W. A. Bassett (1987) Melting of carbon at 50 to 300 kbar. *Phys. Chem. Minerals, 15,* 105-112.

42. Williams, Q., and R. Jeanloz, J. Bass, B. Svendsen and T. J. Ahrens (1987) The melting curve of iron to 250 gigapascals: A constraint on the temperature at the Earth's center. *Science, 236,* 181-182.

43. Williams, Q., and R. Jeanloz (1988) Spectroscopic evidence for pressure-induced coordination changes in silicate glasses and melts. *Science, 239,* 902-905.

44. Xu, J. A., H. K. Mao and P. M. Bell (1986) High-pressure ruby and diamond fluorescence: Observations at 0.21 to 0.55 Terapascal. *Science, 232,* 1404-1406.

45. Yagi, T., P. M. Bell and H. K. Mao (1979) Phase relations in the system $MgO-FeO-SiO_2$ between 150 and 700 kbar at 1000°C. *Carnegie Inst. Wash. Yearbook, 78,* 614-615.

SUBJECT INDEX